Springer Series in
CHEMICAL PHYSICS 91

Springer Series in

CHEMICAL PHYSICS

The purpose of this series is to provide comprehensive up-to-date monographs in both well established disciplines and emerging research areas within the broad fields of chemical physics and physical chemistry. The books deal with both fundamental science and applications, and may have either a theoretical or an experimental emphasis. They are aimed primarily at researchers and graduate students in chemical physics and related fields.

75 **Basic Principles in Applied Catalysis**
By M. Baerns

76 **The Chemical Bond**
A Fundamental Quantum-Mechanical Picture
By T. Shida

77 **Heterogeneous Kinetics**
Theory of Ziegler-Natta-Kaminsky Polymerization
By T. Keii

78 **Nuclear Fusion Research**
Understanding Plasma-Surface Interactions
Editors: R.E.H. Clark and D.H. Reiter

79 **Ultrafast Phenomena XIV**
Editors: T. Kobayashi, T. Okada, T. Kobayashi, K.A. Nelson, S. De Silvestri

80 **X-Ray Diffraction by Macromolecules**
By N. Kasai and M. Kakudo

81 **Advanced Time-Correlated Single Photon Counting Techniques**
By W. Becker

82 **Transport Coefficients of Fluids**
By B.C. Eu

83 **Quantum Dynamics of Complex Molecular Systems**
Editors: D.A. Micha and I. Burghardt

84 **Progress in Ultrafast Intense Laser Science I**
Editors: K. Yamanouchi, S.L. Chin, P. Agostini, and G. Ferrante

85 **Progress in Ultrafast Intense Laser Science II**
Editors: K. Yamanouchi, S.L. Chin, P. Agostini, and G. Ferrante

86 **Free Energy Calculations**
Theory and Applications in Chemistry and Biology
Editors: Ch. Chipot and A. Pohorille

87 **Analysis and Control of Ultrafast Photoinduced Reactions**
Editors: O. Kühn and L. Wöste

88 **Ultrafast Phenomena XV**
Editors: P. Corkum, D. Jonas, D. Miller, and A.M. Weiner

89 **Progress in Ultrafast Intense Laser Science III**
Editors: K. Yamanouchi, S.L. Chin, P. Agostini, and F. Ferrante

90 **Thermodynamics and Fluctuations far from Equilibrium**
By J. Ross

91 **Progress in Ultrafast Intense Laser Science IV**
Editors: K. Yamanouchi, A. Becker, R. Li, and S.L. Chin

Kaoru Yamanouchi
Andreas Becker
Ruxin Li
See Leang Chin
(Eds.)

Progress in Ultrafast Intense Laser Science

Volume IV

With 133 Figures

Springer

Professor Dr. Kaoru Yamanouchi
The University of Tokyo
Department of Chemistry
Hongo 7-3-1, 113-0033 Tokyo, Japan

Dr. Andreas Becker
MPI für Physik komplexer Systeme
Nöthnitzer Str. 38, 01187 Dresden, Germany
Department of Physics and JILA
University of Colorado, 440 UCB,
Boulder, CO 80309-0440, USA

Professor Dr. Ruxin Li
Chinese Academy of Sciences
Shanghai Institute of Optics and Fine Mechanics
P.O.Box 800-21, 201800 Shanghai, China

Professor See Leang Chin
Laval University
Department for Physics
Engineering Physics and Optics & Center for Optics
Photonics and Laser
Quebec City, Quebec G1V 0A6 Canada

Springer Series in Chemical Physics ISSN 0172-6218

ISBN 978-3-642-08864-3 e-ISBN 978-3-540-69143-3

Cover design: eStudio Calamar Steinen

Printed on acid-free paper

9 8 7 6 5 4 3 2 1

springer.com

Preface

It is a great pleasure that we are now publishing the fourth volume of the series on PUILS, through which we have been introducing the progress in ultrafast intense laser science, the frontiers of which are rapidly expanding, thanks to the progress in ultrashort and high-power laser technologies. The interdisciplinary nature of this research field is attracting researchers with different expertise and backgrounds.

As in the previous volumes on PUILS, each chapter in the present volume, which is in the range of 15–25 pages, begins with an introduction in which a clear and concise account of the significance of the topic is given, followed by a description of the authors' most recent research results. All the chapters are peer-reviewed. The articles of this fourth volume cover a diverse range of the interdisciplinary research field, and the topics may be grouped into four categories: strong field ionization of atoms (Chaps. 1–2), excitation, ionization and fragmentation of molecules (Chaps. 3–5), nonlinear intense optical phenomena and attosecond pulses (Chaps. 6–8), and laser solid interactions and photoemissions (Chaps. 9–11).

From the third volume, the PUILS series has been edited in liaison with the activities of the JSPS Core-to-Core Program on Ultrafast Intense Laser Science (FY2004) and JILS (Japan Intense Light Field Science Society), which is now a sponsor organization responsible for the regular publication of the PUILS series. As described in the prefaces of the previous volumes, the PUILS series also collaborates with the annual symposium series of ISUILS (http://www.isuils.jp), designed to stimulate interdisciplinary discussion at the forefront of ultrafast intense laser science.

We would like to take this opportunity to thank all the authors who have kindly contributed to the PUILS series by describing their most recent work at the frontiers of ultrafast intense laser science. We also thank the reviewers who have read the submitted manuscripts carefully. One of the co-editors (KY) thanks Ms. Chie Sakuta and Maki Oyamada for their help with the editing processes. Last but not least, our gratitude goes out to Dr. Claus Ascheron, Physics Editor of Springer Verlag at Heidelberg, for his kind support.

We hope this volume will convey the excitement of Ultrafast Intense Laser Science to the readers, and stimulate interdisciplinary interactions among researchers, thus paving the way to explorations of new frontiers.

University of Tokyo — *Kaoru Yamanouchi*
University of Colorado — *Andreas Becker*
Chinese Academy of Sciences — *Ruxin Li*
Laval University — *See Leang Chin*
October, 2008

Contents

1 A Gauge-Invariant Theory of Intense-Field Coulomb Approximations to All Orders
Farhad H.M. Faisal 1

2 Strong-Field Photoionization by Few-Cycle Laser Pulses
Gerhard G. Paulus 17

3 Wavepacket Dynamics of Molecules in Intense Laser Fields
Hirohiko Kono, Katsunori Nakai, Manabu Kanno, Yukio Sato, Shiro Koseki, Tsuyoshi Kato, and Yuichi Fujimura 41

4 Molecular Ion Beams Interrogated with Ultrashort Intense Laser Pulses
Itzik Ben-Itzhak 67

5 Dynamical Reaction Theory for Vibrationally Highly Excited Molecules
Mikito Toda 91

6 Relativistic Quantum Dynamics in Intense Laser Fields
Guido R. Mocken, Yousef I. Salamin, and Christoph H. Keitel 113

7 Attosecond Scale Multi-XUV-Photon Processes
Dimitris Charalambidis, Paris Tzallas, Emmanouil P. Benis, and George D. Tsakiris 133

8 Ultrashort Pulse Collapse in Quadratic Media
Jian Wu and Heping Zeng 159

9 Femtosecond Laser Induced Various Luminescent Phenomena in Solid Materials
Jianrong Qiu, Li Wang, and Bin Zhu 185

10 Prospect of Laser-Driven X-Ray Lasers for Extension to Shorter Wavelengths
Yoshiaki Kato and Tetsuya Kawachi 215

11 Femtosecond Laser Applications in Micro/Nano Science and Technology: Nonlinear Effects in Photonic Crystal Fibers, Femtosecond Laser-Induced Forward Transfer, and Femtosecond Laser Manipulation System for Biology
Ching-Yue Wang, Qi-Rong Xing, Yan-Feng Li, Ming-Lie Hu, Li Yang, Ji-Xian Gong, Wei Jia, and Lu Chai 233

Index 261

List of Contributors

Emmanouil P. Benis
Foundation for Research
and Technology – Hellas
Institute of Electronic Structure
and Laser
P.O. Box 1527
71110 Heraklion (Crete)
Greece

Itzik Ben-Itzhak
J.R. Macdonald Laboratory
Department of Physics
Kansas State University
Manhattan, KS 66506-2601
USA
ibi@phys.ksu.edu

Lu Chai
Ultrafast Laser Laboratory,
College of Precision Instrument and
Optoelectronics Engineering, and
Key Laboratory of Opto-electronics
Information and Technical Science
(Ministry of Education)
Tianjin University
Tianjin 300072
People's Republic of China
lu_chai@tju.edu.cn

Dimitris Charalambidis
Foundation for Research
and Technology – Hellas,
Institute of Electronic Structure
and Laser
P.O. Box 1527
71110 Heraklion (Crete), Greece

Department of Physics, University
of Crete
P.O. Box 2208
71003 Heraklion (Crete), Greece

Max-Planck-Institut für
Quantenoptik
85748 Garching, Germany

Farhad H.M. Faisal
Fakultät für Physik
Universität Bielefeld
Postfach 100131
33501 Bielefeld, Germany
ITAMP, Harvard-Smithsonian
Center for Astrophysics
60 Garden Street, MS 14
Cambridge, MA 02138, USA

Yuichi Fujimura
Department of Chemistry
Graduate School of Science
Tohoku University
Sendai 980-8578, Japan

Ji-Xian Gong
Ultrafast Laser Laboratory
College of Precision Instrument
and Optoelectronics Engineering
and Key Laboratory
of Opto-electronics Information
and Technical Science (Ministry of
Education), Tianjin University
Tianjin 300072
People's Republic of China
gongjixian@126.com

Ming-Lie Hu
Ultrafast Laser Laboratory
College of Precision Instrument
and Optoelectronics Engineering
and Key Laboratory of
Opto-electronics Information
and Technical Science (Ministry
of Education), Tianjin University
Tianjin 300072
People's Republic of China
huminglie@tju.edu.cn

Wei Jia
Ultrafast Laser Laboratory,
College of Precision Instrument
and Optoelectronics
Engineering, and Key Laboratory
of Opto-electronics Information
and Technical Science
(Ministry of Education)
Tianjin University
Tianjin 300072
People's Republic of China
jiaw@tju.edu.cn

Manabu Kanno
Department of Chemistry
Graduate School of Science
Tohoku University
Sendai 980-8578
Japan

Tsuyoshi Kato
Department of Chemistry
School of Science
The University of Tokyo
7-3-1 Hongo, Bunkyo-ku
Tokyo 113-0033
Japan

Yoshiaki Kato
The Graduate School for the Creation
of New Photonics Industries
1955-1 Kurematsu-cho
Nishi-ku, Hamamatsu
Sizuoka 431-1202
Japan
y.kato@gpi.ac.jp

Tetsuya Kawachi
Advanced Photon Research Center
Japan Atomic Energy Agency
8-1 Umemidai
Kizugawa Kyoto 619-0215
Japan
kawachi.tetsuya@jaea.go.jp

Christoph H. Keitel
Max-Planck-Institut für Kernphysik
Heidelberg, Germany
keitel@mpi-hd.mpg.de

Hirohiko Kono
Department of Chemistry
Graduate School of Science
Tohoku University
Sendai 980-8578
Japan
hirohiko-kono@mail.tains.
tohoku.ac.jp

Shiro Koseki
Department of Chemistry
Graduate School of Science
Osaka Prefecture University
Osaka 599-8531
Japan

Yan-Feng Li
Ultrafast Laser Laboratory
College of Precision Instrument
and Optoelectronics
Engineering, and Key Laboratory
of Opto-electronics Information
and Technical Science
(Ministry of Education)
Tianjin University
Tianjin 300072
People's Republic of China
yanfengli@tju.edu.cn

Guido R. Mocken
Max-Planck-Institut für Kernphysik
Heidelberg, Germany
mocken@mpi-hd.mpg.de

Katsunori Nakai
Department of Chemistry
School of Science
The University of Tokyo
7-3-1 Hongo, Bunkyo-ku
Tokyo 113-0033
Japan

Gerhard G. Paulus
Institute of Optics and Quantum
Electronics
Max-Wien-Platz 1
07743 Jena Germany
and
Department of Physics
Texas A&M University
College Station
TX 77843-4242, USA

Jianrong Qiu
Department of Materials Science
and Engineering
Zhejiang University
Hangzhou 310027, China
qjr@zju.edu.cn

Yousef I. Salamin
Max-Planck-Institut für Kernphysik
Heidelberg, Germany
Physics Department
American University of Sharjah
Sharjah, United Arab Emirates
ysalamin@aus.edu

Yukio Sato
Department of Chemistry
Graduate School of Science
Tohoku University
Sendai 980-8578, Japan

Mikito Toda
Physics Department
Faculty of Science
Nara Women's University
Nara 630-8506, Japan
toda@ki-rin.phys.nara-wu.ac.jp

George D. Tsakiris
Max-Planck-Institut für
Quantenoptik 85748 Garching
Germany

Paris Tzallas
Foundation for Research
and Technology – Hellas
Institute of Electronic
Structure and Laser
P.O. Box 1527
71110 Heraklion (Crete)
Greece

Ching-Yue Wang
Ultrafast Laser Laboratory
College of Precision Instrument
and Optoelectronics Engineering
and Key Laboratory of
Opto-electronics Information
and Technical Science
(Ministry of Education)
Tianjin University
Tianjin 300072
People's Republic of China
chywang@tju.edu.cn

Li Wang
Department of Materials Science and Engineering
Zhejiang University
Hangzhou 310027, China
grace083635@yahoo.com.cn

Jian Wu
State Key Laboratory of Precision Spectroscopy
East China Normal University
Shanghai 200062
China
jwu@phy.ecnu.edu.cn

Qi-Rong Xing
Ultrafast Laser Laboratory
College of Precision Instrument and Optoelectronics Engineering
and Key Laboratory of Opto-electronics Information and Technical Science
(Ministry of Education)
Tianjin University
Tianjin 300072
People's Republic of China
xingqr@yahoo.com

Li Yang
Ultrafast Laser Laboratory
College of Precision Instrument and Optoelectronics Engineering
and Key Laboratory of Opto-electronics Information and Technical Science
(Ministry of Education)
Tianjin University
Tianjin 300072
People's Republic of China
y_lyang@yahoo.com.cn

Heping Zeng
State Key Laboratory of Precision Spectroscopy
East China Normal University
Shanghai 200062, China
hpzeng@phy.ecnu.edu.cn

Bin Zhu
Department of Materials Science and Engineering
Zhejiang University
Hangzhou 310027, China
zhubinaa@126.com

1

A Gauge-Invariant Theory of Intense-Field Coulomb Approximations to All Orders

Farhad H.M. Faisal

Abstract. We present a gauge invariant theory of intense-field Coulomb approximations to *all* orders. The present theory, overcomes a longstanding discrepancy between the strong-field velocity and the length gauge approximations, as well as accounts for the long-range Coulomb interaction in the final-state.

1.1 Introduction

Processes in intense laser fields are currently among the most vigorously pursued research problems in the domain of atomic, molecular and optical physics. The so-called strong-field KFR-approximations [1–4] have provided fruitful insights into a wide range of highly non-perturbative processes in intense laser fields (e.g. *Topical Review* [5]). However, the strong-field approximations in the "velocity" and "length" gauges differ and thus constitute two distinct approximations. In the last several decades many authors have discussed the problem of gauge invariance of the strong-field approximations, and various arguments have been advanced to justify the use of one or the other model in practice, but no derivation of the equivalence of the velocity and of the length gauge approximations has been achieved, until very recently [6, 7]. Another aspect that deserves attention in this context is the usual presence of the long-range Coulomb potential that can affect the electron motion in the field-free final states. In this report, explicit expressions of the intense-field transition amplitudes are derived, which are gauge invariant to *all* orders of approximation, and account for the presence of the long-range Coulomb interaction as well. The method of derivation of the present result is closely analogous to the one applied by us earlier [6, 7].

1.2 Quantum Mechanical Gauge Transformation

The Schrödinger equation of an electron interacting with an electromagnetic field, in the *minimal* coupling scheme, is given by

$$\left(i\hbar\frac{\partial}{\partial t} - eA_o\right)|\Psi(t)>= \frac{(\boldsymbol{p}_{\rm op} - \frac{e}{c}\boldsymbol{A})^2}{2m}|\Psi(t)>, \tag{1.1}$$

where the four-potential $(A_0, \boldsymbol{A})$ is defined by the scalar potential $A_0 \equiv A_0(\boldsymbol{r}, t)$ and the vector potential $\boldsymbol{A} \equiv \boldsymbol{A}(\boldsymbol{r}, t)$. The quantum mechanical gauge transformation consists in changing the potentials by a scalar function $\xi(\boldsymbol{r}, t)$, which leaves the electromagnetic field strengths unchanged, *and* simultaneously changing the wave-function by the gauge factor $\mathrm{e}^{-\mathrm{i}\frac{e}{\hbar c}\xi}$, where $\xi \equiv \xi(\boldsymbol{r}, t)$ is a scalar function of space and time:

$$\begin{aligned} A_0 \to \tilde{A}_0 &= A_0 + \frac{1}{c}\frac{\partial \xi}{\partial t}, \\ \boldsymbol{A} \to \tilde{\boldsymbol{A}} &= \boldsymbol{A} - \frac{\partial \xi}{\partial \boldsymbol{r}}, \\ |\Psi(t)>\to |\tilde{\Psi}(t)> &= \mathrm{e}^{-\mathrm{i}\frac{e}{\hbar c}\xi}|\Psi(t)>. \end{aligned} \tag{1.2}$$

Substituting (1.2) in (1.1) and performing the simple calculations, one obtains in the new gauge,

$$\left(i\hbar\frac{\partial}{\partial t} - e\tilde{A}_o\right)|\tilde{\Psi}(t)>= \frac{(\boldsymbol{p}_{\rm op} - \frac{e}{c}\tilde{\boldsymbol{A}})^2}{2m}|\tilde{\Psi}(t)>. \tag{1.3}$$

The essence of the gauge transformation in quantum mechanics is thus the invariance of the *form* of the Schrödinger equation with respect to the potentials in the old and in the new (tilde) gauges, as it is immediately apparent from the form of (1.1) and (1.3). This ensures the requirement of physical invariance of the transition probabilities (and the expectation values of Hermitian observables) in the old and the new gauges. Furthermore, the gauge invariance of the (transition) *probabilities* implies that the transition amplitudes must also be the same, or may differ at most by a *constant* phase factor, in the two gauges.

1.3 S-Matrix Series for Transition Amplitudes

Let the Schrödinger equation of the interacting system with a total Hamiltonian $H(t)$ be,

$$\left(i\hbar\frac{\partial}{\partial t} - H(t)\right)|\Psi(t)>= 0. \tag{1.4}$$

We define the Green's function (or propagator), $G(t, t_0)$, associated with $H(t)$, by the inhomogeneous equation

$$\left(i\hbar\frac{\partial}{\partial t} - H(t)\right) G(t, t_0) = \delta(t - t_0). \tag{1.5}$$

The solution of the Schrödinger equation (1.4) which satisfies the general initial condition,

$$\begin{aligned} |\Psi(t)> &= 0, \qquad t < t_0, \\ |\Psi(t)> &= |\phi_{\rm i}(t_0)>, \qquad t = t_0^{(+)}, \end{aligned} \tag{1.6}$$

where $|\phi_{\rm i}(t_0)>$ is the initial state of the system prepared at $t = t_0$, is given by

$$|\Psi(t)> = i\hbar G(t, t_0)|\phi_{\rm i}(t_0)> . \tag{1.7}$$

The solution (1.7) of (1.4) is readily verified by operating on (1.7) with $(i\hbar\frac{\partial}{\partial t} - H(t))$, using (1.5), and noting that due to the resulting delta function, (1.4) is fulfilled for all t not equal to t_0. For $t = t_0$, an integration in the immediate neighborhood of the singularity reproduces the initial condition (1.6).

The probability amplitude of a transition from a (non-interacting) initial state $|\phi_{\rm i}(t_0)>$, prepared at $t = t_0$, to a final state $|\Phi_{\rm f}(t_{\rm f})>$ at $t = t_{\rm f}$, is given by,

$$\begin{aligned} S_{\rm if} &= < \Phi_{\rm f}(t_{\rm f})|\Psi(t_{\rm f})> \\ &= i\hbar < \Phi_{\rm f}(t_{\rm f})|G(t_{\rm f}, t_0)|\phi_{\rm i}(t_0)>, \end{aligned} \tag{1.8}$$

where in the second line we have used (1.7). To obtain a systematic expansion of transition amplitudes of interest we use a convenient expression of the total propagator G of the system, in terms of the partial propagators, G_0 and G_i, associated with the partial Hamiltonians, H_0 and H_i, defined by the two partitions of the total Hamiltonian, $H(t) = H_0(t) + V_0(t)$ and $H(t) = H_i(t) + V_i(t)$, respectively. We define the partial propagators $G_0(t, t')$ and $G_i(t, t')$, associated respectively, with $H_0(t)$ and $H_i(t)$, by

$$\left(i\hbar\frac{\partial}{\partial t} - H_0(t)\right) G_0(t, t') = \delta(t - t'), \tag{1.9}$$

and,

$$\left(i\hbar\frac{\partial}{\partial t} - H_i(t)\right) G_i(t, t') = \delta(t - t'). \tag{1.10}$$

For any t between the initial-time t_0, and the final-time $t_{\rm f}$, we can obtain a convenient series expansion of $G(t, t_0)$ in terms of the partial propagator $G_0(t, t_0)$ and $G_i(t, t')$ as follows. First, we have the relation,

$$G(t, t_0) = G_0(t, t_0) + \int_{t_0}^{t_{\rm f}} {\rm d}t_1 G(t, t_1) V_0(t_1) G_0(t_1, t_0). \tag{1.11}$$

This relation can be easily verified by operating with $(\mathrm{i}\frac{\partial}{\partial t} - H_0(t))$ on (1.11) and using (1.5). Alternatively, we also have

$$G(t,t_0) = G_0(t,t_0) + \int_{t_0}^{t_\mathrm{f}} \mathrm{d}t_1 G_0(t,t_1)V_0(t_1)G(t_1,t_0). \tag{1.12}$$

This can be verified by operating it with $(\mathrm{i}\hbar\frac{\partial}{\partial t} - H_0(t))$, using (1.9) and simplifying. Then, substituting (1.12) in (1.11), we get,

$$\begin{aligned} G(t,t_0) = G_0(t,t_0) &+ \int_{t_0}^{t_\mathrm{f}} \mathrm{d}t_1 G_0(t,t_1)V_0(t_1)G_0(t_1,t_0) \\ &+ \int_{t_0}^{t_\mathrm{f}}\int_{t_0}^{t_\mathrm{f}} \mathrm{d}t_2\mathrm{d}t_1 G_0(t,t_2)V_0(t_2)G(t_2,t_1)V_0(t_1)G_0(t_1,t_0). \end{aligned} \tag{1.13}$$

Next, we expand G in terms of the intermediate propagator G_i, as

$$\begin{aligned} G(t_2,t_1) = G_i(t_2,t_1) &+ \int_{t_0}^{t_\mathrm{f}} \mathrm{d}t_3 G_i(t_2,t_3)V_i(t_3)G_i(t_3,t_1) \\ &+ \int_{t_0}^{t_\mathrm{f}}\int_{t_0}^{t_\mathrm{f}} \mathrm{d}t_4\mathrm{d}t_3 G_i(t_2,t_4)V_i(t_4)G_i(t_4,t_3)V_i(t_3)G_0(t_3,t_1) + \cdots . \end{aligned} \tag{1.14}$$

Finally, substituting (1.14) for $G(t_2,t_1)$ on the right-hand side of (1.13), inserting the resulting series expansion in (1.8), and relabelling the integration variables, we obtain the desired form of the S-matrix series (c.f. [5,8]) for the transition between the field-free initial state, $|\phi_\mathrm{i}(t_0)>$, and

$$S_\mathrm{if} = \sum_{n=0}^{\infty} S_\mathrm{if}^{(n)}, \tag{1.15}$$

with,

$$S_\mathrm{if}^{(0)} = i\hbar < \phi_\mathrm{f}(t_\mathrm{f})|G_0(t_\mathrm{f},t_0)|\phi_\mathrm{i}(t_0)> \tag{1.16}$$

$$S_\mathrm{if}^{(1)} = i\hbar \int \mathrm{d}t_1 < \phi_\mathrm{f}(t_\mathrm{f})|G_0(t_\mathrm{f},t_1)V_0(t_1)G_0(t_1,t_0)|\phi_\mathrm{i}(t_0)> \tag{1.17}$$

$$\begin{aligned} S_\mathrm{if}^{(2)} = i\hbar \int \mathrm{d}t_2\mathrm{d}t_1 &< \phi_\mathrm{f}(t_\mathrm{f})|G_0(t_\mathrm{f},t_2)V_0(t_2)G_i(t_2,t_1) \\ &\times V_0(t_1)G_0(t_1,t_0)|\phi_\mathrm{i}(t_0)> \end{aligned} \tag{1.18}$$

$$\cdots\cdots\cdots = \cdots\cdots\cdots\cdots$$

$$\begin{aligned} S_\mathrm{if}^{(n)} = i\hbar \int &\mathrm{d}t_n\mathrm{d}t_{n-1}\cdots\mathrm{d}t_2\mathrm{d}t_1 < \phi_\mathrm{f}(t_\mathrm{f})|G_0(t_\mathrm{f},t_n) \\ &\times V_0(t_n)G_i(t_n,t_{n-1})V_i(t_{n-1}) \times \cdots \\ &\cdots \times V_i(t_2)G_i(t_2,t_1)V_0(t_1)G_0(t_1,t_0)|\phi_\mathrm{i}(t_0)>, \end{aligned} \tag{1.19}$$

where $\int$ stands for the (multiple) integrations in the same range, t_0 to t_f.

For the sake of simplicity, in the rest of the paper we shall assume the usual dipole (or long-wavelength) approximation in which the vector potential, $\boldsymbol{A}(t)$, and the electric field strength, $\boldsymbol{E}(t) \equiv -\frac{1}{c}\dot{\boldsymbol{A}}(t)$, become functions of t only.

1.4 All Order Transition Amplitudes in the Velocity Gauge

The *minimal* coupling Schrödinger equation (1.1) in the dipole approximation defines the total Hamiltonian, $H(t)$, in the so-called velocity gauge,

$$H(t) = \frac{(\boldsymbol{p}_{\mathrm{op}} - \frac{e}{c}\boldsymbol{A}(t))^2}{2m} + V, \tag{1.20}$$

where we have set $eA_0 \equiv V$ for the binding potential. We rewrite the total Hamiltonian (1.20) by adding zero, i.e. by adding and subtracting $V_0(t) = \frac{e}{m}\dot{\boldsymbol{A}} \cdot \boldsymbol{r}$, and introduce the initial partition of $H(t)$ as $H(t) = H_0(t) + V_0(t)$, with,

$$H_0(t) = \left(\frac{(\boldsymbol{p}_{\mathrm{op}} - \frac{e}{c}\boldsymbol{A}(t))^2}{2m} + V - \frac{e}{c}\frac{\partial \boldsymbol{A}(t)}{\partial t} \cdot \boldsymbol{r}\right), \tag{1.21}$$

$$V_0(t) = \frac{e}{c}\frac{\partial \boldsymbol{A}(t)}{\partial t} \cdot \boldsymbol{r}. \tag{1.22}$$

Consider the solution of the Schrödinger equation,

$$i\hbar\frac{\partial}{\partial t}|\psi_j(t) >= H_0(t)|\psi_j(t) >, \tag{1.23}$$

where $H_0(t)$ is defined by (1.21). The exact solutions of interest are,

$$|\psi_j(t) >= \mathrm{e}^{\mathrm{i}\frac{e}{\hbar c}\boldsymbol{A}(t)\cdot\boldsymbol{r}}\mathrm{e}^{-\frac{i}{\hbar}H_{\mathrm{a}}t}|\phi_j >, \tag{1.24}$$

where $|\phi_j >$ form a complete set of orthonormal eigen-functions of the "atomic" Schrödinger equation

$$\begin{aligned} H_{\mathrm{a}}|\phi_j > &= \left(\frac{\boldsymbol{p}_{\mathrm{op}}^2}{2m} + V\right)|\phi_j > \\ &= E_j|\phi_j >, \end{aligned} \tag{1.25}$$

for all index j (discrete and continuous). The validity of our solutions (1.24), can be readily verified by direct substitution in (1.23), and simplifying the algebra using,

$$\begin{aligned} \left(\boldsymbol{p}_{\mathrm{op}} - \frac{e}{c}\boldsymbol{A}(t)\right)|\psi_j(t) > &= \left(\boldsymbol{p}_{\mathrm{op}} - \frac{e}{c}\boldsymbol{A}(t)\right)(\mathrm{e}^{\mathrm{i}\frac{e}{\hbar c}\boldsymbol{A}(t)\cdot\boldsymbol{r}})|\phi_j(t) > \\ &= (\mathrm{e}^{\mathrm{i}\frac{e}{\hbar c}\boldsymbol{A}(t)\cdot\boldsymbol{r}})(\boldsymbol{p}_{\mathrm{op}})|\phi_j(t) >, \end{aligned} \tag{1.26}$$

where $|\phi_j(t) >= \mathrm{e}^{-\frac{i}{\hbar}H_{\mathrm{a}}t}|\phi_j >= \mathrm{e}^{-\frac{i}{\hbar}E_jt}|\phi_j >$. Note that the $|\psi_j(t) >$ s' are associated *one-to-one* with the $|\phi_j >$ s', and thus they too satisfy the orthonormal condition,

$$< \psi_{j'}(t)|\psi_j(t) >= \delta_{j',j}, \tag{1.27}$$

and form a complete set,

$$\sum_j |\psi_j(t)><\psi_j(t)| = \mathbf{1}. \tag{1.28}$$

Thus, the propagator, $G_0(t,t_0)$, associated with $H_0(t)$, that satisfies an equation analogous to (1.5) with $H(t)$ replaced by $H_0(t)$ (similar equations hold for the other propagators defined below), can be easily written down,

$$G_0(t,t_0) = -\frac{i}{\hbar}\theta(t-t_0)\sum_j |\psi_j(t)><\psi_j(t_0)|, \tag{1.29}$$

where the symbol $\sum_j$ stands both for the sum over the discrete and the integration over the continuous indices.

Next, we introduce the intermediate-state partition, $H(t) = H_i(t) + V_i(t)$, where

$$H_i(t) = \frac{(\boldsymbol{p}_{\mathrm{op}} - \frac{e}{c}\boldsymbol{A}(t))^2}{2m}, \tag{1.30}$$

$$V_i(t) = V. \tag{1.31}$$

The intermediate reference propagator, $G_i(t,t')$, associated with the Hamiltonian (1.30) can be written in terms of the complete set of the well-known Volkov wave-functions in the velocity gauge (e.g. [9]),

$$|\Phi_{\boldsymbol{p}}(t)>= L^{-\frac{3}{2}}|\boldsymbol{p}> \exp\left(-\frac{i}{\hbar}\int^t \frac{(\boldsymbol{p}-\frac{e}{c}\boldsymbol{A}(t'))^2}{2m}\mathrm{d}t'\right), \tag{1.32}$$

where L^3 is the normalization volume, $lim.\ L \to \infty$; $|\boldsymbol{p}>$ is a plane-wave of momentum $\boldsymbol{p}$: $<\boldsymbol{r}|\boldsymbol{p}>= e^{\frac{i}{\hbar}\boldsymbol{p}\cdot\boldsymbol{r}}$. The Volkov states satisfy the orthogonality condition,

$$<\Phi_{\boldsymbol{p}'}(t)|\Phi_{\boldsymbol{p}}(t)>= \delta_{\boldsymbol{p}'\boldsymbol{p}}, \tag{1.33}$$

and, the completeness relation,

$$\sum_{\boldsymbol{p}} |\Phi_{\boldsymbol{p}}(t)><\Phi_{\boldsymbol{p}}(t)| = \mathbf{1}. \tag{1.34}$$

Thus, the Volkov propagator in the velocity gauge can be written as

$$G_i(t,t') = -\frac{i}{\hbar}\theta(t-t')\sum_{\boldsymbol{p}} |\Phi_{\boldsymbol{p}}(t)><\Phi_{\boldsymbol{p}}(t')|. \tag{1.35}$$

It should be noted that a transition process between the interaction-free initial state $|\phi_{\mathrm{i}}(t_0)>$, and the final state $|\phi_{\mathrm{f}}(t_{\mathrm{f}})>$, is well defined in any gauge, provided the *field-free* conditions are well satisfied initially and finally, e.g.

$$\boldsymbol{A}(t_0) \equiv \boldsymbol{A}(t_{\mathrm{f}}) = 0, \tag{1.36}$$

Incidentally, it is not difficult to define such pulses quite generally, e.g.

$$\boldsymbol{A}(t) = \boldsymbol{A}_0 \sin^{2n}\left\{\frac{\pi(t-t_0)}{(t_\mathrm{f}-t_0)}\right\}\cos\left(\omega(t-t_0)+\delta\right), \tag{1.37}$$

where $t_\mathrm{f} - t_0 \equiv \tau_\mathrm{p}$ is an arbitrary pulse duration, and δ is an arbitrary initial phase; $n \geq 1$. Note also that,

$$\begin{aligned} \boldsymbol{E}(t_0) &\equiv -\frac{1}{c}\dot{\boldsymbol{A}}(t_0) = 0, \\ \boldsymbol{E}(t_\mathrm{f}) &\equiv -\frac{1}{c}\dot{\boldsymbol{A}}(t_\mathrm{f}) = 0. \end{aligned} \tag{1.38}$$

To derive the intense-field transition amplitudes in the velocity gauge, we need only to substitute the explicit expressions of V_0, G_0, V_i, and G_i, given by (1.22), (1.29), (1.31) and (1.35), respectively, in the general S-matrix terms, (1.16)–(1.19), and simplify the algebra. To this end, first, we note the effect of operating with $G_0(t_1,t_0)$ on the initial state $|\phi_\mathrm{i}(t_0)> \equiv \mathrm{e}^{-\mathrm{i}\frac{i}{\hbar}E_i t_0}|\phi_\mathrm{i}>$:

$$\begin{aligned} i\hbar G_0(t_1,t_0)|\phi_\mathrm{i}(t_0)> &= \theta(t_1-t_0)\sum_j |\psi_j(t_1)><\psi_j(t_0)|\phi_\mathrm{i}(t_0)> \\ &= \theta(t_1-t_0)\sum_j |\psi_j(t_1)><\phi_j|\phi_\mathrm{i}> \\ &= \theta(t_1-t_0)\sum_j |\psi_j(t_1)>\delta_{i,j} \\ &= |\psi_i(t_1)>, \qquad t_1 > t_0, \end{aligned} \tag{1.39}$$

where we have used (1.24), noted that $<\phi_j|\phi_\mathrm{i}>=\delta_{ij}$, and that $|\psi_j(t_0)>= |\phi_j(t_0)>$, since, $\boldsymbol{A}(t_0)=0$ from (1.36). For the *field-free* final-state at $t=t_\mathrm{f}$, and for an asymptotically long-range Coulomb potential, $lim.r \to \infty, V = -\frac{Ze^2}{r}$, one can project on the ingoing ("minus") Coulomb wave [10], $|\phi_\mathrm{f}(t_\mathrm{f})> \equiv |\phi^{(-)}_{\boldsymbol{p}_\mathrm{f}}(t_\mathrm{f})>$, where

$$\begin{aligned} <\boldsymbol{r}|\phi^{(-)}_{\boldsymbol{p}_\mathrm{f}}(t)> &= N(\eta_\mathrm{f})\mathrm{e}^{\frac{i}{\hbar}\boldsymbol{p}_\mathrm{f}\cdot\boldsymbol{r}}\,{}_1F_1\Big(-\mathrm{i}\eta_\mathrm{f},1,-\frac{i}{\hbar}(p_\mathrm{f}r+\boldsymbol{p}_\mathrm{f}\cdot\boldsymbol{r})\Big)\,\mathrm{e}^{-\frac{i}{\hbar}\frac{p_f^2}{2m}t_\mathrm{f}}, \\ N(\eta_\mathrm{f}) &= \mathrm{e}^{\frac{\pi\eta_\mathrm{f}}{2}}\Gamma(1+\mathrm{i}\eta_\mathrm{f}), \\ \eta_\mathrm{f} &= \frac{Z\hbar}{p_\mathrm{f}a_0}, \end{aligned} \tag{1.40}$$

where ${}_1F_1(a,c;x)$ is a confluent hypergeometric function. Using (1.29) and (1.33), we deduce also,

$$\begin{aligned} <\phi^{(-)}_{\boldsymbol{p}_\mathrm{f}}(t_\mathrm{f})|G_0(t_\mathrm{f},t)| &= -\frac{i}{\hbar}\theta(t_\mathrm{f}-t)<\psi^{(-)}_{\boldsymbol{p}_\mathrm{f}}(t)| \\ &= -\frac{i}{\hbar}<\psi^{(-)}_{\boldsymbol{p}_\mathrm{f}}(t)|, \qquad t<t_\mathrm{f}, \end{aligned} \tag{1.41}$$

where

$$|\psi_{\boldsymbol{p}_\mathrm{f}}^{(-)}(t)> \equiv \mathrm{e}^{\mathrm{i}\frac{e}{\hbar c}\boldsymbol{A}(t)\cdot\boldsymbol{r}}|\phi_{\boldsymbol{p}_\mathrm{f}}^{(-)}(t)> \tag{1.42}$$

Thus, using (1.39) in the zeroth order amplitude (c.f. (1.16)) we find,

$$\begin{aligned}
S_\mathrm{if}^{(0)} &= i\hbar < \phi_{\boldsymbol{p}_\mathrm{f}}^{(-)}(t_\mathrm{f})|G_0(t_\mathrm{f},t_0)|\phi_\mathrm{i}(t_0)>, \\
&= < \phi_{\boldsymbol{p}_\mathrm{f}}^{(-)}(t_\mathrm{f})|\psi_i(t_\mathrm{f})>, \\
&= < \phi_{\boldsymbol{p}_\mathrm{f}}^{(-)}(t_\mathrm{f})|\phi_\mathrm{i}(t_\mathrm{f})> \\
&= \delta_{i,f},
\end{aligned} \tag{1.43}$$

since, from (1.36), $\boldsymbol{A}(t_\mathrm{f}) = 0$. In a similar way (using (1.39), (1.41), (1.22), and (1.36)) we get for the first order amplitude (c.f. (1.17)),

$$\begin{aligned}
S_\mathrm{if}^{(1)} &= \left(-\frac{i}{\hbar}\right)\int_{t_0}^{t_\mathrm{f}} \mathrm{d}t_1 < \psi_{\boldsymbol{p}_\mathrm{f}}^{(-)}(t_1)|V_0(t_1)|\psi_i(t_1)> \\
&= \left(-\frac{i}{\hbar}\right)\int_{t_0}^{t_\mathrm{f}} \mathrm{d}t_1 < \phi_{\boldsymbol{p}_\mathrm{f}}^{(-)}(t_1)|\left(\frac{e}{c}\dot{\boldsymbol{A}}(t_1)\cdot\boldsymbol{r}\right)|\phi_\mathrm{i}(t_1)>,
\end{aligned} \tag{1.44}$$

where in the last line we have used (1.24). Similarly, for the second order amplitude (c.f. (1.18)) we find,

$$\begin{aligned}
S_\mathrm{if}^{(2)} &= \left(-\frac{i}{\hbar}\right)\int_{t_0}^{t_\mathrm{f}}\int_{t_0}^{t_\mathrm{f}} \mathrm{d}t_2\mathrm{d}t_1 < \psi_{\boldsymbol{p}_\mathrm{f}}^{(-)}(t_2)|V_0(t_2)G_i(t_2,t_1)V_0(t_1)|\psi_i(t_1)> \\
&= \left(-\frac{i}{\hbar}\right)^2\int_{t_0}^{t_\mathrm{f}}\int_{t_0}^{t_\mathrm{f}} \mathrm{d}t_2\mathrm{d}t_1\theta(t_2-t_1)\sum_{\boldsymbol{p}_1} < \phi_{\boldsymbol{p}_\mathrm{f}}^{(-)}(t_2)| \\
&\quad \times \mathrm{e}^{-\mathrm{i}\frac{e}{\hbar c}\boldsymbol{A}(t_2)\cdot\boldsymbol{r}}V_0(t_2)|\Phi_{\boldsymbol{p}_1}(t_2)> \\
&\quad \times < \Phi_{\boldsymbol{p}_1}(t_1)|V_0(t_1)\mathrm{e}^{\mathrm{i}\frac{e}{\hbar c}\boldsymbol{A}(t_1)\cdot\boldsymbol{r}}|\phi_\mathrm{i}(t_1)> \\
&= \left(-\frac{i}{\hbar}\right)^2\int_{t_0}^{t_\mathrm{f}}\mathrm{d}t_2\int_{t_0}^{t_2} \mathrm{d}t_1\sum_{\boldsymbol{p}_1} < \phi_{\boldsymbol{p}_\mathrm{f}}^{(-)}(t_2)| \\
&\quad \times \mathrm{e}^{-\mathrm{i}\frac{e}{\hbar c}\boldsymbol{A}(t_2)\cdot\boldsymbol{r}}\left(\frac{e}{c}\dot{\boldsymbol{A}}(t_2)\cdot\boldsymbol{r}\right)|\Phi_{\boldsymbol{p}_1}(t_2)> \\
&\quad \times < \Phi_{\boldsymbol{p}_1}(t_1)|\left(\frac{e}{c}\dot{\boldsymbol{A}}(t_1)\cdot\boldsymbol{r}\right)\mathrm{e}^{\mathrm{i}\frac{e}{\hbar c}\boldsymbol{A}(t_1)\cdot\boldsymbol{r}}|\phi_\mathrm{i}(t_1)>,
\end{aligned} \tag{1.45}$$

where $\sum_{\boldsymbol{p}} \equiv L^3\int\frac{\mathrm{d}^3p}{(2\pi\hbar)^3}$.

Continuing in an analogous manner, we get the general result for the nth order amplitude in the velocity gauge (c.f. (1.19)),

$$\begin{aligned}
S_\mathrm{if}^{(n)} &= \left(-\frac{i}{\hbar}\right)\int \mathrm{d}_n\mathrm{d}_{n-1}\cdots\mathrm{d}t_2\mathrm{d}t_1 < \psi_{\boldsymbol{p}_\mathrm{f}}^{(-)}(t_n)| \\
&\quad \times \left(\frac{e}{c}\dot{\boldsymbol{A}}(t_n)\cdot\boldsymbol{r}\right)G_i(t_n,t_{n-1})V_i(t_{n-1})\cdots
\end{aligned}$$

$$
\begin{aligned}
&\times \cdots \times V_i(t_2)G_i(t_2,t_1)\left(\frac{e}{c}\dot{\boldsymbol{A}}(t_1)\cdot\boldsymbol{r}\right)|\psi_i(t_1)>,\\
&=\left(-\frac{i}{\hbar}\right)^n \sum_{\boldsymbol{p}_{n-1},\cdots\boldsymbol{p}_1}\int_{t_0}^{t_f}\mathrm{d}t_n\int_{t_0}^{t_n}\mathrm{d}t_{n-1}\cdots\int_{t_0}^{t_3}\mathrm{d}t_2\int_{t_0}^{t_2}\mathrm{d}t_1\\
&\times<\phi_{\boldsymbol{p}_f}^{(-)}(t_n)|\mathrm{e}^{-\mathrm{i}\frac{e}{\hbar c}\boldsymbol{A}(t_n)\cdot\boldsymbol{r}}\left(\frac{e}{c}\dot{\boldsymbol{A}}(t_n)\cdot\boldsymbol{r}\right)|\\
&\times\Phi_{\boldsymbol{p}_{n-1}}(t_n)><\Phi_{\boldsymbol{p}_{n-1}}(t_{n-1})|V|\Phi_{\boldsymbol{p}_{n-2}}(t_{n-1})>\\
&\times<\Phi_{\boldsymbol{p}_{n-2}}(t_{n-2})|V|\Phi_{\boldsymbol{p}_{n-3}}(t_{n-2})>V\cdots V|\Phi_{\boldsymbol{p}_1}(t_2)>\\
&\times<\Phi_{\boldsymbol{p}_1}(t_1)|\left(\frac{e}{c}\dot{\boldsymbol{A}}(t_1)\cdot\boldsymbol{r}\right)\mathrm{e}^{\mathrm{i}\frac{e}{c}\dot{\boldsymbol{A}}(t_1)\cdot\boldsymbol{r}}|\phi_i(t_1)>.
\end{aligned}
\tag{1.46}
$$

1.5 All Order Transition Amplitudes in Length Gauge

Using the gauge function in the dipole approximation, $\xi(\boldsymbol{r},t)=\boldsymbol{A}(t)\cdot\boldsymbol{r}$, and the gauge transformation equations (1.2), one immediately calculates,

$$
\begin{aligned}
\tilde{A}_0 &= A_0+\frac{1}{c}\dot{\boldsymbol{A}}(t)\cdot\boldsymbol{r},\\
\tilde{\boldsymbol{A}} &= \boldsymbol{A}-\boldsymbol{A},\\
&= 0.
\end{aligned}
\tag{1.47}
$$

Thus, from (1.3), we get the total Hamiltonian in the new (the so-called length) gauge,

$$
\tilde{H}(t)=\frac{\boldsymbol{p}_{\mathrm{op}}^2}{2m}+V+\frac{e}{c}\dot{\boldsymbol{A}}(t)\cdot\boldsymbol{r},
\tag{1.48}
$$

where $eA_0\equiv V$. We now partition $\tilde{H}(t)$ as $\tilde{H}(t)=\tilde{H}_0+\tilde{V}_0(t)$, where

$$
\tilde{H}_0\equiv H_{\mathrm{a}}=\frac{\boldsymbol{p}_{\mathrm{op}}^2}{2m}+V,
\tag{1.49}
$$

$$
\begin{aligned}
\tilde{V}_0(t) &= \frac{e}{c}\dot{\boldsymbol{A}}(t)\cdot\boldsymbol{r},\\
&= -e\boldsymbol{E}(t)\cdot\boldsymbol{r},
\end{aligned}
\tag{1.50}
$$

since, by definition $\boldsymbol{E}(t)=-\frac{1}{c}\dot{\boldsymbol{A}}(t)$. The eigen-functions and eigen-energies of $\tilde{H}_0$ are given by (1.25). The propagator associated with $\tilde{H}_0(t)$, therefore, can be easily written as,

$$
\begin{aligned}
\tilde{G}_0(t,t_0) &= -\frac{i}{\hbar}\theta(t-t_0)\sum_j|\tilde{\phi}_j(t)><\tilde{\phi}_j(t_0)|\\
&= -\frac{i}{\hbar}\theta(t-t_0)\sum_j|\phi_j>\mathrm{e}^{-\frac{i}{\hbar}H_{\mathrm{a}}(t-t_0)}<\phi_j|,\\
&= G_{\mathrm{a}}(t,t_0)
\end{aligned}
\tag{1.51}
$$

We note here, in particular, the field-free initial state,

$$|\tilde{\phi}_i(t_0)>\equiv|\phi_{\rm i}(t_0)> = {\rm e}^{-\frac{i}{\hbar}E_i t_0}|\phi_{\rm i}>, \tag{1.52}$$

and the field-free final state,

$$|\tilde{\phi}^{(-)}_{\boldsymbol{p}_{\rm f}}(t_{\rm f})>\equiv|\phi^{(-)}_{\boldsymbol{p}_{\rm f}}(t_{\rm f})> = {\rm e}^{-\frac{i}{\hbar}\frac{p_f^2}{2m}t_{\rm f}}|\phi^{(-)}_{\boldsymbol{p}_{\rm f}}> . \tag{1.53}$$

Next, we introduce the intermediate-state partition, $\tilde{H}(t) = \tilde{H}_i(t) + \tilde{V}_i(t)$, where

$$\tilde{H}_i(t) = \frac{\boldsymbol{p}^2_{\rm op}}{2m} + \frac{e}{c}\dot{\boldsymbol{A}}(t)\cdot\boldsymbol{r}, \tag{1.54}$$

$$\tilde{V}_i(t) = V. \tag{1.55}$$

The well-known Volkov states in the length gauge (e.g. [9]), associated with the intermediate Volkov Hamiltonian in length gauge, $\tilde{H}_i(t)$, are

$$\begin{aligned}|\tilde{\Phi}_{\boldsymbol{p}}(t)> &= L^{-\frac{3}{2}}|\left(\boldsymbol{p}-\frac{e}{c}\boldsymbol{A}(t)\right)> \exp\left(-\frac{i}{\hbar}\int^t\frac{(\boldsymbol{p}-\frac{e}{c}\boldsymbol{A}(t'))^2}{2m}{\rm d}t'\right)\\ &= {\rm e}^{-{\rm i}\frac{e}{\hbar c}\boldsymbol{A}(t)\cdot\boldsymbol{r}}|\Phi_{\boldsymbol{p}}(t)>,\end{aligned} \tag{1.56}$$

where $|\Phi_{\boldsymbol{p}}(t)>$ is the Volkov solution in the velocity gauge, (1.32). Therefore, the Volkov propagator in the length gauge, $\tilde{G}_i(t,t')$, can be written in terms of the Volkov propagator in the velocity gauge, $G_i(t,t')$, (1.35), as:

$$\begin{aligned}\tilde{G}_i(t,t') &= -\frac{i}{\hbar}\theta(t-t')|\tilde{\Phi}_{\boldsymbol{p}}(t)><\tilde{\Phi}_{\boldsymbol{p}}(t')|,\\ &= {\rm e}^{-{\rm i}\frac{e}{\hbar c}\boldsymbol{A}(t)\cdot\boldsymbol{r}}G_i(t,t'){\rm e}^{+i\frac{e}{\hbar c}\boldsymbol{A}(t')\cdot\boldsymbol{r}'}.\end{aligned} \tag{1.57}$$

To obtain the transition amplitude in the length gauge, we need simply to substitute the tilde-quantities (length gauge), $\tilde{G}_0$, $\tilde{V}_0$, $\tilde{G}_i$ and $\tilde{V}_i$, defined above, in place of the respective non-tilde quantities, in the formal expressions for the amplitudes, (1.16)–(1.19), and simplify the algebra. To this end, first, we consider the effect of operating with $\tilde{G}_0(t_1,t_0)$ on the initial state $|\tilde{\phi}_i(t_0)>$. using (1.51), and (1.52), we calculate,

$$\begin{aligned}i\hbar\tilde{G}_0(t_1,t_0)|\tilde{\phi}_i(t_0) &= i\hbar G_0(t_1,t_0)|\phi_i(t_0)\\ &= \theta(t_1-t_0)\sum_j|\phi_j(t_1)><\phi_j(t_0)|\phi_{\rm i}(t_0)>\\ &= \theta(t_1-t_0)\sum_j|\phi_j(t_1)>\delta_{ij}\\ &= |\phi_{\rm i}(t_1)>, \qquad t_1>t_0.\end{aligned} \tag{1.58}$$

Similarly, using (1.51) and (1.53), we get,

$$
\begin{aligned}
< \tilde{\phi}^{(-)}_{\boldsymbol{p}_\mathrm{f}}(t_\mathrm{f})|\tilde{G}_0(t_\mathrm{f},t) &= < \tilde{\phi}^{(-)}_{\boldsymbol{p}_\mathrm{f}}(t_\mathrm{f})|G_0(t_\mathrm{f},t),\\
&= -\frac{\mathrm{i}}{\hbar}\theta(t_\mathrm{f}-t) < \phi^{(-)}_{\boldsymbol{p}_\mathrm{f}}(t)|,\\
&= -\frac{i}{\hbar} < \tilde{\phi}^{(-)}_{\boldsymbol{p}_\mathrm{f}}(t)|, \qquad t<t_\mathrm{f}, \qquad (1.59)
\end{aligned}
$$

since from (1.36), $\boldsymbol{A}(t_\mathrm{f})=0$. Next, using the equal-time orthonormality of the length gauge Volkov states, (1.56), and the Volkov propagator, we calculate for the intermediate projections (1.57), we get,

$$
\begin{aligned}
< \tilde{\Phi}_{\boldsymbol{p}}(t)|\tilde{G}_i(t,t')| &= -\frac{i}{\hbar}\theta(t-t') < \tilde{\Phi}_{\boldsymbol{p}}(t')|\\
&= -\frac{i}{\hbar} < \Phi_{\boldsymbol{p}}(t')|(\mathrm{e}^{+i\frac{e}{\hbar c}\boldsymbol{A}(t')\cdot\boldsymbol{r}'}), \qquad t'<t<t_\mathrm{f}. \quad (1.60)
\end{aligned}
$$

To find the zeroth order amplitude (1.16), we use (1.58) and (1.36), and get,

$$
\begin{aligned}
\tilde{S}^{(0)}_\mathrm{if} &= i\hbar < \tilde{\phi}^{(-)}_{\boldsymbol{p}_\mathrm{f}}(t_\mathrm{f})|\tilde{G}_0(t_\mathrm{f},t_0)|\tilde{\phi}_i(t_0) >\\
&= < \phi^{(-)}_{\boldsymbol{p}_\mathrm{f}}(t_\mathrm{f})|\mathrm{e}^{\mathrm{i}\frac{e}{\hbar c}\boldsymbol{A}(t_\mathrm{f})\cdot\boldsymbol{r}}|\phi_\mathrm{i}(t_\mathrm{f}) >\\
&= < \phi^{(-)}_{\boldsymbol{p}_\mathrm{f}}(t_\mathrm{f})|\phi_\mathrm{i}(t_\mathrm{f}) >\\
&= \delta_{f,i}\\
&= S^{(0)}_\mathrm{if}, \qquad (1.61)
\end{aligned}
$$

where we have used $\boldsymbol{A}(t_\mathrm{f})=0$ from (1.36); the last line follows from a comparison with the velocity gauge result, (1.43). Next, we may recall ((1.22), and (1.50)) that, $\tilde{V}_0(t)=V_0(t)=\frac{e}{c}\boldsymbol{A}(t)\cdot\boldsymbol{r}$, and ((1.31) and (1.55)) that, $\tilde{V}_i(t)=V_i(t)=V$. Then, using (1.59) and (1.58) in the definition of the first order amplitude (1.17), we get,

$$
\begin{aligned}
\tilde{S}^{(1)}_\mathrm{if} &= \int_{t_0}^{t_\mathrm{f}} \mathrm{d}t_1 < \tilde{\phi}^{(-)}_{\boldsymbol{p}_\mathrm{f}}(t_1)\tilde{V}_0(t_1)\tilde{G}_0(t_1,t_0)|\tilde{\phi}_i(t_0) >\\
&= \left(-\frac{i}{\hbar}\right)\int_{t_0}^{t_\mathrm{f}} \mathrm{d}t_1 < \phi^{(-)}_{\boldsymbol{p}_\mathrm{f}}(t_1)|\left(\frac{e}{c}\dot{\boldsymbol{A}}(t_1)\cdot\boldsymbol{r}\right)|\phi_\mathrm{i}(t_1) >\\
&= S^{(1)}_\mathrm{if}, \qquad (1.62)
\end{aligned}
$$

where the last line follows from a comparison with (1.44). Continuing in the same way, using (1.58) and (1.59), for the second order amplitude (1.18) we get,

$$
\begin{aligned}
\tilde{S}^{(2)}_\mathrm{if} &= \left(-\frac{i}{\hbar}\right)\int \mathrm{d}t_2\mathrm{d}t_1 < \tilde{\phi}^{(-)}_{\boldsymbol{p}_\mathrm{f}}(t_2)|\tilde{V}_0(t_2)\tilde{G}_i(t_2,t_1)\tilde{V}_0(t_1)|\tilde{\phi}_i(t_1) >\\
&= \left(-\frac{i}{\hbar}\right)^2\int_{t_0}^{t_\mathrm{f}}\int_{t_0}^{t_\mathrm{f}} \mathrm{d}t_2\mathrm{d}t_1\theta(t_2-t_1)\sum_{\boldsymbol{p}_1} < \phi^{(-)}_{\boldsymbol{p}_\mathrm{f}}(t_2)|V_0(t_2)|
\end{aligned}
$$

$$\times \, \mathrm{e}^{-\mathrm{i}\frac{e}{\hbar c}\boldsymbol{A}(t_2)\cdot\boldsymbol{r}} \Phi_{\boldsymbol{p}_1}(t_2) >$$
$$\times < \Phi_{\boldsymbol{p}_1}(t_1)|\mathrm{e}^{\mathrm{i}\frac{e}{\hbar c}\boldsymbol{A}(t_1)\cdot\boldsymbol{r}} V_0(t_1)|\phi_{\mathrm{i}}(t_1) >$$
$$= \left(-\frac{i}{\hbar}\right)^2 \int_{t_0}^{t_\mathrm{f}} \mathrm{d}t_2 \int_{t_0}^{t_2} \mathrm{d}t_1 \sum_{\boldsymbol{p}_1} < \phi^{(-)}_{\boldsymbol{p}_\mathrm{f}}(t_2)$$
$$\times \,|\left(\frac{e}{c}\dot{\boldsymbol{A}}(t_2)\cdot\boldsymbol{r}\right) \mathrm{e}^{-\mathrm{i}\frac{e}{\hbar c}\boldsymbol{A}(t_2)\cdot\boldsymbol{r}} |\Phi_{\boldsymbol{p}_1}(t_2) >$$
$$\times < \Phi_{\boldsymbol{p}_1}(t_1)|\mathrm{e}^{\mathrm{i}\frac{e}{\hbar c}\boldsymbol{A}(t_1)\cdot\boldsymbol{r}} \left(\frac{e}{c}\dot{\boldsymbol{A}}(t_1)\cdot\boldsymbol{r}\right)|\phi_{\mathrm{i}}(t_1) >$$
$$= S^{(2)}_{\mathrm{if}}, \tag{1.63}$$

where the last line follows from a comparison with (1.45). Using (1.58), (1.59) and (1.60) in the nth order amplitude (1.19), we get, first,

$$\tilde{S}^{(n)}_{\mathrm{if}} = \left(-\frac{i}{\hbar}\right) \int \mathrm{d}t_n \mathrm{d}t_{n-1} \cdots \mathrm{d}t_2 \mathrm{d}t_1 < \tilde{\phi}_{\boldsymbol{p}_\mathrm{f}}(t_n)|\tilde{V}_0(t_n)\tilde{G}_i(t_n, t_{n-1})\tilde{V}_i(t_{n-1}) \cdots$$
$$\times \, \tilde{G}_i(t_2, t_1)\tilde{V}_0(t_1)|\tilde{\phi}_i(t_1) > . \tag{1.64}$$

This expression can be further reduced using successively the identity,

$$< \tilde{\Phi}_{p_j}(t_j)|\tilde{V}_i(t_j)\tilde{G}_i(t_j, t_{j-1}) = -\frac{i}{\hbar}\theta(t_j - t_{j-1}) \sum_{\boldsymbol{p}_{j-1}} < \Phi_{\boldsymbol{p}_j}(t_j)|V|\Phi_{\boldsymbol{p}_{j-1}}(t_j) >$$
$$\times < \Phi_{\boldsymbol{p}_{j-1}}(t_{j-1})|(\mathrm{e}^{\mathrm{i}\frac{e}{\hbar c}\boldsymbol{A}(t_{j-1})\cdot\boldsymbol{r}}), \tag{1.65}$$

where $j = 2, 3, \cdots n$, and $\boldsymbol{p}_n \equiv \boldsymbol{p}_\mathrm{f}$, which is readily established from (1.56) and (1.57). Thus, recalling that $\tilde{V}_i(t) = V$, and $\tilde{V}_0(t) = (\frac{e}{c}\dot{\boldsymbol{A}}(t)\cdot\boldsymbol{r})$, noting the presence of the theta-functions (associated with the propagators) which determine the integration intervals, and applying the identity, (1.65), to the nth order amplitude in the length gauge, (1.64), we finally deduce,

$$\tilde{S}^{(n)}_{\mathrm{if}} = \left(-\frac{i}{\hbar}\right)^n \sum_{\boldsymbol{p}_{n-1}\cdots\boldsymbol{p}_1} \int_{t_0}^{t_\mathrm{f}} \mathrm{d}t_n \int_{t_0}^{t_n} \mathrm{d}t_{n-1} \cdots \int_{t_0}^{t_3} \mathrm{d}t_2 \int_{t_0}^{t_2} \mathrm{d}t_1$$
$$\times < \phi_{\boldsymbol{p}_\mathrm{f}}(t_n)|\mathrm{e}^{-\mathrm{i}\frac{e}{\hbar c}\boldsymbol{A}(t_n)\cdot\boldsymbol{r}} \left(\frac{e}{c}\dot{\boldsymbol{A}}(t_n)\cdot\boldsymbol{r}\right)|\Phi_{\boldsymbol{p}_{n-1}}(t_n) >$$
$$\times < \Phi_{\boldsymbol{p}_{n-1}}(t_{n-1})|V|\Phi_{n-2}(t_{n-1}) > \cdots\cdots < \Phi_{\boldsymbol{p}_2}(t_2)|V|\Phi_{\boldsymbol{p}_1}(t_2) >$$
$$\times < \Phi_{\boldsymbol{p}_1}(t_1)|\mathrm{e}^{\mathrm{i}\frac{e}{\hbar c}\boldsymbol{A}(t_1)\cdot\boldsymbol{r}} \left(\frac{e}{c}\dot{\boldsymbol{A}}(t_1)\cdot\boldsymbol{r}\right)|\phi_{\mathrm{i}}(t_1) >$$
$$= S^{(n)}_{\mathrm{if}}. \tag{1.66}$$

The last equality, which follows from a comparison with (1.46), establishes the desired gauge invariance of the intense-field approximations, in the velocity and length gauges, in *all* orders, exactly.

1.5.1 Simultaneous Coulomb Cum Field Effect

It should be noted that often the simultaneous effect of Coulomb-potential and the laser-field becomes rather dominant, e.g. in the adiabatic and/or tunneling regime which we define by the two conditions:

$$\begin{aligned} \hbar\omega &< eF_0 a_\mathrm{B} < E_\mathrm{B} \\ \gamma &= p_\mathrm{B}\omega / eF_0 < 1, \end{aligned} \tag{1.67}$$

where F_0 is the peak field strength, $\frac{p_\mathrm{B}^2}{2m} = E_\mathrm{B}$ is the binding energy of the electron equal to the ionization potential I_p, and $a_\mathrm{B} = \frac{\hbar}{p_\mathrm{B}}$ is the characteristic dimension of the initial state of the electron.

The present expansion, if carried out to all orders can, of course, account for it exactly. However, such summations are not usually practicable. Therefore, it is useful to account for the same, even approximately, in the leading orders of the expansion. In fact, the general scheme of the intense-field S-matrix theory [5] (as also in the present case) permits the use of alternative intermediate Green's functions, in place of the plane-wave Volkov Green's function, $\tilde{G}_i(t,t')$, employed above. Thus, we may conveniently use an approximate Coulomb–Volkov (CV) Green's function in length gauge, $\tilde{G}_i^{\mathrm{CV}}(t,t')$, that is based on the rudimentary Coulomb–Volkov wave function that was used fruitfully earlier to interpret the observed momentum distributions of the intense-field ionization of atoms [11, 12]. The approximate CV wave function of interest is given by [11, 12],

$$\tilde{\Phi}_{\boldsymbol{p}}^{(-)}(\boldsymbol{r},t) = \frac{1}{L^{3/2}} \mathrm{e}^{-\frac{i}{\hbar 2m}\int^t p_{t'}^2 \mathrm{d}t'} \mathrm{e}^{\frac{i}{\hbar}\boldsymbol{p}_t\cdot\boldsymbol{r} + i\eta_t \ln(2(p_t/\hbar)r)}, \tag{1.68}$$

where $\boldsymbol{p}_t \equiv \boldsymbol{p} - \frac{e}{c}\boldsymbol{A}(t)$ and $\eta_t \equiv \frac{Z\hbar}{p_t a_0}$. It takes account of the Coulomb phase distortion in the classically forbidden domain, $k_\mathrm{B} r_0 >> k_\mathrm{B} r >> 1$, $\hat{\boldsymbol{r}} \parallel \hat{\boldsymbol{p}}$, when $\hbar\omega/E_\mathrm{B} << 1$, $F/F_\mathrm{B} << 1$, where $k_\mathrm{B} = p_\mathrm{B}/\hbar$, and $E_\mathrm{B} = \frac{p_\mathrm{B}^2}{2m}$ is the binding energy of the active electron, and F_B is the corresponding binding field-strength; r_0 is of the order of the radius of the outer turning point of the classically forbidden region. The accuracy of this approximate Coulomb–Volkov wavefunction may be estimated by direct substitution into the Schrödinger equation, with the laser interaction in the length gauge + the Coulomb potential, and comparing the smallness of the rest-terms in the Hamiltonian with respect to the initial total energy E_B. Using (1.68), we may write the approximate CV Green's function in length gauge as,

$$\begin{aligned} \tilde{G}_i^{\mathrm{CV}}(\boldsymbol{r},t;\boldsymbol{r}',t') = &-\frac{i}{\hbar}\theta(t-t')\sum_{\boldsymbol{p}} \mathrm{e}^{-\frac{i}{\hbar 2m}\int_{t'}^{t} p_\tau^2 \mathrm{d}\tau} \mathrm{e}^{\frac{i}{\hbar}(\boldsymbol{p}_t\cdot\boldsymbol{r} - \boldsymbol{p}_{t'}\cdot\boldsymbol{r}')} \\ &\times \mathrm{e}^{i\eta_t \ln(2(p_t/\hbar)r) - i\eta_{t'} \ln(2(p_{t'}/\hbar)r')} \end{aligned} \tag{1.69}$$

Note, in passing, that (1.69) reduces to the exact plane-wave Volkov case in the same gauge in the absence of the Coulomb potential ($Z = 0$). Equation (1.69) can be used conveniently in the adiabatic and tunnel regime in place of the plane-wave Volkov Green's function, $\tilde{G}_i(t,t')$, e.g. starting with the second order amplitude, (1.63); in the same regime a stationary phase approximation is often applicable which is most suitably carried out in the complex t-plane near the vicinity of the (complex) poles of the integrand (cf. [11–13]).

1.6 Brief Comments and Summary

Before concluding this report, we may observe the following:

(a) It is clear from (1.43), that the present formulation correctly reproduces the desirable orthogonality between the initial and the final field-free states of the system, at the 0th order.

(b) The first order amplitude, (c.f. 1.44) only contributes to a single-photon transition process, e.g. for a high enough incident photon energy, equal to, or greater than the threshold energy of the process. For sub-threshold incident photon energies, the leading term of the series is given by the second order term, (1.44), or (1.62).

(c) For a *long* (or adiabatic) laser pulse, it is often convenient to Fourier transform the *periodic* part of the integrand with respect to t_n, in any order $n = 1, 2, 3, \cdots$, and to carry out the final $\mathrm{d}t_n$-integration *analytically* between the limits $t_0 \to -\infty$ and $t_\mathrm{f} \to \infty$, to obtain:

$$S_{\mathrm{if}}^{(n)} = -\frac{2\pi i}{L^{\frac{3}{2}}} \sum_s T_s^{(n)}(\boldsymbol{p}_\mathrm{f}) \times \delta\left(\frac{p_f^2}{2m} + |E_i| + U_\mathrm{p} - s\hbar\omega\right), \quad s = 0, \pm 1, \pm 2, \pm 3, \cdots, \tag{1.70}$$

where $T_s^{(n)}(\boldsymbol{p}_\mathrm{f})$ is the sth Fourier component of the periodic part of the nth order amplitude. This allows one immediately to determine the bound-free transition probability per unit time, or the fundamental *rate* of the process, from the (generalized Fermi golden) rule:

$$d\Gamma(\boldsymbol{p}_\mathrm{f}) = \frac{2\pi}{\hbar} \sum_{s \geq s_0} \left| \sum_n T_s^{(n)}(\boldsymbol{p}_\mathrm{f}) \right|^2 \times \delta\left(\frac{p_f^2}{2m} + |E_i| + U_\mathrm{p} - s\hbar\omega\right) \frac{d^3 p_\mathrm{f}}{(2\pi\hbar)^3}, \tag{1.71}$$

where $s_0 \equiv \left[\frac{(p_f^2/2m + |E_i| + U_\mathrm{p})}{\hbar\omega}\right]_{\mathrm{int.}} + 1$, $U_\mathrm{p} \equiv \frac{e^2 F_0^2}{4m\omega^2}$ is the so-called ponderomotive energy, ω is the carrier frequency, and F_0 is the peak field strength.

(d) For an *ultrashort* laser pulse, when a steady rate of the transition ($i \to f, i \neq f$) might be absent, one can simply take the absolute square of the sum of the time-dependent amplitudes, (1.44)–(1.46), to obtain the transition probability of interest, directly.

(e) Finally, we note that for more complex *many-body* atomic and molecular systems as well as for other transition processes (e.g. for molecular high-harmonic generation signals), explicitly gauge invariant expressions can be obtained by employing (*mutatis mutandis*) the same method as used above.

To summarize: a gauge invariant theory of Coulomb-modified intense-field approximations is presented which explicitly demonstrates the equivalence of the velocity- and the length-gauge transition probabilities in *all* orders. The present theory thus not only overcomes a longstanding discrepancy between the strong-field approximations in the two gauges (that is resolved only recently [6, 7]), but also takes exact account of the long-range Coulomb interaction for ionization.

Acknowledgments

This work was partially supported by the National Science Foundation through a grant for the Institute for Theoretical Atomic, Molecular and Optical Physics at Harvard University and Smithsonian Astrophysical Observatory.

References

1. L.V. Keldysh, Sov. Phys. JETP **20**, 1307 (1965)
2. F.H.M. Faisal, J. Phys. B **6**, L89 (1973)
3. H.R. Reiss, Phys. Rev. A **22**, 1786 (1980)
4. F.H.M. Faisal, Nuovo Cimento B **33**, 775 (1976)
5. A. Becker, F.H.M. Faisal, J. Phys. B **38**, R1 (2005)
6. F.H.M. Faisal, J. Phys. B **40**, F145 (2007)
7. F.H.M. Faisal, Phys. Rev. A **75**, 063412 (2007)
8. F.H.M. Faisal, A. Becker, in *Selected Topics in Electron Physics*, ed. by D.M. Campbell, H. Kleinpoppen (Plenum, New York, 1996), p. 397
9. F.H.M. Faisal, *Theory of Multiphoton Processes* (Plenum, New York, 1987), p. 1.7
10. L.D. Landau, *Quantum Mechanics*, 3rd edn. (Pergamon, New York, 1981)
11. F.H.M. Faisal, G. Schlegel, J. Phys. B **38**, L223 (2005)
12. F.H.M. Faisal, G. Schlegel, J. Mod. Opt. **53**, 207 (2006)
13. G.F. Gribakin, M.Yu. Kuchiev, Phys. Rev. A **55**, 3760 (1997)

2

Strong-Field Photoionization by Few-Cycle Laser Pulses

Gerhard G. Paulus

Abstract. The electric field of few-cycle laser pulses can be precisely controlled by controlling their "absolute" phase. Intense phase-stabilized few-cycle laser pulses give rise to novel effects in strong-field and above-threshold ionization (ATI). The phase dependence of the photoelectron spectra is a sensitive probe of the dynamics of the strong-field ionization processes underlying attosecond laser physics. Conversely, the phase-dependence of ATI spectra can be used for the measurement of the absolute phase. This, a summary of the experimental status of phase-dependent strong-field ionization, and a description of unsolved problems are the main points of this contribution.

2.1 Introduction

Attosecond XUV- and electron pulses can be generated by interaction of intense ultra-short laser pulses with atoms and also with molecules. This is remarkable as the optical period of the commonly used lasers is 2.5 fs long which means that the processes leading to the attosecond pulses not only must evolve within an optical cycle, but also be precisely synchronized to the driving optical field. The key for understanding attosecond laser physics is understanding strong-field photoionization. It has been shown in the last decade of the twentieth century that all of the characteristic effects observed in strong-field laser–atom interaction – namely the plateaus in high-harmonic generation (HHG) [1] and above-threshold ionization (ATI) [2] as well as the knee in non-sequential double-ionization [3] – are due to a class of trajectories of photo-ionizing electrons that are driven back to the vicinity of the core by the oscillating laser field [4–6], for reviews see [7–9]. Depending on the instant of ionization and return, these electrons may hit the ion core with attosecond timing precision and with energies up to $3.17\,U_\mathrm{P}$, where U_P is the ponderomotive energy [4]. Even at the nowadays fairly modest intensity of $10^{14}\,\mathrm{W\,cm^{-2}}$, the ponderomotive energy is 6 eV (again assuming 800 nm wavelength). Therefore, high-harmonics, in fact XUV attosecond pulses, can be generated in the

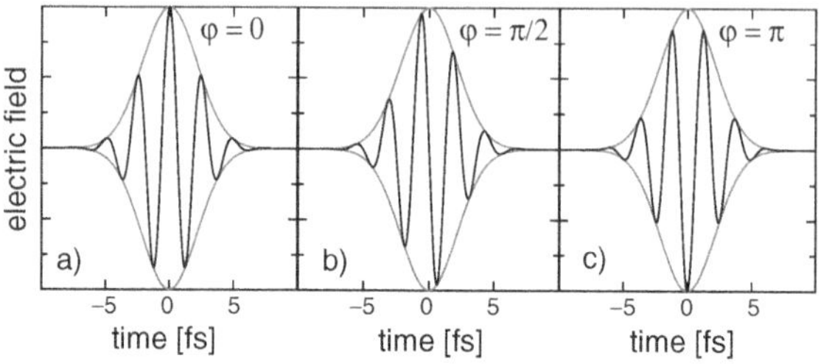

Fig. 2.1. 5-fs laser pulses consist of less than two optical cycles (few-cycle pulses). The temporal evolution of the electric field of such pulses depends on the phase of the carrier with respect to the envelope, the so-called carrier-envelope (CE) phase or "absolute" phase. By convention, the origin of the time scale is chosen to be at the maximum of the envelope of the pulses. Then an absolute phase $\varphi = 0$ corresponds to cosine-like pulses and $\varphi = -\pi/2$ to sine-like pulses. [20]

case the returning electron makes a transition to the ground state. Impact ionization leads to non-sequential double-ionization [5], and elastic scattering to the ATI plateau [6,10]. Although numerous experiments have revealed some serious deficits of this essentially classical picture of strong-field laser–atom interaction, the model has proven to be a very useful guiding principle. This also holds with respect to synchronization of the attosecond pulses with the driving field.

The very fact that the dynamics of strong-field ionization and attosecond pulse generation proceed within and are phase-locked to single optical cycles suggests that it is highly desirable to use as few optical cycles for the driving field as possible [11,12]. One immediate benefit are isolated attosecond pulses whereas longer pulses lead to trains of, in general, not identical attosecond pulses. If the duration (FWHM) of the driving pulses can be reduced to less than three optical cycles (few-cycle pulses), it is in addition possible to tailor the optical cycles by controlling the phase ("absolute" phase) of the carrier oscillation with respect to the maximum of the pulse envelope. This facilitates direct control of the electron trajectories responsible for attosecond pulse generation and works the better the shorter the driving pulse. The absolute phase φ obviously plays a key role in attosecond laser physics. For an unambiguous definition, the field $\mathcal{E}(t)$ of the laser pulse is written as a product of envelope $\mathcal{E}_0(t)$ and carrier wave:

$$\mathcal{E}(t) = \mathcal{E}_0(t) \cdot \cos(\omega t + \varphi). \tag{2.1}$$

This definition also explains the frequently used jargon of sine- and cosine-like pulses which are just special cases of few-cycle laser pulses.

The pivotal importance of the absolute phase for attosecond laser physics and other modern branches of laser physics explains the many proposals and attempts to devise a scheme for its measurement [13–18]. In this paper we discuss strong-field photoionization, i.e., above-threshold ionization, by few-cycle pulses. Some of the results are directly relevant for attosecond science as they address the dynamics of strong-field photoionization, others for the application of ATI for phase measurement. This paper concentrates on experimental aspects, unsolved problems, and phase measurement. A review on the entire scope of ATI with few-cycle laser pulses can be found in [19]. ATI offers numerous possibilities for phase-measurement and investigation of few-cycle strong-field ionization if the number of electrons emitted in positive ("right") and negative ("left") direction is compared.[1] For phase measurement it is important to realize that the requirements are manifold:

- The effect needs to exhibit a strong asymmetry, ideally in an energy range where the count rate is high.
- The dependence of the effect on the absolute phase should be known with high accuracy.
- The phase dependence (e.g., the phase at which there is equal count rate on the left and the right detector) should not depend on intensity, pulse duration, etc. Otherwise re-calibration is necessary as conditions change or errors occur if changes go unnoticed.
- Ideally the degree of asymmetry can be used for a measurement of the pulse duration. This, however, requires a particularly good understanding of the underlying physical mechanisms.

2.2 Detection of the "Absolute" Phase

The most conspicuous feature of few-cycle pulses is their asymmetry. Assuming the conditions implied by (2.1) a cosine-like pulse exhibits mirror-symmetry at $t = 0$ while sine-like pulses exhibit inversion symmetry. It is important to realize that the symmetry or lack of thereof refers to a single optical cycle: Shaping pulses is not new unless the envelope is shaped such that it changes considerably within one optical cycle. In this sense all long pulses exhibit mirror and inversion symmetry. As a consequence of the Curie principle, the photoelectron angular distribution exhibits at least inversion symmetry as long as we accept the approximation that strong-field ionization is essentially a single-cycle process. This symmetry can be broken by few-cycle laser pulses, provided ionization is a highly non-linear process. (Linear photoionization cannot lead to asymmetric photoionization because the integral of the field of a propagating electromagnetic pulse must be zero.)

The broken inversion symmetry of photoelectron angular distributions is equivalent to a negative correlation (or anti-corrleation) of photoelectrons emitted in opposite directions. The significance of this insight is that phase-stabilized amplified few-cycle pulses are not necessary for the detection of phase effects. Rather, correlations can be detected – actually with very high

[1] Throughout this paper it is assumed that only electrons emitted parallel to the polarization are detected unless otherwise noted.

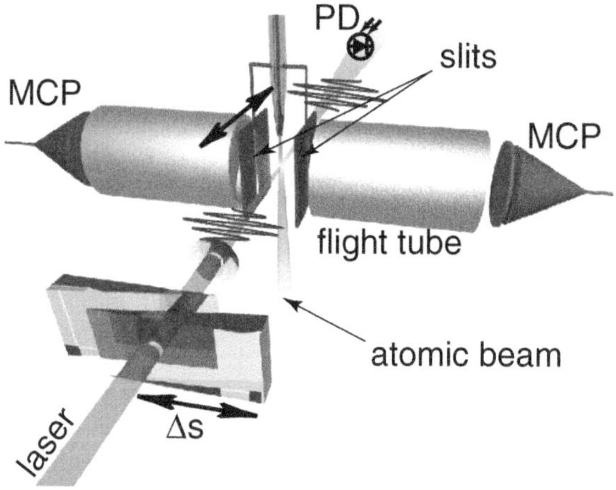

Fig. 2.2. *Stereo-TOF spectrometer.* Two opposing flight tubes are mounted inside a ultra-high vacuum apparatus. Atoms or molecules can be fed in through a glass nozzle and are ionized by a focussed few-cycle laser pulse (in reality, reflective optics is used). A pair of slits, which can be moved with a vacuum manipulator, transmits only electrons ionized in a small fraction of the focus. The absolute phase can be changed either by moving the pair of slits (Gouy effect) of by moving one of the glass wedges thus changing glass dispersion. [42]

sensitivity – by analyzing shot for shot the electron yield recorded by two electron detectors positioned at the left and the right of the laser focus. Due to its characteristic scheme, the instrument is referred to as an stereo-ATI spectrometer, see Fig. 2.2. Each laser pulse results in a pair of integers representing the number of electrons detected for this pulse on the left and the right electron detector. After accumulating data for 10^5–10^6 laser pulses a two-dimensional histogram, which represents the probability distribution of the pairs of integers, i.e., the electrons detected simultaneously on the left and the right detector, can be produced, typically presented in false-color coding. In such a contingency (or correlation) map, anti-correlations show up as structures perpendicular to the diagonal: If a certain pulse produces many electrons emitted to the left, then there is a high probability that only a few electrons will be emitted to the right, and vice versa. This exactly has been measured for Krypton atoms exposed to 780-nm circularly polarized laser pulses of 6–7 fs pulse duration (FWHM) and $5 \times 10^{13}\,\mathrm{W\,cm^{-2}}$ intensity [20], see Fig. 2.3. A weaker anti-correlation has been found for linear polarization.

The correlation method was also investigated theoretically [21]. For this calculation a Keldysh-type model [22] was adapted to few-cycle pulses. One result has been the dependence of the strength of anti-correlation as a function of pulse duration. An earlier prediction [23] that circular polarization leads to stronger inversion asymmetries was confirmed qualitatively. The theoretical modeling also addressed the central problem of the correlation approach:

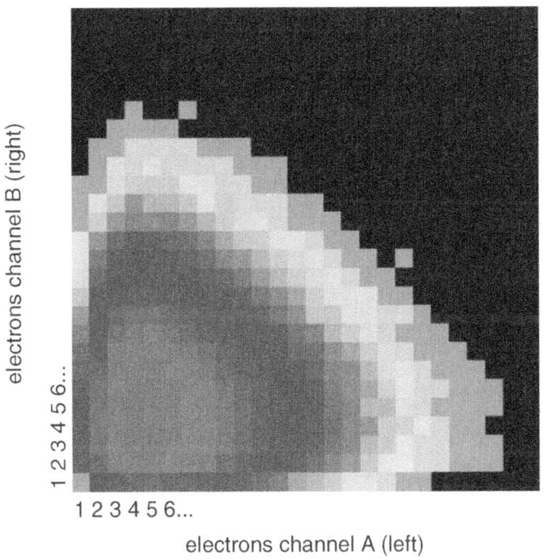

Fig. 2.3. Evidence of absolute-phase effects from few-cycle laser pulses. In this contingency map, every laser shot is recorded according to the number of photoelectrons measured in the left and the right arm of the stereo-ATI spectrometer. The number of laser shots with electron numbers according to the coordinates of the pixel is coded in grey shades. The signature of the absolute phase is an anticorrelation in the number of electrons recorded with the left and the right detector. In the contingency map they form a structure perpendicular to the diagonal [20]. Shown here is a measurement with krypton atoms for circular laser polarization, a pulse duration of 6 fs, and an intensity of $5 \times 10^{13}\,\mathrm{W\,cm^{-2}}$. [20]

Laser pulse fluctuations create positive correlations. Due to the non-linear dependence of multi-photon ionization probability on intensity, even modest laser pulse fluctuations create positive correlations that can mask the effects of the absolute phase. Also this finding agrees well with the experimental evidence.

2.3 Asymmetric Ionization

We now proceed with a quantitative analysis of the energy-integrated photoelectron yield in opposite directions and parallel to the laser polarization. It may be instructive to start the discussion with a naive and in fact essentially wrong model: The asymmetry $R(\varphi)$ defined as the ratio of the energy-integrated ("total") electron yield to the left and to the right is estimated using the quasi-static tunneling ionization probability [24, 25]

$$P_{\mathrm{ion}}(t) \propto \left(\frac{2\,\kappa^3}{|\mathcal{E}(t)|}\right)^{2/\kappa-1} \exp\left[-\frac{2\,\kappa^3}{3|\mathcal{E}(t)|}\right], \tag{2.2}$$

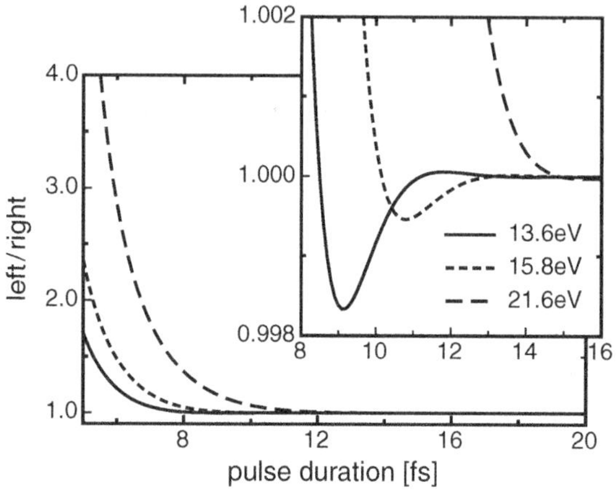

Fig. 2.4. Asymmetry of photoelectron yield calculated by integrating static tunneling ionization probabilities for positive and negative sign of the field. The result of this oversimplified model is shown as a function of pulse duration and ionization potential. The inset shows that the function is not quite monotonic. The intensity used was $10^{14}\,\mathrm{W\,cm^{-2}}$

where $\kappa^2 = 2E_{\mathrm{Ion}}$ with E_{Ion} being the ionization threshold. With the suggestive (but invalid!) assumption that electrons generated at instants where the field had opposite sign would travel to opposite directions, the yield on the left and on the right detector would be estimated by integrating $P_{\mathrm{ion}}(t)$ for positive and negative sign of the field separately. The ratio $R_{\mathrm{max}} = \max(R(\varphi))$ of the yield for left and right direction ("asymmetry") calculated in this way is shown in Fig. 2.4 in dependence of the pulse duration. The asymmetry is a steep function of pulse duration and ionization threshold, as expected. This suggests that $R(\varphi)$ can be used not only for phase measurement, but also for determining the pulse duration with the remarkable feature of increasing sensitivity with decreasing pulse duration. It will turn out that this promise is not kept.

The oversimplification of the above model comes from the fact that an electron that tunnels at an instant $t = t_0$ will be deflected by the subsequent evolution of the field. Invoking the strong-field approximation which assumes that the field does not affect the ground state and the atomic potential does not affect electrons in the continuum, it is straightforward to predict the emission direction of the electron. From conservation of canonical momentum $\boldsymbol{p}_{\mathrm{can}} = \boldsymbol{p} - e\boldsymbol{A}(t)$, one finds that the electron will be emitted in the opposite direction of the vector potential $\boldsymbol{A}$ at the instant of tunneling t_0. An assumption used is that the electron momentum at $t = t_0$ is negligible. The fact that the vector potential of a sine-like pulse is essentially cosine-like and vice versa, together with the fact that the tunneling probability is of course still be given by the field strenth $\mathcal{E}(t_0)$, has two consequences: (a) the most asymmetric pulse shape, i.e., cosine-like pulses, will lead to symmetric yield, i.e., the same number of electrons on the left and the right detector:

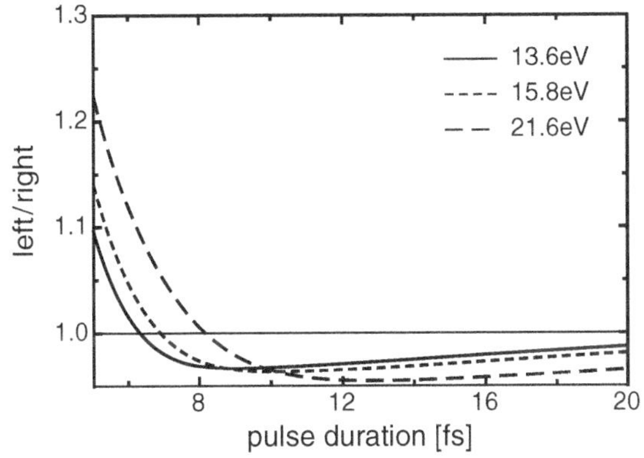

Fig. 2.5. Asymmetry of photoelectron yield calculated by the classical model of strong-field ionization for the same conditions as in Fig. 2.4

$R_{\max} = R(\pm\pi/2), z = R(n\pi), n \in Z$ (b) The strength of the asymmetry will be greatly reduced, see Fig. 2.5. An illustration of the latter effect can be found in [26].

It is interesting to note that the asymmetry as a function of pulse duration changes sign for both models considered with the effect being more pronounced for the "correct" model, see Figs. 2.4 and 2.5. The experimental evidence (see Fig. 2.7) reveals however that the classical model is wrong in predicting equal electron yield to the left and to the right for cosine-like pulses. This result was confirmed by theoretical investigations based on numerical solutions of the time-dependent Schrödinger equation (TDSE) [27]. For typical conditions (6 fs, $0.7 \times 10^{14}\,\mathrm{W\,cm^{-2}}$) symmetric electron yield is generated at a phase $\varphi \approx -0.3\pi$. However, other phase dependencies of R are also observed. The attempt to establish a relationship between left–right contrast and pulse duration by solving the TDSE for various pulse lengths leads to a surprisingly complex picture, see Fig. 2.6. If we try to model the logarithm of $R(\varphi)$ by a sine-like function, we would make the ansatz

$$\log R(\varphi) = \log R_{\max} \cdot \sin(\varphi + \bar{\varphi}). \tag{2.3}$$

If the classical model were correct, $\bar{\varphi} = 0$. However, Fig. 2.6 indicates that $\bar{\varphi}$ can assume any value and $R_{\max}$ depends on intensity and, as more detailed calculations reveal, pulse duration in a non-trivial way. It appears that the asymmetry $R(\varphi)$ of the energy-integrated photoelectron yield is not a reliable indicator for the absolute phase. In addition, $R(\varphi)$ cannot be used for estimating or measuring the pulse duration. This conclusion is drawn from calculations using a 1D model atom. 3D calculations can certainly be expected to deliver a quantitatively different result. Qualitatively, however, the above conclusion can be expected to be valid in consideration that results of 3D calculations do also result in a complicated dependence of R on intensity, pulse duration, etc. [28–30]. Although it is obvious that the strong-field

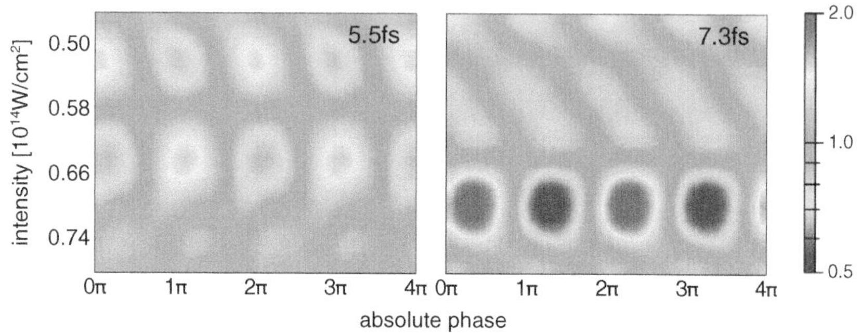

Fig. 2.6. The left–right ratio of the energy-integrated photoelectron yield as a function of the phase φ and intensity. The figure displays the result for a 1D model atom for 5.5 and 7.3-fs pulses. As the pulse duration is increased, the pattern changes gradually and moves in direction of lower intensities. For comparison to experiment, focal averaging needs to be taken into account, i.e., averaging over a range of intensities

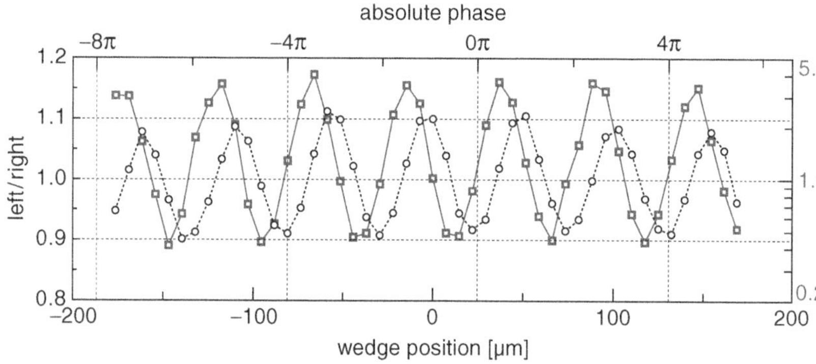

Fig. 2.7. Measurement of $R(\varphi)$ (*open dots*). The phase φ was changed by shifting one of the glass wedges (see Fig. 2.2). The *open-square* data points correspond to ATI plateau electrons that will be discussed in Sect. 2.4.2

approximation, i.e., neglecting the atomic potential, is responsible for the discrepancies of analytical and numerical results, the underlying mechanisms of the intricate dependencies of the simplest of all quantities that can be used to characterize asymmetric ionization are not at all clear. It seems, however, that the non-monotonic behavior that can be observed in the simple classical model contributes to the effects.

2.4 Above-Threshold Ionization Spectra

So far only the energy-integrated (or total) yield has been considered. We now turn our attention to energy-resolved spectra. A characteristic of strong-field ionization is that more photons than necessary for ionization may be absorbed by an atom which leads to correspondingly high electron kinetic energies (above-threshold ionization, ATI) [31].

Phase-dependent photoelectron spectra can only be measured if the absolute phase can be kept constant for many laser pulses, unless the spectrometer is constructed such that a complete spectrum can be measured for a single laser pulse. Single-shot operation is possible, however, it compromises dynamic range and energy resolution. An important development has been the invention of a technique (self-referencing or f-to-$2f$ technique) allowing the use of femtosecond lasers as frequency rulers in precision metrology [13, 14]. This innovation had profound consequences also for time-domain applications like strong-field ionization, high-harmonic generation, and in particular attosecond science. The reason is that stabilizing a femtosecond laser's frequency comb implies that the rate with which the absolute phase changes is stabilized, too. For a review see [32]. In 2003 the self-referencing technique was implemented in a femtosecond laser amplifier [33, 34] thus opening the full potential of tailored few-cycle pulses for time-domain applications. It should be noted that phase stabilization by the f-to-$2f$ technique may lead to small drifts of the absolute phase due to dispersion [35].

Figure 2.8 displays a series of ATI spectra measured for different absolute phases. Although the absolute phases (and thus also the temporal evolution of the electric fields) of the respective pulses are displayed in this set of figures, it should be noted that they are not known a priori. What is known, however, is the difference in absolute phase for each measurement relative to, e.g., the first one. This is because the phase is changed by changing the amount of glass in the laser beam, c.f. Fig. 2.2. From the dispersion properties of the inserted glass found in the literature, it is straightforward to calculate the phase change: A piece of glass of thickness d induces a change $\Delta\varphi$ of the absolute phase given by

$$\Delta\varphi = \omega d \left(\frac{1}{v_{\mathrm{ph}}} - \frac{1}{v_{\mathrm{g}}} \right), \tag{2.4}$$

where v_{ph} and v_{g} are phase and group velocity, respectively.

$$v_{\mathrm{ph}} = \frac{c}{n}, \tag{2.5}$$

$$v_{\mathrm{g}} = \frac{c}{n} + \frac{c\lambda}{n^2} \frac{\partial n}{\partial \lambda}, \tag{2.6}$$

where c is the vacuum speed-of-light and n the refractive index. This means that, at the central wavelength $\lambda_0 = 760\,\mathrm{nm}$ of the laser used in our experiments, a fused silica glass plate with a thickness $d = 26\,\mu\mathrm{m}$ changes the absolute phase by $\Delta\varphi = \pi$. A cosine-like pulse would be converted into a minus-cosine-like pulse.

The ATI spectra displayed in Fig. 2.8 exhibit a typical structure that is well-known from longer pulses: The electron yield decreases steeply with increasing energy. For intensities exceeding $0.5 \times 10^{14}\,\mathrm{W\,cm}$ this drop comes

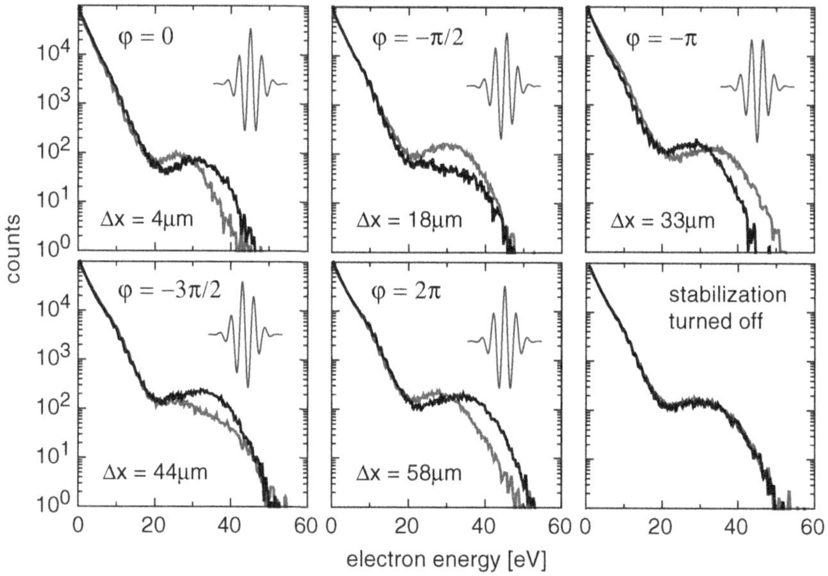

Fig. 2.8. Xenon photoelectron spectra for different absolute phases controlled by fine movement of one of the wedges. Δx indicates the glass added hereby. The *black curves* correspond to emission to the right (positive direction), the *gray ones* to the opposite direction. The inset shows the corresponding real time variation of the electric field, as deduced from comparison of the phase-dependence of the plateau electrons with theory. Only without phase stabilization were identical spectra measured left and right

to a sudden hold at 20–30 eV and a plateau-like structure develops [2]. This means an approximately constant electron yield up to the cut-off energy where the yield decreases in an essentially exponential way again. The phenomenon has already been mentioned in the introduction. It is due to electrons that re-collide elastically with the ion core. In the case of backscattering, the electron velocity is altered such that the electron will be further accelerated in the oscillating electric field of the laser after the re-collision event. For long pulses classical and quantum calculations predict a cutoff energy of $10\,U_P$ in good agreement with experimental evidence [6, 36]. An ATI spectrum therefore consists of high-energy (plateau) electrons that underwent rescattering and low-energy electrons that left the atom directly, i.e., without rescattering. Consequently the latter are often referred to as "direct" electrons.

There are a few obvious consistency tests for these measurements: Changing the absolute phase π should result in identical spectra. This is fulfilled to a large extent, but not completely. The reason is that adding glass into the beam not only changes the absolute phase, but also higher-order dispersion that affects the pulse duration. Similarly, changing the phase by π reverses the roles of left and right. From the experimental data it is obvious that the strongest phase effects occur for high-energy (>20 eV) photoelectrons.

However, there are also noticeable phase effects for the low-energy electrons which are difficult to see in Fig. 2.8 due to the semi-logarithmic scale used. A comparison of the asymmetry for direct and rescattered electrons can be seen in Fig. 2.7. In fact, Figs. 2.8 and 2.7 display data from the same experimental run. The curves of Fig. 2.7 correspond to energy-integrated yields where the integration extended from 0 to 50 eV for the open dots and from 20 to 50 eV for the open squares.

A non-trivial observation is that there is no phase for which the spectra measured on the left and the right would be identical – at least for high-energy electrons. This immediately suggests that the processes underlying the ATI plateau electrons are not instantaneous but extent over an appreciable fraction of the laser pulse. Obviously, this conclusion is in perfect agreement with the present understanding of re-collission effects like the ATI plateau.

2.4.1 Low-Energy (Direct) ATI Electrons

The direct electrons are particularly simple to model, at least in the framework of the classical model [37]. From conservation of canonical momentum (cf. Sect. 2.3) and the assumption that the electron has zero momentum at the exit of the tunnel, i.e., $\boldsymbol{p}_0 := \boldsymbol{p}(t_0) = 0$, we immediately arrive at the conclusion that the maximum electron energy for a few-cycle laser pulse is $2U_\mathrm{P}$ and is obtained for sine-like pulses which have cosine-like vector potentials.[2] It is of course the hallmark of few-cycle laser pulses that the cutoff energy becomes phase-dependent. Its variation with phase and pulse duration is shown in Figs. 2.9 and 2.10.

Although the existence of the 2-U_P cutoff has been impressively demonstrated by Gallagher in microwave ionization experiments of Rydberg atoms [38][3], it seems to be hardly known that the effect can also play a role at 800 nm. In fact, there are hardly any measured spectra in the literature that proof its significance for, e.g., rare gas atoms. The reasons may be the following: The ionization yield between 0 and $2U_\mathrm{P}$ drops quickly. This can even be understood with the classical model because the higher the electron energy, the smaller the electric field at $t = t_0$, the instant at which the electron enters the continuum. The sharp classical cutoff at $2U_\mathrm{P}$ is replaced by an exponential roll-off due to quantum (essentially tunneling) effects. The slope of the spectrum in

[2] The ponderomotive energy is the cycle-averaged kinetic energy of an otherwise free electron in an oscillating electric field and is given by $U_\mathrm{P} = I/(4\omega^2)$ in atomic units. I is the laser intensity and ω the laser angular frequency. The definition of U_P is ambiguous for laser pulses, in particular for few-cycle pulses. Using the maximum of the pulse envelope for I results in a definition that connects seamlessly to long pulses and therefore is used here.

[3] The microwave experiments were long-pulse experiments. Due to ponderomotive acceleration the 2-U_P cutoff then is shifted to $3U_\mathrm{P}$.

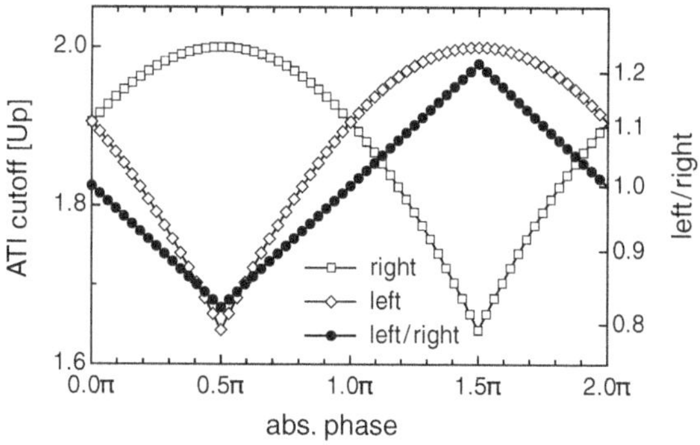

Fig. 2.9. Phase-dependence of the cutoff energy of direct electrons for 2-cycle pulses, i.e., 5.3 fs pulse width (FWHM) at 800 nm. The curves with the open points show the cutoff energy in units of U_P for the spectra measured with the left (*diamonds*) and the right (*squares*) detector. The curve with the solid points displays the ratio of the afore-mentioned curves. Note the logarithmic scale for this curve

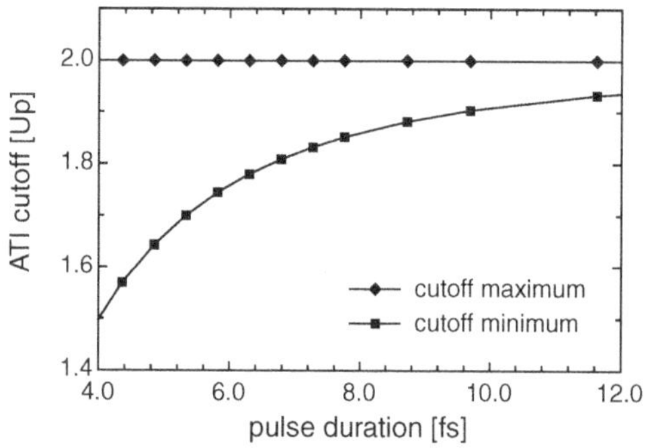

Fig. 2.10. The maximum and minimum cutoff energy as a function of pulse duration. The maximum for electrons emitted to the right is obtained at $\varphi = 0.5\pi$ and the minimum at $\varphi = 1.5\pi$ as shown in Fig. 2.9 for the example of 2-cycle pulses. The 2-U_P maximum for the cutoff as well as the phase for which this maximum is obtained are immediate consequences of the strong-field approximation

the classically forbidden region is the steeper the higher the intensity, i.e., the more the classical the situation.[4]

If the slopes of the spectrum for energies smaller and larger than $2\,U_P$ happen to be similar, the 2-U_P cutoff cannot be recognized. This is generally the case for intensities smaller than $10^{14}\,\mathrm{W\,cm^{-2}}$ because the classically allowed energy region ($E < 2\,U_P$) is small and the slope in the classically forbidden region ($E > 2\,U_P$) is not very steep. Exposing atoms to significantly higher intensities, on the other hand, often leads to spectra that are difficult to

[4] It is helpful to use an energy scale in units of the ponderomotive potential for the spectra in order to clearly see the effect.

interpret due to saturation of ionization and focal averaging. Few-cycle laser pulses help avoiding these problems to a certain degree and, in fact, the effect can be seen for ATI spectra generated by sub-10-fs pulses [39][5] and obviously also in computer experiments, see e.g., Fig. 5 in [36].

For phase-stabilized few-cycle pulses, the phase-dependence of the 2-U_P cutoff has also been observed [35, 40]. Considering the potential of the effect for measuring the absolute phase, it should be noted that the left–right asymmetry can be very significant, much greater than the one for the total yield. Plateau electrons, which will be discussed in the following section, have a comparable if not higher asymmetry. However, their yield is lower by about a factor of 100. Another very important advantage is that the phase dependence of direct electrons beyond the 2-U_P cutoff follows closely the classical model as shown in Figs. 2.11 and 2.12. It may sound ironic if not contradictory that the classical model fails badly in reproducing the phase dependence in the classically allowed energy region but does have some predictive power for electrons of classically forbidden energy. Their relatively high energy suggests however that their physics is much less involved than for low-energy electrons. This is confirmed by computer simulations. It should be mentioned that the asymmetry is not only a function of pulse duration but certainly at least also of the energy interval selected and, considering the discussion in the previous paragraph, the laser intensity (higher intensities lead to stepper cutoffs). Therefore, spectral resolution for a phasemeter based on this effect is necessary.

It is apparent from Figs. 2.11 and 2.12 that the classically forbidden electrons can also serve as an excellent probe for measuring the duration of pulses of less than 7 fs duration (FWHM) if the spectra are sufficiently well characterized. While the contrast for longer pulses is still sufficient, the method breaks down due to effects reminiscent of those discussed in Sect. 2.3. Although 3D simulations yield results in general agreement with the ones presented here [28–30], quantitative discrepancies are possible. The experimental results suggest in fact a less complicated behavior than the simulations. This may be due to focal averaging or deficiencies even in the 3D calculations: the only atom for which exact simulations are possible is Hydrogen. However, for Hydrogen no experimental data exists.

2.4.2 High-Energy (Plateau) ATI Electrons

The phase-dependent spectra of high-energy or plateau electrons are fundamentally different from low-energy electrons. An indication of that is the already mentioned fact that there is no phase for which spectra measured

[5] It is worthwhile to mention that the 2-U_P cutoff was observed for Argon. Apparently, the reason is that the slope of Argon spectra is steeper for energies above $2\,U_P$ than for other rare gases. This fact is known since the discovery of the ATI plateau [2] but has remained unexplained to the present day.

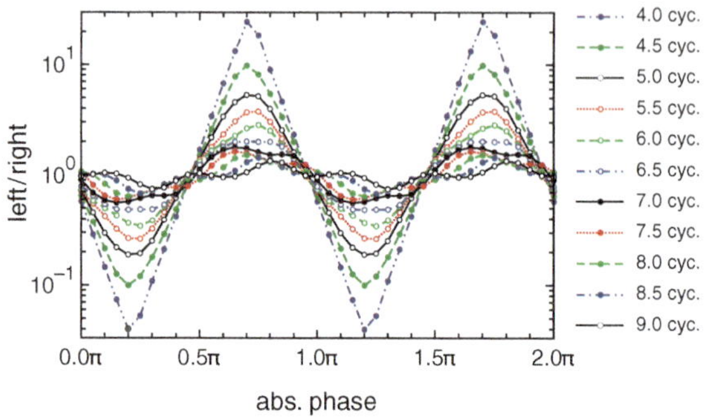

Fig. 2.11. Asymmetry of ATI electrons beyond the 2-U_P cutoff (30–40 eV) for various pulse durations. The data was obtained by solving the TDSE for a 1D model atom with an ionization threshold corresponding to Argon. The intensity is $1.8 \times 10^{14}\,\mathrm{W\,cm^{-2}}$ ($2U_\mathrm{P} \approx 21\,\mathrm{eV}$)

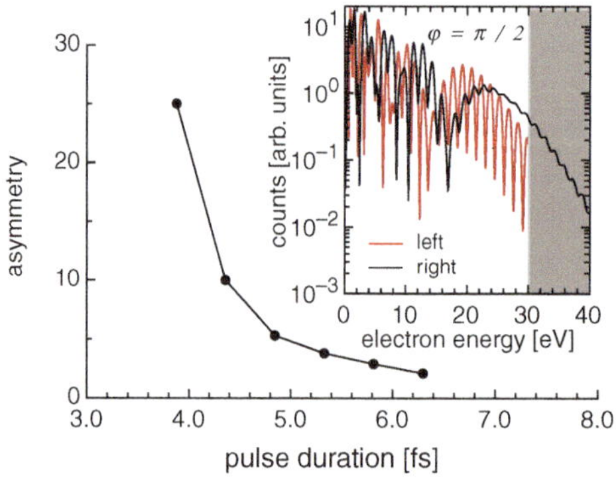

Fig. 2.12. The amplitude of the curves of Fig. 2.11, i.e., the maximum asymmetry, as a function of pulse duration. The inset displays an example of the pairs of spectra from which the other data displayed in Fig. 2.11 and this figure were derived

in opposite directions would be identical. This as well as the high contrast of phase effects for ATI-plateau electrons can be understood in the following way: First of all, high-energy electrons returning to the ion core can only be created in sub-femtosecond time intervals close to peaks of the electric field of the laser pulse. However, the probability that they tunnel through the atomic potential at t_0 depends exponentially on the field strength $\mathcal{E}(t_0)$ and – as few-cycle pulses are involved – it is likely only for those very few optical half-cycles close to the pulse maximum. Generally, the highest kinetic energies are reached for electrons returning to the core at $t = t_1$ such that the electric field becomes nearly zero ($\mathcal{E}(t_1) = 0$). For few-cycle pulses, in addition, the field amplitude $\mathcal{E}_0$ needs to be as high as possible for $t > t_1$ in order to allow efficient acceleration after rescattering. Since the start time t_0 and return time t_1 differ by about 75% of one optical cycle, both conditions – namely the highest possible field strength at t_0 and highest possible amplitude after t_1 – are hard to meet and result in a strong dependence of photoionization on the absolute phase. Number, strength, and timing of the wave-packets lead to distinctive structures in the ATI spectra. Their analysis therefore provides detailed information about the key processes of attosecond science. Generally, quantum mechanical calculations are in good qualitative agreement with this classical treatment [41].[6]

The ATI plateau is interesting with respect to phase measurement because the phase-dependence in this part of the spectrum is particularly strong. In addition, in some sense similarly to classically forbidden direct electrons, the influence of the atomic potential on the ATI plateau electrons is small. Therefore even a classical simulation can lead to a meaningful calibration for the absolute phase [42]. The fact that the count rates in the plateau region are 100–1,000 times lower than for direct electrons is not a real problem. An ATI phasemeter can be built in such a way that even a 30-μJ pulse will produce hundreds of plateau electrons.

In contrast to direct electrons, however, it does not appear to be a good idea to use only electrons beyond the 10-U_P cutoff because of the low abundance. Nevertheless, the ATI plateau cutoff energy is the most obvious feature to investigate. Figure 2.13 displays its dependence on absolute phase for 2-cycle pulses. In Fig. 2.14 the dependence of the maximum and and minimum cutoff on pulse duration is shown. While these figures certainly have some

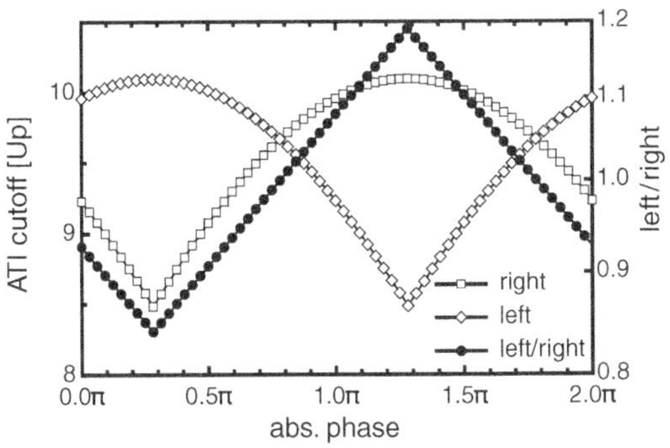

Fig. 2.13. Phase-dependence of the ATI plateau cutoff energy for 2-cycle pulses, i.e., 5.3 fs pulse width (FWHM) at 800 nm. The curves with the open points show the cutoff energy in units of U_P for the spectra measured with the left (*diamonds*) and the right (*squares*) detector. The curve with the solid points displays the ratio of the afore-mentioned curves. Note the logarithmic scale for this curve

[6] It should be mentioned at this point that for few-cycle conditions the ATI plateau is well-pronounced for Xenon but not so for Argon. Krypton displays the typical plateau only at low intensities [39]. For Helium and Neon experimental data does not seem to be available.

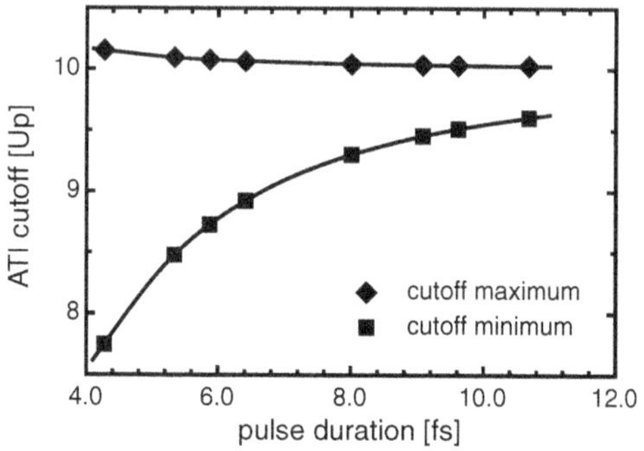

Fig. 2.14. The maximum and minimum cutoff energy as a function of pulse duration. The maximum for electrons emitted to the left is obtained at $\varphi = 0.3\pi$ and the minimum at $\varphi = 1.3\pi$ as shown in Fig. 2.13 for the example of 2-cycle pulses

significance they do not convey the complexity of the entire picture. In fact, each of the strongest cycles creates its own plateau with its own cutoff. The heights of the plateaus depend on the field strength at the time when the respective electron trajectories started. Figure 2.15 gives an impression of the behavior of these cutoffs. Quantum calculations do not exhibit sharp cutoffs (see Fig. 2.16) and it may be difficult to recognize them in cases where the intensity or the ionization probability is low.

For a reasonably complete analysis, interference effects also need to be considered. Interference of trajectories launched by subsequent cycles lead to small-scale fringes with a spacing being given by approximately the photon energy. They are reminiscent and in fact they are closely related to the well-known ATI peak structure and will be discussed in the next paragraph. Large-scale interferences, i.e., interference maxima and minima with an energy separation much bigger than the photon energy, arise from the interference of short and long trajectories: All electron energies below the plateau cutoff can be realized by at least two trajectories of different travel time. The trajectory that is launched later will re-collide earlier and vice versa. These facts are of course well-known from longer pulses. For few-cycle pulses all of the effects cited depend on the absolute phase and many also on pulse duration which makes it difficult to present a concise overview.

Naturally, this has consequences if the phase-dependent ATI plateau is considered for being used to measure the absolute phase. Here one has to distinguish two cases: When the pulse is just to be characterized it is possible to scan the absolute phase from 0 to 2π. In this case the comparison to even classical or semiclassical calculations results in a reliable calibration [42] that is confirmed by all tests that can be performed using other phase-dependent features of ATI spectra. If an ATI phasemeter is to be used in a servo loop, a much faster approach is necessary and the phase cannot be scanned. The safest approach would be to measure the ATI spectra and compare to corresponding

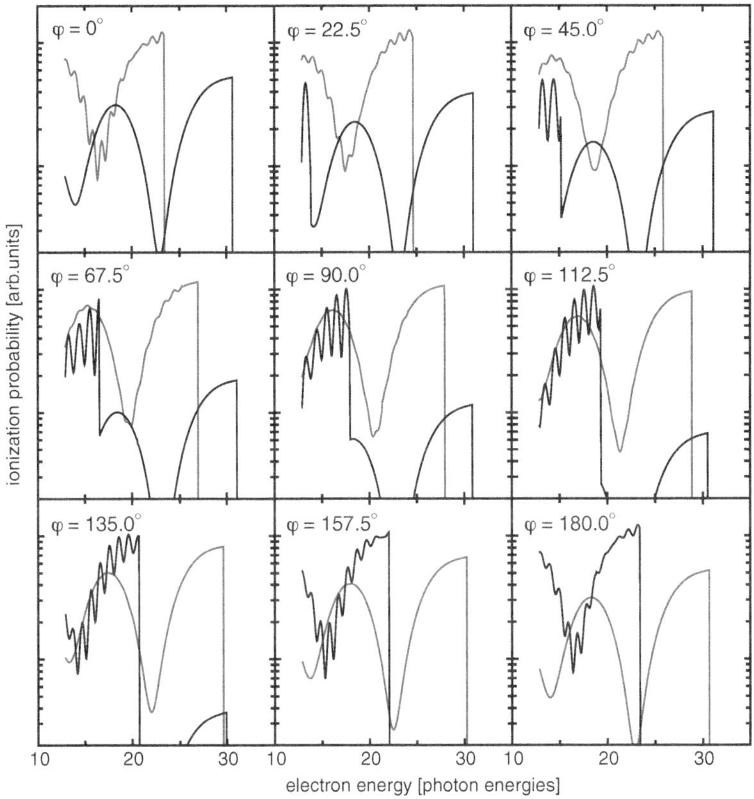

Fig. 2.15. ATI spectra (rescattered electrons only) for various absolute phases calculated with the classical model for a 2-cycle pulse. Interference effects are reproduced by calculating the classical action for the electrons and adding electron trajectories with the same final momentum coherently. One recognizes two types of interferences: Fringes with close to one photon energy separation which arise from wave packets emitted in subsequent optical cycles. In case only one optical cycle leads to significant amplitude for a given electron momentum, the spectrum is smoot. The other type of interference with a fringe spacing of almost 10 photon energies results from interference of long and short electron trajectories. *Black gray* curves correspond to spectra recorded in positive (negative) direction. The intensity is 0.8×10^{14} W cm^{-2} corresponding to $10\,U_\mathrm{P} \approx 31\,\hbar\omega$. A logarithmic scale is used. [26]

solutions of the TDSE. This, however, is not practical because it seems that Xenon is the preferred atom presently (cf. footnote 6) for which accurate calculations are not available. In addition, focal averaging would need to be included.[7]

[7] The 1D calculations on which some of the present analysis is based are certainly not suitable for calibrating a phasemeter that takes advantage of ATI plateau electrons. The major reason is that the strength of the plateau for a 1D model atom

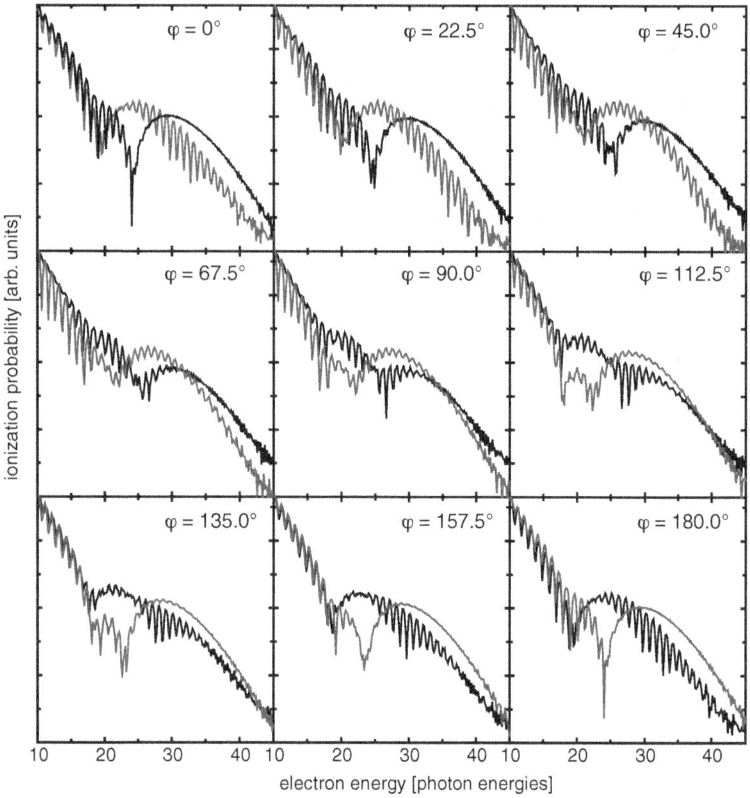

Fig. 2.16. ATI spectra (rescattered electrons only) for various absolute phases calculated with the classical model for a 2-cycle pulse. Interference effects are reproduced by calculating the classical action for the electrons and adding electron trajectories with the same final momentum coherently. One recognizes two types of interferences: Fringes with close to one photon energy separation which arise from wave packets emitted in subsequent optical cycles. In case only one optical cycle leads to significant amplitude for a given electron momentum, the spectrum is smooth. The other type of interference with a fringe spacing of almost 10 photon energies results from interference of long and short electron trajectories. *Black gray curves* correspond to spectra recorded in positive (negative) direction. The intensity is $0.8 \times 10^{14}\,\mathrm{W\,cm^{-2}}$ corresponding to $10\,U_\mathrm{P} \approx 31\,\hbar\omega$. A logarithmic scale is used

2.5 Phase-Dependent Interferences: The Attosecond Double Slit

ATI spectra generated by many-cycle pulses are characterized by a series of peaks separated by the photon energy. The position of these peaks is given by $(N + s)\hbar\omega + E_\mathrm{Ion} - U_\mathrm{P}$, where N is the number of photons necessary

is considerably smaller than for Xenon. Therefore, the multiple cutoff energies for different optical cycles play a much more pronounced role.

for ionization at zero intensity, s the order of the peak, E_{Ion} the ionization potential, and U_{P} the ponderomotive energy. Dynamically induced resonances may cause additional peaks. For few-cycle pulses the picture changes in many respects. Figure 2.17 displays measured electron spectra. In Fig. 2.17a the spectra recorded at the left and the right detectors are shown for ±cosine-like and ±sine-like pulses as defined in Fig. 2.1. A problem in depicting such spectra is that they quickly roll off with increasing electron energy. This roll-off was eliminated by dividing the spectra by the average of all spectra over the pulse's phase. Peak position, peak contrast, and even peak separation depend on the absolute phase. It seems to be virtually impossible to account for all of these effects within the framework of the conventional theory of ATI.

A better interpretation is based on an analogy to the double-slit experiment. Such an analogy has also been used to interpret multiphoton excitation experiments with Rydberg atoms exposed to microwaves with a variable number of cycles [43]. Here we take advantage of the fact that photoionization of atoms with an ionization threshold much greater than the photon energy is a highly nonlinear process. For intense fields, the first step can be described by optical field ionization, as already noted. This immediately explains that ionization can take place in one attosecond window (or slit) in time per half-cycle close to its extremum. By using phase-controlled few-cycle laser pulses, it is possible to manipulate the temporal evolution of the field, thus gradually opening or closing the slits. In this way which-way information about the instants of ionization is controlled. Depending on the field, one or two half-cycles (or anything in between) contribute to the electron amplitude for a given direction and electron energy. This corresponds to a varying degree of which-way information and, accordingly, to varying contrast of the interference fringes. The temporal slits leading to electrons of given final momentum are spaced by approximately the optical period. This results in a fringe spacing close to the photon energy.

Indeed, clear interference fringes with varying visibility are observed as expected from the discussion above. The highest visibility is observed for −sine-like pulses in the positive ("right") direction. For the same pulses, the visibility is very low in the opposite direction. Changing the phase by π interchanges the role of left and right as expected. The most straightforward explanation is to assume that, for −sine-like pulses, there are two slits and no which-way information for the positive direction and just one slit and (almost) complete which-way information in the negative direction. The fact that the interference pattern does not entirely disappear is caused by the pulse duration, which is still slightly too long to create a perfect single slit. A semi-classical analysis of the problem qualitatively reproduces the effect. However, the fringes are observed in the opposite direction as compared to the prediction of the simple model. Numerical simulations on the other hand faithfully reproduce the observed pattern [19, 44]. Using the effect for phase measurement is questionable as long as it is not completely understood. A more promising application seems to be to use this type of time-domain

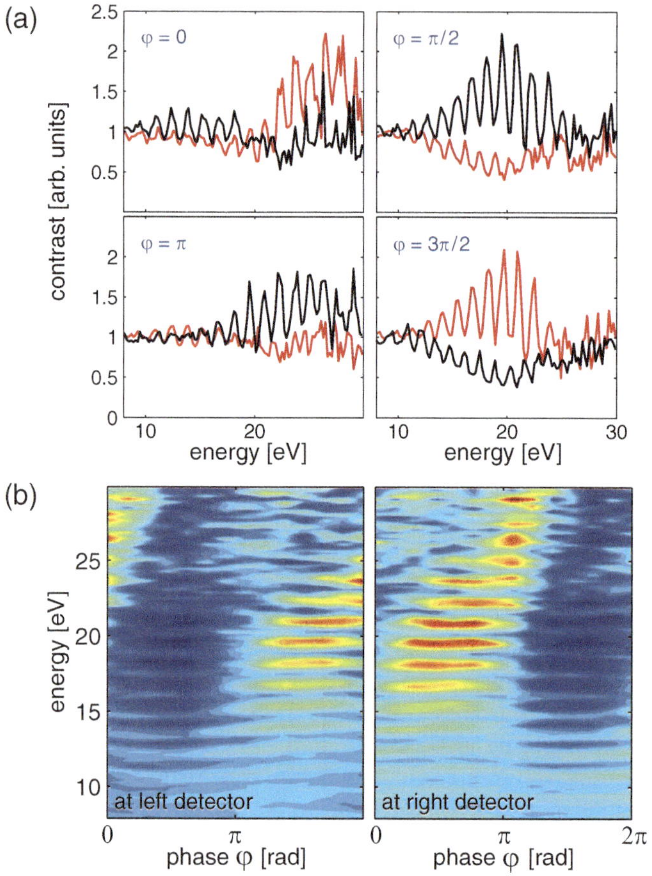

Fig. 2.17. Photoelectron spectra of argon measured with 6-fs laser pulses for intensity $1 \times 10^{14}\,\mathrm{W\,cm}^2$ as a function of the phase. Panel (**a**) displays the spectra for ±sine- and ±cosine-like laser fields. The *gray* curves are spectra recorded with the left detector (negative direction), while the black curves relate to the positive direction. For $\varphi = \pi/2$ the fringes exhibit maximum visibility for electron emission to the right, while in the opposite direction minimum fringe visibility is observed. In addition, the fringe positions are shifted. Panel (**b**) displays the entire measurement where the fringe visibility is coded in shades of *gray*. The fringe positions vary as the phase φ of the pulse is changed. This causes the wave-like bending of the stripes in these figures. Both panels, in principle, show the same information because a phase shift of π mirrors the pulse field in space and thus reverses the role of positive and negative direction. However, the data shown were recorded simultaneously but independently were the phase φ was varied between 0 and 2π. [44]

interferometry for learning more about the dynamics of the electron wave packet as the ionization process proceeds.

2.6 Conclusions

ATI spectra consist of a low-energy ("direct") and a high-energy (plateau) part. The latter is due to recolliding electrons, whereas the former is created by electrons that leave the atom directly without a second encounter with the atom. Both parts of the spectrum depend on the absolute phase in a characteristic way. The behavior of plateau electrons is simple in the sense that it can be reproduced by simple classical and semi-classical models. However, there are many different mechanisms that determine the shape of the spectra ranging from the varying amplitude of subsequent optical cycles to interference effects. Surprisingly complicated is the energy-integrated electron yield which is dominated by electrons with close to zero energy. The obvious reason for the difficulties may be that low-energy electrons are more strongly affected by the intra-atomic field which has a strength comparable to the laser field in the vicinity of the atom. The simplest phase-dependence is observed in the cutoff regions for direct and rescattered electrons.

Acknowledgments

I would like to thank the many coworkers and colleagues who helped establishing the, in my opinion, fascinating field of few-cycle quantum optics. In particular, I would like to mention W. Becker, F. Grasbon, F. Krausz, F. Lindner, D. B. Milošević, M. Nisoli, P. Villoresi, and H. Walther. This work was supported by The Welch Foundation (grant A-1562) and the National Science Foundation (grant PHY-0555568).

References

1. M. Ferray, A. L'Huillier, X.F. Li, L.A. Lompré, G. Mainfray, C. Manus, Multiple-harmonic conversion of 1064 nm radiation in rare gases. J. Phys. B: At. Mol. Opt. Phys. **21**, L31, 1988
2. G.G. Paulus, W. Nicklich, H. Xu, P. Lambropoulos, H. Walther, Plateau in above threshod ionization spectra. Phys. Rev. Lett. **72**, 2851, 1994
3. A. L'Huillier, L.A. Lompre, G. Mainfray, C. Manus, Multiply charged ions induced by multiphoton absorption in rare gases at 0.53 μm. Phys. Rev. A **27**, 2503, 1983
4. K.C. Kulander, K.J. Schafer, K.L. Krause, in *Dynamics of Short-Pulse Excitation, Ionization, and Harmonic Conversion*, ed. by B. Piraux, A. l'Huillier, K. Rzazewski. Super-Intense Laser-Atom Physics. (Plenum Press, New York, 1993), p. 95

5. P.B. Corkum, Plasma perspective on strong-field multiphoton ionization. Phys. Rev. Lett. **71**, 1994, 1993
6. G.G. Paulus, W. Becker, W. Nicklich, H. Walther, Rescattering effects in above-threshold-ionization: a classical model. J. Phys. B: At. Mol. Opt. Phys. **27**, L703, 1994
7. P. Salières, A. l'Huillier, P. Antoine, M. Lewenstein, Study of the spatial and temporal coherence of high-order harmonics. Adv. At. Mol. Opt. Phys. **141**, 83, 1999
8. W. Becker, F. Grasbon, R. Kopold, D.B. Milošević, G.G. Paulus, H. Walther, Above-threshold ionization: from classical features to quantum effects. Adv. At. Mol. Opt. Phys. **48**, 35, 2002
9. R. Dörner, Th. Weber, M. Weckenbrock, A. Staudte, M. Hattass, H. Schmidt-Böcking, R. Moshammer, J. Ullrich, Multiple ionization in strong laser fields. Adv At. Mol. Opt. Phys. **48**, 1, 2002
10. B. Yang, K.J. Schafer, B. Walker, K.C. Kulander, P. Agostini, L.F. DiMauro, Intensity-dependent scattering rings in high order above-threshold ionization. Phys. Rev. Lett. **71**, 3770, 1993
11. I.P. Christov, M.M. Murnane, H.C. Kapteyn, High-harmonic generation of attosecond pulses in the "single-cycle" regime. Phys. Rev. Lett. **78**, 1251, 1997
12. M. Hentschel, R. Kienberger, Ch. Spielmann, G.A. Reider, N. Milosevic, T. Brabec, P. Corkum, U. Heinzmann, M. Drescher, F. Krausz, Attosecond metrology. Nature **414**, 509, 2001
13. J. Reichert, R. Holzwarth, Th. Udem, T.W. Hänsch, Measuring the frequency of light with mode-locked lasers. Opt. Commun. **172**, 59, 1999
14. D.J. Jones, S.A. Diddams, J.K. Ranka, A. Stentz, R.S. Windeler, J.L. Hall, S.T. Cundiff, Carrier-envelope phase control of femtosecond mode-locked lasers and direct optical frequency synthesis. Science **288**, 635, 2000
15. T.M. Fortier, P.A. Roos, D.J. Jones, S.T. Cundiff, R.D.R. Bhat, J.E. Sipe, Carrier-envelope phase-controlled quantum interference of injected photocurrents in semiconductors. Phys. Rev. Lett. **92**, 147403, 2004
16. O.D. Mücke, T. Tritschler, M. Wegener, U. Morgner, F.X. Kärtner, G. Khitrova, H.M. Gibbs, Carrier-wave Rabi flopping: role of the carrier-envelope phase. Opt. Lett. **29**, 2160, 2004
17. A. Apolonski, P. Dombi, G.G. Paulus, M. Kakehata, R. Holzwarth, Th. Udem, Ch. Lemell, K. Torizuka, J. Burgdörfer, T.W. Hänsch, F. Krausz, Observation of light-phase-sensitive photoemission from a metal. Phys. Rev. Lett. **92**, 073902, 2004
18. M. Kreß, T. Löffler, M.D. Thomson, R. Dörner, H. Gimpel, K. Zrost, T. Ergler, R. Moshammer, U. Morgner, J. Ullrich, H.G. Roskos, Determination of the carrier-envelope phase of few-cycle laser pulses with terahertz-emission spectroscopy. Nature Physics **2**, 327–331, 2006
19. D.B. Milošević, G.G. Paulus, D. Bauer, W. Becker, Above-threshold ionization by few-cycle pulses. J. Phys. B: At. Mol. Opt. Phys. **39**(14), R203–R262, 2006
20. G.G. Paulus, F. Grasbon, H. Walther, P. Villoresi, M. Nisoli, S. Stagira, E. Priori, S. De Silvestri, Absolute-phase phenomena in photoionization with few-cycle laser pulses. Nature **414**, 182, 2001
21. D.B. Milošević, G.G. Paulus, W. Becker, Phase-dependent effects of a few-cycle laser pulse. Phys. Rev. Lett. **89**, 153001, 2002
22. W. Becker, R.R. Schlicher, M.O. Scully, Final-state effects in above-threshold ionisation. J. Phys. B: At. Mol. Opt. Phys. **19**, L785, 1986

23. P. Dietrich, F. Krausz, P.B. Corkum, Determining the absolute carrier phase of a few-cycle laser pulse. Opt. Lett. **25**, 16, 2000
24. A.M. Perelomov, V.S. Popov, M.V. Terent'ev. Sov. Phys. JETP **24**, 207–217, 1967
25. C.Z. Bisgaard, L.B. Madsen, Tunneling ionization of atoms. Am. J. Phys. **72**(2), 249–254, 2003
26. G.G. Paulus, F. Lindner, D.B. Milošević, W. Becker, Phase-controlled single-cycle strong-field photoionization. Phys. Scr. **T110**, 120–125, 2004.
27. S. Chelkowski, A.D. Bandrauk, A. Apolonski, Phase-dependent asymmetries in strong-field photoionization by few-cycle laser pulses. Phys. Rev. A **70**, 013815, 2004
28. S. Chelkowski, A.D. Bandrauk, Sensitivity of spatial photoelectron distributions to the absolute phase of an ultrashort intense laser pulse. Phys. Rev. A **65**, 061802(R), 2002
29. J. Tate, Ph. Colosimo, L.F. DiMauro, G.G. Paulus, Phase-dependence of strong-field photoionization with few-cycle laser pulses, 2008 (unpublished)
30. D. Bauer. Private communication, 2006
31. P. Agostini, F. Fabre, G. Mainfray, G. Petite, N.K. Rahman, Free-free transitions following six-photon ionization of xenon atoms. Phys. Rev. Lett. **42**, 1127, 1979
32. S.T. Cundiff, Phase stabilization of ultrashort optical pulses. J. Phys. D: Appl. Phys. **35**, R43, 2002
33. A. Baltuška, Th. Udem, M. Uiberacker, M. Hentschel, E. Goulielmakis, Ch. Gohle, R. Holzwarth, V.S. Yakovlev, A. Scrinzi, T.W. Hänsch, F. Krausz, Attosecond control of electronic processes by intense light fields. Nature **421**, 611, 2003
34. A. Baltuška, G.G. Paulus, F. Lindner, R. Kienberger, F. Krausz. in *Generation and Measurement of Intense Phase-Controlled Few-Cycle Laser Pulses*, ed, by J. Ye, S.T. Cundiff. Femtosecond Optical Frequency Comb Technology. (Springer, New York, 2005), pp. 263–313
35. G.G. Paulus, F. Lindner, H. Walther, A. Baltuška, F. Krausz, Measurement of the phase of few-cycle laser pulses. J. Mod. Opt. **52**, 221, 2005
36. G.G. Paulus, W. Becker, H. Walther, Classical rescattering effects in two-color above-threshold ionization. Phys. Rev. A **52**, 4043, 1995
37. H.B. van Linden van den Heuvell, H.G. Muller, *Multiphoton Processes, Cambridge Studies in Modern Optics.* (Cambridge University Press, Cambridge, 1988), p. 25
38. T.F. Gallagher, Above-threshold ionization in low-frequency limit. Phys. Rev. Lett. **61**, 2304, 1988
39. F. Grasbon, G.G. Paulus, H. Walther, P. Villoresi, G. Sansone, S. Stagira, M. Nisoli, S.De Silvestri, Above-threshold ionization at the few-cycle limit. Phys. Rev. Lett. **91**, 173003, 2003
40. F. Lindner, G.G. Paulus, H. Walther, A. Baltuška, E. Goulielmakis, M. Lezius, F. Krausz, Gouy phase shift for few-cycle laser pulses. Phys. Rev. Lett. **92**, 113001, 2004
41. D.B. Milošević, G.G. Paulus, W. Becker, High-order above-threshold ionization with few-cycle pulse: a meter of the absolute phase. Opt. Express **11**, 1418, 2003
42. G.G. Paulus, F. Lindner, H. Walther, A. Baltuška, E. Goulielmakis, M. Lezius, F. Krausz, Measurement of the phase of few-cycle laser pulses. Phys. Rev. Lett. **91**, 253004, 2003

43. R.B. Watkins, W.M. Griffith, M.A. Gatzke, T.F. Gallagher, Multiphoton resonance with one to many cycles. Phys. Rev. Lett. **77**, 2424, 1996
44. F. Lindner, M.G. Schätzel, H. Walther, A. Baltuška, E. Goulielmakis, F. Krausz, D.B. Milošević, D. Bauer, W. Becker, G.G. Paulus, Attosecond double-slit experiment. Phys. Rev. Lett. **95**, 040401, 2005

3

Wavepacket Dynamics of Molecules in Intense Laser Fields

Hirohiko Kono, Katsunori Nakai, Manabu Kanno, Yukio Sato, Shiro Koseki, Tsuyoshi Kato, and Yuichi Fujimura

Abstract. The dynamical behavior of H_2^+ in near-infrared, intense laser fields ($I > 10^{13}$ W cm^{-2} and $\lambda > 700$ nm) was discussed on the basis of the results of accurate electronic and nuclear wave packet propagation obtained by the application of the dual transformation method. Using "field-following" time-dependent (TD) adiabatic states defined as the eigenfunctions of the "instantaneous" electronic Hamiltonian including the dipole interaction with laser fields, we clarified the dynamics of the bound electron, ionization processes, Coulomb explosion processes, and field-induced molecular vibration of H_2^+. The analyses indicate that the electron dynamics and nuclear (reaction) dynamics of *polyatomic* molecules in intense fields can be described by using the potential surfaces of TD adiabatic states and the nonadiabatic coupling elements between those states. To obtain the TD adiabatic states of a molecule, one can diagonalize the electronic Hamiltonian including the interaction with the instantaneous laser electric field by ab initio electronic structure calculations. We then present the results of simulation as to how much vibrational energy is acquired by C_{60} (or C_{60}^{z+}) through the interaction with an ultrashort intense pulse of $\lambda = 1{,}800$ nm. This type of simulation was carried out by incorporating an ab initio classical molecular dynamics method into the framework of the TD adiabatic state approach. The results indicate that large-amplitude vibration with energy of >20 eV is induced in the $h_g(1)$ prolate-oblate mode of C_{60} or C_{60}^{z+}. We found that the acquired vibrational energy is maximized at $T_p \sim T_{vib}/2$, where T_p is the pulse length and T_{vib} is the vibrational period of the $h_g(1)$ mode.

3.1 Introduction

The frontier of light-matter interaction has been expanded into new areas by the development of high-power lasers [1]. Novel nonperturbative electronic dynamics of atoms in intense laser fields, such as tunnel ionization [2, 3] and higher-order harmonic generation of emission (HHG) [4, 5], have been discovered. An epoch is marked by the advent of attosecond light pulses generated by using HHG in the soft X-ray regime [6]. Its application has become a hot topic in recent years [7, 8]. Nowadays, the intensity of focused light generated by near-infrared (IR) Ti:Sapphire laser systems can easily exceed

$I = 10^{16}\,\mathrm{W\,cm^{-2}}$. A laser field of light intensity of $3.5 \times 10^{16}\,\mathrm{W\,cm^{-2}}$, which corresponds to $5.1 \times 10^{9}\,\mathrm{V\,cm^{-1}}$, exerts on an electron a force that is as strong as the electron–nucleus interaction in a hydrogen atom.

The introduction of handy high-power laser systems into the field of chemistry is innovative and inevitable. Compared to atoms in intense fields, molecules exhibit complex phenomena arising from their additional (vibrational and rotational) degrees of freedom, such as charge-symmetric and -asymmetric dissociation [9], alignment with an external field [10,11], creation of electronically excited fragments [12], and ionization rates that depend on the internuclear distance (known as enhanced ionization) [13–15]. In intense laser fields ($I > 10^{13}\,\mathrm{W\,cm^{-2}}$), the Coulombic potentials that the electrons are placed in are greatly distorted [16–20]: A large part of the electron density is transferred among nuclei within a half optical cycle (~1.3 fs for wavelength $\lambda = 800\,\mathrm{nm}$ light) [21–25]. Intramolecular electron transfer induced by such an intense field triggers structural deformation or nuclear motion of molecules. For example, near-IR intense laser pulses can induce structural deformations, such as bond stretching and bond angle bending [26], and even chemical reactions [27,28].

The resultant structural deformations in turn change the electronic response to the field, e.g., the efficiency of intramolecular electron transfer and the probability of a tunnel type of ionization. As the field becomes stronger and its optical period ($=2\pi/\omega$, where ω is the frequency of light) becomes longer, an electron penetrates (or goes beyond) the distorted "quasistatic" barrier(s) for ionization before the sign of the laser electric field changes [2,3,16–20]. The condition for quasistatic tunnel ionization is characterized by Keldysh as follows: the so-called Keldysh parameter $\gamma \equiv \omega\sqrt{2I_\mathrm{P}}/f(t) \ll 1$ (in atomic units) [16], where I_P is the ionization potential of the system and $f(t)$ is the electric field envelope (field amplitude) at time t. Chin and his coworkers have experimentally found out that the condition for tunnel ionization is $\gamma < 1/2$ [19,20]. This condition means that the picture of tunnel ionization is valid as long as the tunneling time for an electron to go through the distorted potential at classical speed is less than a quarter of the laser field's period.

One characteristic of ionization of a molecule in a near-IR intense laser field is enhanced ionization; ionization is greatly enhanced at a critical internuclear distance, R_c, which is twice as long as or much longer than the equilibrium internuclear distance, R_e [13–15]. Enhanced ionization is a universal phenomenon; it has been experimentally observed for various diatomic and polyatomic molecules [29] such as CO_2 [26]. Detailed theoretical analyses [15,21,25,30,31] have revealed that enhanced ionization of H_2^+ is due to electron localization originating from the transition between the ground and first excited electronic states characterized by its large transition moment proportional to the internuclear distance. This type of transition is associated with a transition between a bonding and a corresponding antibonding molecular orbital and it is named "charge resonance transition" by Mulliken [32].

Charge resonance ionization is viewed as a typical mechanism of enhanced ionization for general molecules of an odd number of electrons.

The correlation between ionization and structural deformation plays a key role in fragmentation or reaction processes [33–35]. In other words, the competition between ionization and fragmentation governs the fate of the molecule in an intense filed. One interesting example is C_{60}. It is known that C_{60} in a laser field of $\lambda \leq 800\,\text{nm}$ decomposes into fragments, e.g., through the loss of C_2 units [36–40]. On the other hand, for the case of intense ultrashort 70-fs laser pulses of long wavelength $\lambda \sim 1{,}600\,\text{nm}$, highly charged C_{60} cations up to C_{60}^{12+} are produced almost without fragmentation [41]. This is just an example to indicate that the development of laser systems generating intense femtosecond pulses has made a new tool for manipulating the dynamics of molecules on an ultrashort time scale available to chemists.

Our research on the dynamical behavior of molecules in intense fields started from an investigation of the dynamics of a very basic example, namely, H_2^+. To quantitatively understand the ultrafast electronic dynamics in intense fields, it is necessary to solve the time-dependent Schrödinger equation for the electronic degrees of freedom of a molecule. We developed an efficient grid point method, the dual transformation method [42, 43], for accurate propagation of an electronic wavepacket. We have applied this method to H_2^+ [21,25] and H_2 [22,23]. The vibrational degree of freedom is incorporated in the calculation of H_2^+ without resorting to the Born–Oppenheimer (B–O) approximation.

The calculated electronic and nuclear dynamics of H_2^+ is analyzed by using "field-following" time-dependent adiabatic electronic states $\{|n\rangle\}$ defined as the eigenfunctions of the "instantaneous" electronic Hamiltonian $H_{\text{el}}(t)$ including the interaction of the electric dipole with a laser electric field $\varepsilon(t)$ [21,23,25,44–47]. Field-free adiabatic bound states can be used to diagonalize $H_{\text{el}}(t)$. In a high-intensity regime, "field-following" TD eigenvalues interpreted as TD adiabatic potential surfaces can cross each other in energy so that nonadiabatic transitions between adiabatic states occur [21–23,25,44–47]. We have demonstrated how "doorway states" to ionization or to structural deformation of H_2^+ and H_2 can be identified by population analysis of TD adiabatic states. The analysis has shown that the nuclear dynamics in a near-IR field can be described in terms of wavepacket propagation on TD adiabatic potentials and nonadiabatic transitions due to temporal change in $\varepsilon(t)$. It also turns out that tunnel ionization to Volkov states [18,48] (quantum states of a free electron in a laser field) proceeds mainly from an adiabatic state or from a small number of adiabatic states.

The above description of dynamics in terms of TD adiabatic states constructed from bound electronic states, i.e., the TD adiabatic state approach, is applicable to the investigation of electronic and nuclear (reaction) dynamics of *polyatomic* molecules in intense laser fields [49–51]. As in the case of H_2^+, we expand the total wavefunction of a polyatomic molecule in terms of the eigenfunctions of the instantaneous electronic Hamiltonian. In this representation, while intramolecular *multielectron* dynamics of a polyatomic molecule in intense fields is described by TD adiabatic states, resultant nuclear dynamics is described by multichannel wavepacket propagation on TD adiabatic potentials with *nonadiabatic transitions* between them due to temporal change in $\varepsilon(t)$. This approach has an advantage that the properties of the adiabatic states of polyatomic molecules can be evaluated by ab initio molecular orbital methods intended to calculate bound electronic states [49, 52]. We have applied the approach to investigate the dynamics of CO_2 [49] and ethanol C_2H_5OH in intense laser fields [50, 51].

In this review, we first present the results of our theoretical investigation of electronic and nuclear dynamics of H_2^+ in near-IR intense fields. Population analysis in terms of TD adiabatic states is presented. The TD adiabatic state approach developed for polyatomic molecules is next outlined. We report the results of applications of the TD adiabatic state approach to the field-induced vibrational dynamics of a large molecule, namely, C_{60}.

3.2 Electronic and Nuclear Dynamics of H_2^+ in Intense Laser Fields

In this section, we show how the electronic and nuclear wavepacket of H_2^+ in an intense laser field evolves temporally and spatially and present the results of TD adiabatic state analysis of the wavepacket dynamics. We introduce the important concept of "doorway state to tunnel ionization" by using a basis set of TD adiabatic states.

3.2.1 Numerical Solution of the TD Schrödinger Equation for a 3D H_2^+ Hamiltonian

We first present a three-dimensional (3D) Hamiltonian for the electronic and nuclear degrees of freedom of H_2^+ in a laser electric field. The dynamics of one electron is in general described by three electronic coordinates; e.g., three cylindrical coordinates z, ρ, and φ ($x = \rho\cos\varphi, y = \rho\sin\varphi, z = z$). In this study, molecular vibration is taken into account quantum mechanically. On the other hand, the molecular axis is assumed to be fixed parallel to the polarization direction of the applied laser electric field $\varepsilon(t)$. Then, the molecular axis component of the electronic angular momentum, $m\hbar$, is conserved (the magnetic quantum number m is conserved). It is thus convenient to set the electronic coordinate z parallel to the molecular axis; after separation of the center-of-mass of a molecule, the total wavefunction can be written as a product form of $\Phi(z, \rho, R)\mathrm{e}^{\mathrm{i}m\varphi}/\sqrt{2\pi}$.

Choosing the midpoint between the two protons as the origin of the electronic coordinates, we then reduce the TD Schrödinger equation for the four degrees of freedom (z, ρ, φ, R) to that for $\Phi(z, \rho, R)$ as

$$\mathrm{i}\frac{\partial}{\partial t}\Phi(z, \rho, R) = \left[-\frac{1}{m_{\mathrm{p}}}\frac{\partial^2}{\partial R^2} + H_{\mathrm{el}}(t)\right]\Phi(z, \rho, R), \tag{3.1}$$

where m_{p} is the proton mass and the instantaneous electronic Hamiltonian $H_{\mathrm{el}}(t)$ is the sum of the Born–Oppenheimer (B–O) electronic Hamiltonian at zero field strengths, H_{el}^0, and the dipole interaction with light, $V_\varepsilon(z, t)$:

$$H_{\mathrm{el}}(t) = H_{\mathrm{el}}^0 + V_\varepsilon(z, t). \tag{3.2}$$

Throughout this paper atomic units are used unless otherwise noted (i.e., the electron mass m_{e}, Bohr radius a_0, elementary electric charge e, hartree E_{h}, and $\hbar$ are all set to unity). The explicit form of H_{el}^0 is given by [53]

$$H_{\mathrm{el}}^0 = -\frac{1}{2\mu_{\mathrm{e}}}\left(\frac{\partial^2}{\partial\rho^2} + \frac{1}{\rho}\frac{\partial}{\partial\rho} + \frac{\partial^2}{\partial z^2}\right) + \frac{m^2}{2\rho^2} + V(z, \rho, R) + \frac{1}{R}, \tag{3.3}$$

where $\mu_{\mathrm{e}} = 2m_{\mathrm{p}}m_{\mathrm{e}}/(2m_{\mathrm{p}} + m_{\mathrm{e}}) \approx m_{\mathrm{e}}$ and $V(z, \rho, R)$ is the Coulomb attractive potential exerted on the electron

$$V(z, \rho, R) = -\frac{1}{\sqrt{\rho^2 + (z - R/2)^2}} - \frac{1}{\sqrt{\rho^2 + (z + R/2)^2}}. \tag{3.4}$$

The nuclear repulsion energy $1/R$ is included in (3). The dipole interaction between the molecule and the electric field $\varepsilon(t)$ is expressed as

$$V_\varepsilon(z, t) = [1 + m_{\mathrm{e}}/(2m_{\mathrm{p}} + m_{\mathrm{e}})]\, z\varepsilon(t) \approx z\varepsilon(t). \tag{3.5}$$

Equation (3.1), which contains both electronic and vibrational degrees of freedom, can be solved "exactly" by an efficient grid point method for accurate propagation of an electronic and nuclear wavepacket of a Coulombic system without resorting to the Born–Oppenheimer approximation [21, 25]. In this method, called "dual transformation," both the wavefunction and the Hamiltonian are transformed consistently to overcome the numerical difficulties arising from the divergence of the Coulomb potentials involved. The wavefunction is transformed so that it is zero at the Coulomb singular points and the choice of new scaled coordinates introduced for the variable transformation in the Hamiltonian is also crucial. The details of the application of the dual transformation method to H_2^+ are given in [42,43]. We have also applied the method to the investigation of the dynamics of the two-electron system H_2 [22, 23].

Numerical results of the dynamics of H_2^+ in an intense laser field are presented in Sect. 3.2.3.

3.2.2 Field-Following Adiabatic States and Field-Induced Nonadiabatic Coupling

We show in the next subsection that intramolecular electronic motion in H_2^+ which triggers nuclear dynamics and tunnel ionization to H_2^{2+} can be understood in terms of field-following TD adiabatic states. The TD adiabatic states are obtained by diagonalizing the instantaneous H_2^+ electronic Hamiltonian $H_{\mathrm{el}}(t)$ in terms of bound states of H_{el}^0, $|\psi_1\rangle, |\psi_2\rangle, \ldots, |\psi_n\rangle$. The matrix elements of $H_{\mathrm{el}}(t)$ are functions of adiabatic parameters R and t. Diagonalization of $H_{\mathrm{el}}(t)$ yields the TD adiabatic states $\{|j(t)\rangle\}$ and energies $\{E_j(t)\}$, where j runs from 1 to n in ascending order with respect to energy. To diagonalize $H_{\mathrm{el}}(t)$, we used here the lowest two bound electronic states $1\,s\sigma_g$ and $1\,s\sigma_u$ of H_{el}^0 (abbreviated as $|g\rangle$ and $|u\rangle$, respectively), which are radiatively coupled with each other by the transition dipole moment $\langle g|\, z\, |u\rangle \simeq R/2$. This large transition moment is due to the charge resonance character of the electronic transition between $1\,s\sigma_g$ and $1\,s\sigma_u$, [32].

In this two-state model, the TD eigenfunctions, $|1\rangle$ and $|2\rangle$, which are adiabatically connected with $1\,s\sigma_g$ and $1\,s\sigma_u$ at $\varepsilon(t) = 0$, respectively, are given analytically as [21,25,54,55]

$$|1\rangle = \cos\,\Theta\;|g\rangle - \sin\,\Theta\;|u\rangle\,, \tag{3.6a}$$

$$|2\rangle = \cos\,\Theta\;|u\rangle + \sin\,\Theta\;|g\rangle\,, \tag{3.6b}$$

where Θ is a function of t and R

$$\Theta = \frac{1}{2}\arctan\left[\frac{2\,\langle g|\, z\, |u\rangle\, \varepsilon\,(t)}{\Delta E_{ug}\,(R)}\right], \tag{3.7}$$

with the B–O energy separation $\Delta E_{ug}(R) = E_u(R) - E_g(R)$. Here, $E_g(R)$ and $E_u(R)$ are the eigenvalues of H_{el}^0 for $1\,s\sigma_g$ and $1\,s\sigma_u$, respectively. The corresponding eigenvalues (potential surfaces) are

$$\begin{aligned} E_{2,1}\,(R,t) &= \frac{1}{2}\left[E_g\,(R) + E_u\,(R) \pm \sqrt{\Delta E_{ug}^2 + 4\,|\langle g|\, z\, |u\rangle\;\varepsilon(t)\,|^2}\right] \\ &\simeq \frac{1}{2}\left[E_g\,(R) + E_u\,(R) \pm R\,|\varepsilon\,(t)\,|\right] \text{ for large } R. \end{aligned} \tag{3.8}$$

The instantaneous electrostatic potential $V(z,\rho,R) + V_\varepsilon(z,t) + 1/R$ at $\rho = 0$ and $R = 4a_0$ is plotted in Fig. 3.1 as a function of z, together with the energies E_1 and E_2: (a) $\varepsilon(t) = 0$ and (b) $\varepsilon(t) = 0.053\,E_{\mathrm{h}}/ea_0$. The field strength in SI units is related to the value of $\varepsilon(t)$ in atomic units of field strength, i.e., in units of E_{h}/ea_0, as follows: $5.14 \times 10^{11}\varepsilon(t)\,\mathrm{V m^{-1}}$. If the value of $\varepsilon(t)$ is assumed to be equal to the field envelope (field amplitude), the corresponding light intensity I in SI units is given by $I = 3.5 \times 10^{16}\varepsilon(t)^2\,\mathrm{W\,cm^{-2}}$.

There are two potential wells around the nuclei, i.e., $z = \pm R/2$. As shown in Fig. 3.1a, the two wells are identical to each other with inversion about $z = 0$ when $\varepsilon(t) = 0$. The dipole interaction energy of an electron in a field $\varepsilon(t)$ is $\varepsilon(t)R/2$ at the right nucleus and $-\varepsilon(t)R/2$ at the left nucleus. As a result, the potential well formed around the right nucleus ascends and the well formed around the left nucleus descends when $\varepsilon(t) > 0$, as shown in Fig. 3.1b;

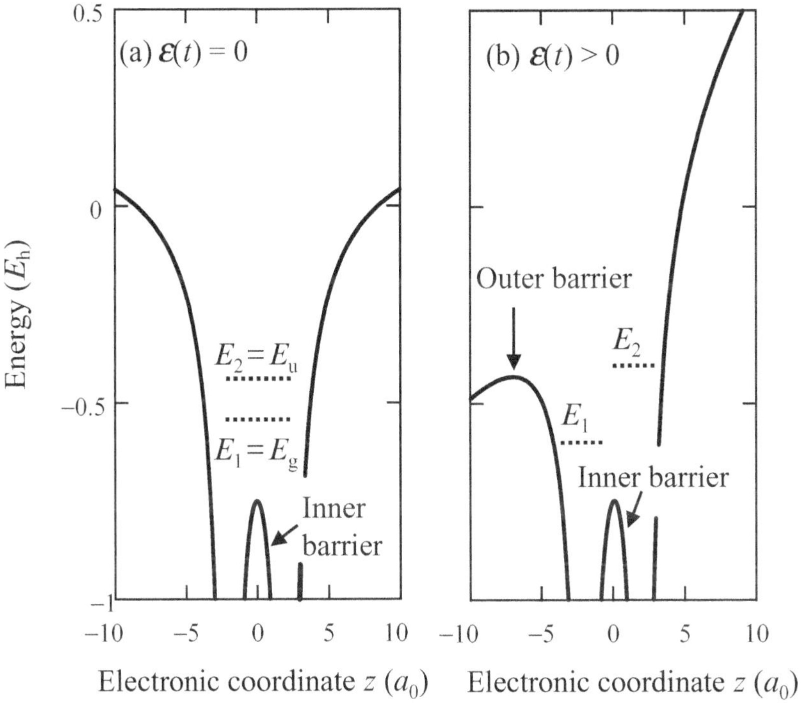

Fig. 3.1. The instantaneous electrostatic potential $V(z, \rho, R) + V_\varepsilon(z, t) + 1/R$ (*solid lines*) of H_2^+ as a function of the electronic coordinate z along the molecular axis at (**a**) $\varepsilon(t) = 0$ and (**b**) $\varepsilon(t) = 0.053\, E_h/ea_0$ ($I = 10^{14}\,\mathrm{W\,cm^{-2}}$). The internuclear distance is fixed at $R = 4a_0$ and the molecular axis is assumed to be parallel to the polarization direction of the laser electric field. The energies E_1 and E_2 of the lowest two TD adiabatic states $|1\rangle$ and $|2\rangle$ are denoted by *horizontal dotted lines.* E_g and E_g in (**a**) are the Born–Oppenheimer adiabatic energies of $1\,s\sigma_g$ and $1\,s\sigma_u$, E_g and E_u, respectively. The electrostatic potential has two wells around the nuclei, i.e., $z = \pm R/2$. As shown in (**b**), a barrier that an electron can tunnel through or go over (outer barrier) is formed when $\varepsilon(t) \neq 0$, in addition to the inner barrier

the positions of the ascending and descending wells are reversed when $\varepsilon(t) < 0$. In addition to the barrier between the two wells (inner barrier), when $|\varepsilon(t)| \neq 0$, a barrier with finite width is formed outside the descending well (outer barrier). An electron can tunnel through or go over the outer barrier.

The electron densities of $|1\rangle$ and $|2\rangle$ are high near the descending and ascending wells, respectively. Therefore, the values of E_1 and E_2 at large $R\,|\varepsilon(t)|$ are dominated by the energy shifts $\mp R\,|\varepsilon(t)|/2$ of the descending and ascending wells, respectively, as shown in (3.8). The adiabatic energies $E_1\,(R,t)$ and $E_2\,(R,t)$ at $\varepsilon(t) = 0.053\,E_h/ea_0$ are plotted in Fig. 3.2 against R, together with $E_g(R)$ and $E_u(R)$. The heights of the inner and outer barriers for tunnel or over-barrier ionization also depend on R, which are denoted by

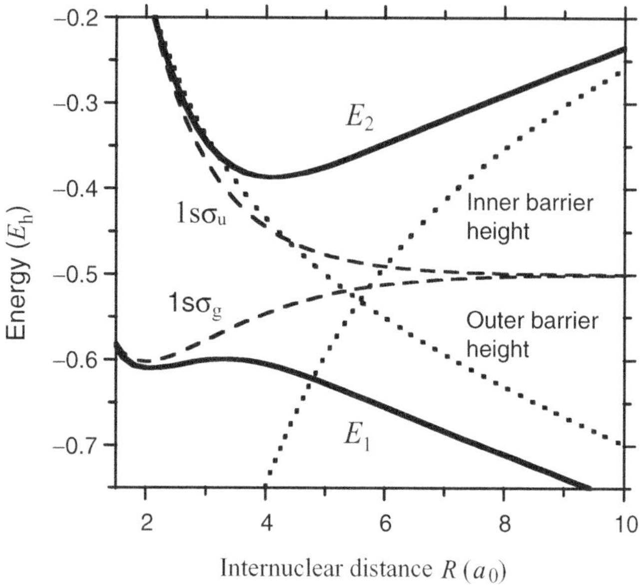

Fig. 3.2. Potential energies $E_1(R,t)$ and $E_2(R,t)$ of the lowest two field-following adiabatic states $|1\rangle$ and $|2\rangle$ of H_2^+ (denoted by *solid lines*) at $\varepsilon(t) = 0.053\,E_h/ea_0$, i.e., at $I = 10^{14}\,\mathrm{W\,cm^{-2}}$ as a function of internuclear distance R. The *broken lines* denote the Born–Oppenheimer potential surfaces $E_g(R)$ and $E_u(R)$ of $1\,s\sigma_g$ and $1\,s\sigma_u$. The *dotted lines* denote the heights of the inner and outer barriers for tunnel or over-barrier ionization. As R increases, the inner barrier height increases while the outer barrier height decreases. The internuclear repulsion $1/R$ is included in the energy, as in Fig. 3.1

dotted lines in Fig. 3.2. We clarify the mechanism of enhanced ionization of H_2^+ in Sect. 3.2.4 by comparing $E_1\,(R,t)$ and $E_2\,(R,t)$ to the barrier heights.

We projected the "exact" electronic and nuclear wavepacket $\Phi(z,\rho,R;t)$ onto the two TD adiabatic states $|1\rangle$ and $|2\rangle$ to investigate the H_2^+ nuclear dynamics and ionization dynamics of H_2^+. The two projected components $\chi_1(R,t) = \langle 1 \mid \Phi(t)\rangle$ and $\chi_2(R,t) = \langle 2 \mid \Phi(t)\rangle$ (integrated over ρ and z), which are regarded as the vibrational wavefunctions propagating on E_1 and E_2, are used for analysis of the mechanism of ionization as well as field-induced nuclear dynamics. The results are presented in Sect. 3.2.3.

We also used an expansion formula for the total wavefunction $|\Phi\rangle$ as

$$|\Psi(z,\rho,R;t)\rangle = \chi_1'(R,t)\,|1\rangle + \chi_2'(R,t)\,|2\rangle\,. \tag{3.9}$$

where $|\Psi\rangle$ is an approximate wavefunction for $|\Phi\rangle$. It will be shown in Sect. 3.2.3 that the key processes of intramolecular electronic motion and nuclear motion in H_2^+ can be described by this expansion while ionization is not directly taken into account. The following approach using the two TD adiabatic states can be straightforwardly extended to cases where the number of TD adiabatic states are more than two. The results of a six-state model are summarized in Sect. 3.2.4.

Inserting (3.9) into (3.1), we derived the coupled equations within the framework of the two-state model as [21, 25, 30, 31]

$$\frac{\partial}{\partial t}\chi_1'(R) = -\mathrm{i}\left[-\frac{1}{m_\mathrm{p}}\frac{\partial^2}{\partial R^2} + E_1(R,t) + \frac{1}{m_\mathrm{p}}\left(\frac{\partial\Theta}{\partial R}\right)^2\right]\chi_1'(R) - \Lambda(R,t)\chi_2'(R), \tag{3.10a}$$

$$\frac{\partial}{\partial t}\chi_2'(R) = -\mathrm{i}\left[-\frac{1}{m_\mathrm{p}}\frac{\partial^2}{\partial R^2} + E_2(R,t) + \frac{1}{m_\mathrm{p}}\left(\frac{\partial\Theta}{\partial R}\right)^2\right]\chi_2'(R) + \Lambda(R,t)\chi_1'(R), \tag{3.10b}$$

where the total coupling $\Lambda(R,t)$ which induces nonadiabatic transitions between the two TD adiabatic states is

$$\Lambda(R,t) = \left\langle 1 \left| \frac{\partial}{\partial t} \right| 2 \right\rangle - \frac{2\mathrm{i}}{m_\mathrm{p}}\frac{\partial\Theta}{\partial R}\frac{\partial}{\partial R} - \frac{\mathrm{i}}{m_\mathrm{p}}\frac{\partial^2\Theta}{\partial R^2} + \cdots . \tag{3.11}$$

Here the small terms $\langle g|\,\partial^2/\partial R^2\,|g\rangle/m_\mathrm{p}$ and $\langle u|\,\partial^2/\partial R^2\,|u\rangle/m_\mathrm{p}$ are not explicitly written in the off-diagonal element (3.11) and in the diagonal elements inside the square brackets of (3.10). The term $\langle 1|\,\partial/\partial t\,|2\rangle = \partial\Theta/\partial t$ is the nonadiabatic coupling due to temporal change in $\varepsilon(t)$ [54,55] and the other coupling terms in (3.11) are due to the joint effect of the electric field and the nuclear motion.

When typical values of the nuclear velocities are assumed, the effect of the first term in (3.11), namely, $\partial\Theta/\partial t$, is much larger than that of the second term for homonuclear diatomic ions (in a near-IR field) [54, 55]. In the case of heteronuclear diatomic ions, the dominant coupling term which causes hopping between TD surfaces is also $\partial\Theta/\partial t$. The second term causes significant hopping at small internuclear distances only when the field-induced Stark shift is, in the domain of small R, comparable to the energy gap between the field-free states.

It is expected from (3.10) and (3.11) that field-induced nonadiabatic transitions occur if the main coupling term $\partial\Theta/\partial t$ is large in comparison with the gap between the two adiabatic potential surfaces, $E_2 - E_1$. The term $\partial\Theta/\partial t$ is expressed as

$$\frac{\partial\Theta}{\partial t} = \frac{\langle g|\,z\,|u\rangle}{\Delta E_{ug}(R)}\frac{\partial\varepsilon(t)}{\partial t}(\cos 2\Theta)^2. \tag{3.12}$$

Since $\varepsilon(t)$ is a sinusoidal function of time, $|\partial\varepsilon(t)/\partial t|$ and $(\cos 2\Theta)^2$ becomes largest when $\varepsilon(t) = 0$, i.e., $t_n = n\pi/\omega$ (n is an integer) for

$$\varepsilon(t) = f(t)\sin(\omega t), \tag{3.13}$$

where ω is the light frequency and $f(t)$ is the envelope (or amplitude) of the laser electric field. When $\varepsilon(t)$ changes its sign, the gap $E_2 - E_1$ becomes smallest ($\sim\Delta E_{ug}$) and an avoided crossing takes place. Thus, a nonadiabatic transition between the two adiabatic states is expected to occur within a very short period $\tau_{\rm tr}$ around $\varepsilon(t) = 0$ [30, 56, 57]. It should be noted that $\langle g|\,z\,|u\rangle/\Delta E_{ug}(R)$ increases as R increases. Larger internuclear distances are thus favorable for nonadiabatic transition. A nonadiabatic transition between $|1\rangle$ and $|2\rangle$ corresponds to suppression of electron transfer between the two wells, i.e., two nuclei; in other words, the electron is localized in a well.

If the nuclear motion is slow in comparison with the time scale of $\tau_{\rm tr}$, the nonadiabatic transition probability H_2^+ *per level crossing* at R, $P_{\rm nonad}$, can be defined and it is given by the Landau–Zener form [58, 59]

$$P_{\rm nonad} = \exp(-2\pi\delta), \tag{3.14}$$

where

$$\delta = \Delta E_{ug}(R)^2/[8\,\langle g|\,z\,|u\rangle\, f(t)\omega]. \tag{3.15}$$

For $\delta >> 1$, the electronic state remains in the same adiabatic state such as $|1\rangle$ or $|2\rangle$. For $\delta << 1$, the main route is the nonadiabatic channel. The probability $P_{\rm nonad}$ can be increased as the intensity, frequency, and R are increased. We have obtained conditions for isolated transition (where $\tau_{\rm tr} << 2\pi/\omega$) [30]; $\omega < \Delta E_{ug}\,(R)/2 < 2\,\langle g|\,z\,|u\rangle\, f(t)$ in the adiabatic case and $\max\,[\omega, \Delta E_{ug}\,(R)/2] < 2\,\langle g|\,z\,|u\rangle\, f(t)$ in the diabatic case. The condition for isolated transition holds in the case of near-IR fields. In the following subsections, we show how the nonadiabatic transition between the adiabatic states is correlated to dissociation and ionization processes.

3.2.3 Analysis of Electronic and Nuclear Dynamics of H_2^+ in an Intense Near-IR Field

We show that for an intense near-IR field, a tunnel type of ionization proceeds in the region $R \geq 2R_{\rm e}$ through the $|2\rangle$ state populated from $|1\rangle$, where $R_{\rm e}$ is the equilibrium internuclear distance. The two adiabatic states $|1\rangle$ and $|2\rangle$ (or $1\,s\sigma_g$ and $1\,s\sigma_u$) are essential states for describing the nuclear dynamics and the electronic dynamics prior to ionization. To demonstrate these points, we first present the projected components $\chi_1(R,t)$ and $\chi_2(R,t)$ of the "exact" total wavepacket $\Phi(\rho, z, R; t)$ onto $|1\rangle$ and $|2\rangle$. In the following, we will also show that the solutions of the coupled (3.10a) and (3.10b) for the two-state model are semiquantitatively consistent with the motion of the exact projected components $\chi_1(R,t)$ and $\chi_2(R,t)$.

Experimentally, H_2^+ is prepared in an intense laser field through one-electron ionization of H_2. We assume that H_2 is vertically ionized from its lowest vibrational state $\chi_{\rm vib=0}(R)$ in the ground electronic state $X^1\Sigma_g^+$ to $|1\rangle$ of H_2^+ [60–64]. In the present 3D simulation, the following form is employed as the total wavefunction of H_2^+ at $t = 0 : \Phi(z, \rho, R; t) = \chi_{\text{vib}=0}(R)\,|1\rangle$. Here, we simply use the $1\,s\sigma_g$ B–O electronic wavefunction of H_2^+ as $|1\rangle$ at $t = 0$. The parameters of the applied laser pulse employed in the present simulation are as follows: $\omega = 0.06E_h/\hbar$ ($\lambda = 760$ nm) and the envelope $f(t)$ is linearly ramped with t so that after two optical cycles $f(t)$ attains its peak value $f_0 = 0.12\,E_h/ea_0$ ($I = 5.0 \times 10^{14}\,\text{W cm}^{-2}$). The total pulse duration is 20 cycles (40 fs). We set absorbing boundaries for the electronic coordinates ρ and z to eliminate the outgoing ionizing flux for the evaluation of the ionization probability.

The vibrational components $\chi_1(R, t)$ and $\chi_2(R, t)$ are shown in Figs. 3.3a and 3.3b, respectively. The wavepacket of H_2^+ starts from the equilibrium internuclear distance $R_e \approx 1.5a_0$ of H_2; $\chi_1(R, t = 0) = \chi_{\text{vib}=0}(R)$ and $\chi_2(R, t = 0) = 0$. Since the equilibrium internuclear distance is longer for H_2^+ than for H_2, $\chi_1(R, t)$ initially moves toward larger internuclear distances. Without an external field, the main component of the packet reflects at the outer turning point in E_g at $R \approx 3a_0$, coming back toward the inner turning point. On the other hand, in an intense field, dissociation takes place as well as ionization. When the instantaneous field strength $|\varepsilon(t)|$ is large, the potentials $E_1(R, t)$ and $E_2(R, t)$ given by (3.8) are distorted as shown in Fig. 3.2. The nuclear wavepacket is moved toward larger internuclear distances by the field-induced barrier suppression in $E_1(R, t)$ (bond softening due to a laser field).

In the early stage, in which $R < 3a_0$, the response to the field is adiabatic: The main component of the wavepacket is still in $|1\rangle$. At larger R ($>3a_0$), nonadiabatic transitions occur between $|1\rangle$ and $|2\rangle$ when $\varepsilon(t)$ changes its sign, i.e., when the two adiabatic potential surfaces come closest to each other. As the packet reaches internuclear distances around $R \sim 3.5a_0$, δ becomes as small as 0.23 for $f(t) = 0.12\,E_h/ea_0$; $P_{\text{nonad}} = 0.24$. Around $t = 5.1$ fs ($=4\pi/\omega$), as shown in Fig. 3.3b, a part of the population in $|1\rangle$ is transferred to $|2\rangle$. The temporal structure in Fig. 3.3b clearly demonstrates that nonadiabatic transitions from $|1\rangle$ to $|2\rangle$ occur at every crossing around $t = j\pi/\omega$ ($j \geq 4$), where $\delta < 1$. Part of the nuclear wavepacket in $|1\rangle$ bifurcates into $\chi_1(R, t)$ and $\chi_2(R, t)$ at $R = 3.5a_0$–$4.5a_0$ when $\varepsilon(t) = 0$, owing to high probabilities of nonadiabatic transition between $|1\rangle$ and $|2\rangle$ for $R > 3.5a_0$. After the subsequent avoided crossing (in the subsequent half optical cycle), $\chi_1(R, t)$ and $\chi_2(R, t)$ that reach a region around $R = 5a_0$ destructively interfere with each other in $|2\rangle$ [46]: $|\chi_2(R, t)|^2$ at $R \sim 5a_0$ turns to be almost zero. The probability $|\chi_2(R, t)|^2$ has therefore a gap around $R = 5a_0$, as shown in Fig. 3.3b; i.e., the spatial distribution of $|\chi_2(R, t)|^2$ has two humps A and B.

The wavefunctions $\chi_1'(R, t)$ and $\chi_2'(R, t)$ in the two-state model obtained by solving the coupled (3.10a) and (3.10b) are shown in Figs. 3.3a' and 3.3b', respectively. It should be pointed out that the two-state model of (3.9) gives nearly the same *nuclear* dynamics as the "exact" projected ones $\chi_1(R, t) = \langle 1 \mid \Phi\rangle$ and $\chi_2(R, t) = \langle 2 \mid \Phi\rangle$, although the decrease in population due to ionization cannot be reproduced. This indicates that *the nuclear motion in intense fields is determined by the potential surfaces of a small number of TD*

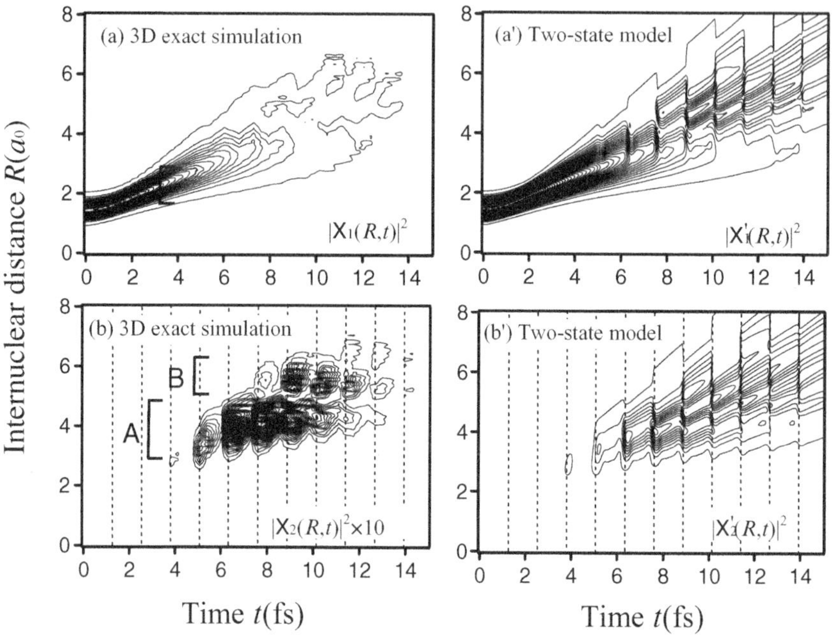

Fig. 3.3. Dynamics of nuclear wavepackets of H_2^+ in an intense 760-nm laser field ($\omega = 0.06\,E_h/\hbar$ and the envelope $f(t)$ is linearly ramped with t so that after two optical cycles $f(t)$ attains its peak value $f_0 = 0.12\,E_h/ea_0$). The nuclear packets $|\chi_1(R,t)|^2 = |\langle 1|\,\Phi\rangle|^2$ and $|\chi_2(R,t)|^2 = |\langle 2|\,\Phi\rangle|^2$ obtained from the full 3D wavepacket $\Phi(z,\rho,R;t)$ are denoted by *solid contour lines* in (**a**) and (**b**), respectively. Panels (**a**') and (**b**') show $|\chi'_1(R,t)|^2$ and $|\chi'_2(R,t)|^2$ in the two-state model defined by (9), respectively. Level crossing moments, at which the two adiabatic energies are closest to each other (i.e., the electric field $\varepsilon(t)$ is zero), are indicated in the lower panels by *vertical dotted lines*. The distribution of $|\chi_2(R,t)|^2$ has two humps in R-space as indicated by the regions A and B

adiabatic states. The concept of TD adiabatic state has already proved useful for designing actual experiments. Using the idea of time-dependent adiabatic state, Niikura et al. have designed a scheme to control the vibrational wave packet on the lowest TD adiabatic state $|1\rangle$ of H_2^+ or D_2^+ [65].

We also performed calculations for the two-state model of H_2^+ on the assumption that all the coupling terms in (3.11) other than the dominant direct field-induced term $\langle 1|\,\partial/\partial t\,|2\rangle = \partial\Theta/\partial t$ are zero (the field-induced coupling model). As expected from the discussion in Sect. 3.2.2, $\chi'_1(R,t)$ and $\chi'_2(R,t)$ obtained by solving the coupled (3.10a) and (3.10b) for the field-induced coupling model are nearly identical with those in Figs. 3.3a and 3.3b. The dominance of field-induced nonadiabatic coupling $\langle 1|\,\partial/\partial t\,|2\rangle$ is the key to develop a practicable way of estimating the nuclear dynamics of general polyatomic molecules in intense laser fields. The details of a proposed method are given in Sect. 3.3.

3.2.4 Adiabatic Tunnel Ionization and Nonadiabatic Ionization

The overall populations $P_1(t) = \int |\chi_1(R,t)|^2 \mathrm{d}R$ and $P_2(t) = \int |\chi_2(R,t)|^2 \mathrm{d}R$ are shown in Fig. 3.4 together with the total norm in the grid space enclosed with the absorbing boundaries, $P_{\mathrm{grid}}(t)$. The increase in $P_2(t)$ *at a crossing point* coincides with the decrease in $P_1(t)$. During the next half cycle after a nonadiabatic transition (as the field strength approaches a local maximum), a reduction in $P_2(t)$ is clearly observed, whereas $P_1(t)$ changes very little. According to nonadiabatic transition theory [57–59], $P_1(t)$ and $P_2(t)$ should be constant between adjacent crossing points if ionization does not take place. The reduction in $P_2(t)$ just after a crossing is hence due to ionization to Volkov states or ionization via the H_2^+ higher TD adiabatic states. This is

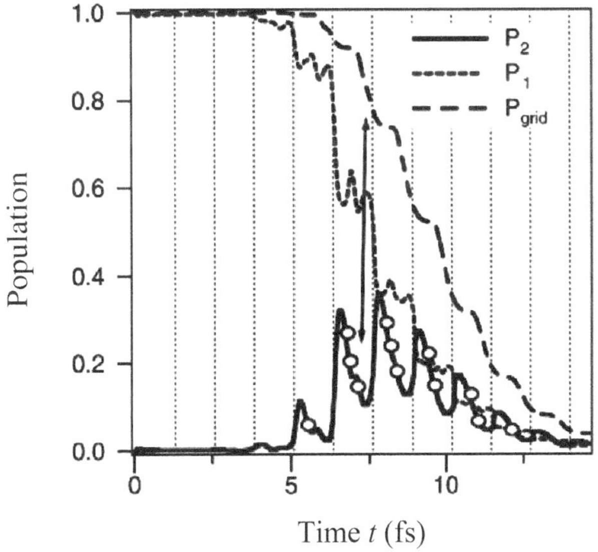

Fig. 3.4. Overall populations $P_1(t)$ and $P_2(t)$ on the lowest two adiabatic states, $|1\rangle$ and $|2\rangle$, of H_2^+ in an intense 760-nm laser field. See, for the parameters of the field, the captions in Fig. 3.3. The populations $P_1(t)$ and $P_2(t)$ integrated over R are denoted by a *dotted line* and a *solid line*, respectively. The total norm in the grid space enclosed with the absorbing boundaries, $P_{\mathrm{grid}}(t)$, is denoted by a *broken line*. Moments of avoided crossing for H_2^+, where $\varepsilon(t) = 0$, are indicated by *vertical dotted lines*. Around $t = 5.1$ fs ($=4\pi/\omega$), the wavepacket approaches $R \sim 3.5a_0$; then, nonadiabatic transitions between $|1\rangle$ and $|2\rangle$ begin. Nonadiabatic transitions occur around level crossing points $t = n\pi/\omega$ for $n \geq 4$ (in the case of the electric field (3.13)). A reduction in $P_2(t)$ denoted by open circles is due to ionization. The upper adiabatic state $|2\rangle$ is regarded as the main doorway state to tunnel ionization. The reduction in $P_2(t)$ corresponds to the subsequent reduction in $P_{\mathrm{grid}}(t)$ (delayed by a time that the ionizing currents need to hit the absorbing boundaries) as indicated by the line with arrows at both ends

also confirmed by the correlation in reduction between $P_{\mathrm{grid}}(t)$ and $P_2(t)$ as indicated by the line with arrows at both ends. We hence conclude that $|2\rangle$ is the main doorway state to ionization. Enhancement of ionization in H_2^+ around $R > 2R_{\mathrm{e}}$ is due to the joint effect of the following three factors: (a) Nonadiabatic transitions from $|1\rangle$ to $|2\rangle$ occur at large R; (b) $|2\rangle$ is easier to ionize than is $|1\rangle$; (c) the ionization probability of $|2\rangle$ has peaks around certain internuclear distances $R > 2R_{\mathrm{e}}$.

The reason why $|2\rangle$ is easier to ionize than is $|1\rangle$ can be explained by comparing the adiabatic energies with the barrier heights for tunnel ionization. As shown in Fig. 3.1, in addition to the inner barrier, the instantaneous electrostatic potential $V(z, \rho, R) + z\varepsilon(t) + 1/R$ distorted in the presence of the field $\varepsilon(t) \neq 0$ has the outer barrier which an electron can tunnel through or go over. The electron must go over or through both the inner and outer barriers in the case of ionization from $|2\rangle$, while the electron goes over or through only the outer barrier in the case of ionization from $|1\rangle$. The barrier heights of the instantaneous electrostatic potential evaluated along the molecular axis are plotted in Fig. 3.2 against R. It is clearly demonstrated that while E_1 is usually below both of the barrier heights, E_2 can be higher than both of the barrier heights in a range around $R_{\mathrm{c}} \approx 6a_0$. The component of $|2\rangle$ which is nonadiabatically created from $|1\rangle$ is easier to ionize than is $|1\rangle$, except at the atomic limit where $R \gg R_{\mathrm{e}}$. If there is a period in which the energy of the adiabatic state is higher than the barrier(s) for ionization, "overbarrier" ionization outmeasures "underbarrier" tunnel ionization. In this paper, "tunnel ionization" simply refers to both types of ionization, though Reiss has pointed out that ionization becomes increasingly distant from tunneling as γ becomes much smaller than unity [66].

We next discuss the ionization of $|2\rangle$ quantitatively. To that end, we calculated the rate of ionization from $|2\rangle$ in a DC field, Γ_2. The ionization rate Γ_2 for constant field strength of $\varepsilon(t) = 0.0533\, E_{\mathrm{h}}/ea_0$ is plotted in Fig. 3.5 as a function of R. The rates Γ_2 were obtained by solving the TD Schrödinger equation of H_2^+ in the DC field. Roughly speaking, Γ_2 is large in a wide range from $R = 5a_0$ to $10a_0$, owing to barrier suppressions favorable for tunnel ionization from $|2\rangle$ (in the region of $R < 10a_0$, the rate of ionization from $|1\rangle$ in a DC field, Γ_1, is one-hundredth as large as Γ_2). Besides, a couple of peaks are observed in the range (Γ_2 also has a peak at $R \approx 0.5a_0$). The maximum ionization rate of Γ_2 is found at $R \approx 9a_0$. For this field strength, at $R \approx 9a_0, E_2$ is however only a little above the inner barrier, as shown in Fig. 3.2. The fast ionization originates from the following facts: At $R \approx 9a_0, |2\rangle$ is resonant with an adiabatic state (originating from states other than $1\,s\sigma_g$ and $1\,s\sigma_u$) in the descending well that is higher than $|1\rangle$; the outer barrier is thin and is much lower than E_2. Another peak is also found at $R \approx 6a_0$, but it is lower than that at $R \approx 9a_0$. The rate of ionization from $|2\rangle$ seems to be more sensitive to the outer barrier than the inner barrier.

We also show in Fig. 3.5 the ionization rates Γ in an alternating intense field [15, 30, 67]. The initial state is the ground state $1\,s\sigma_g$. The parameters for the

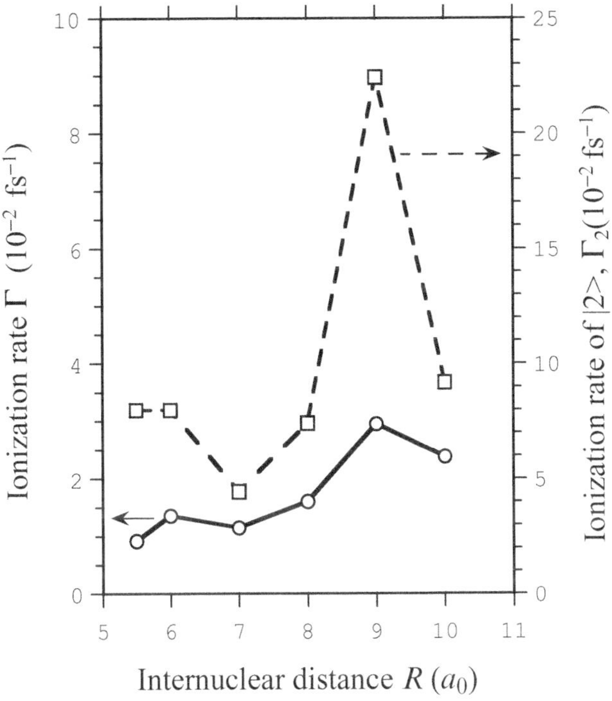

Fig. 3.5. Ionization rates of H_2^+ as a function of internuclear distance R. The open squares denote ionization rates Γ_2 of the upper adiabatic state $|2\rangle$ at a DC constant field strength $\varepsilon(t) = 0.0533\,E_\mathrm{h}/ea_0$ (which corresponds to $I = 10^{14}\,\mathrm{W\,cm^{-2}}$); the open circles denote ionization rates Γ in an alternating intense field of $\lambda = 1{,}064$ nm and $I = 10^{14}\,\mathrm{W\,cm^{-2}}$ under the condition that the initial state is the ground state $1\,s\sigma_g$ (the field envelope has a five-cycle linear ramp). The scales for Γ_2 and Γ are marked on the right and left ordinates, respectively

applied field are as follows: $\omega = 0.0428\,E_\mathrm{h}/\hbar$ ($\lambda = 1{,}064$ nm) and the envelope $f(t)$ attains its constant peak value $f_0 = 0.0533\,E_\mathrm{h}/ea_0$ ($I = 10^{14}\,\mathrm{W\,cm^{-2}}$) after a linear ramp of five cycles. Γ is defined as the stationary ionization rate after the ramp. The ionization rate Γ in an alternating intense field is correlated with Γ_2 (e.g., both have peaks at $R = 6a_0$ and $9a_0$), which also indicates that ionization proceeds via the $|2\rangle$ nonadiabatically created from the initial state $1\,s\sigma_g$. The double peak structure is also observed for the case of $\lambda = 760$ nm; it is universal in the near-IR regime.

The above mechanism of ionization is called charge resonance-enhanced ionization [15, 30] because nonadiabatic transitions occur between the two adiabatic states $|1\rangle$ and $|2\rangle$ arising from a charge resonance pair of $1\,s\sigma_g$ and $1\,s\sigma_u$. The analysis in terms of the populations of the *two* adiabatic states is validated by the fact that in the high-intensity and low-frequency regime only these two states are mainly populated before ionization [45, 68]. By using six states as a basis set for the diagonalization of $H_{\rm el}(t)$, we have confirmed that the bound state component other than $|1\rangle$ and $|2\rangle$ can be just as large as 5–10% of the total bound state component. Moreover, all the intermediates other than $|1\rangle$ and $|2\rangle$ vanish completely within a quarter cycle due to overbarrier ionization. It is concluded that $|1\rangle$ and $|2\rangle$ (or $1\,s\sigma_g$ and $1\,s\sigma_u$) are essential states for describing the electronic dynamics prior to tunnel ionization.

We also calculated three types of ionization rates of H_2^+: (a) the rate of ionization in a field of $\lambda = 760\,\text{nm}$ and $f_0 = 0.1\,E_{\rm h}/ea_0$ ($I = 3.5 \times 10^{14}\,\text{W cm}^{-2}$); (b) the rate of adiabatic tunnel ionization from the lowest TD adiabatic state in a DC field of $\varepsilon(t) = 0.1\,E_{\rm h}/ea_0$; (c) the adiabatic tunnel ionization rate averaged over one optical cycle of a field of $f(t) = 0.1\,E_{\rm h}/ea_0$ (according to the quasiperiodic tunneling theory). We denote these three rates by Γ, Γ_1, and $\bar{\Gamma}_1$, respectively. The initial state for the calculation of Γ is chosen to be the ground electronic state $1\,s\sigma_g$.

The rates Γ, Γ_1, and $\bar{\Gamma}_1$ are plotted in Fig. 3.6 as a function of R. Although Γ_1 is slightly higher than Γ up to $R = 3a_0$, the cycle average of adiabatic tunnel ionization rate, $\bar{\Gamma}_1$, is much smaller than Γ (by a factor of $\sim$4). This

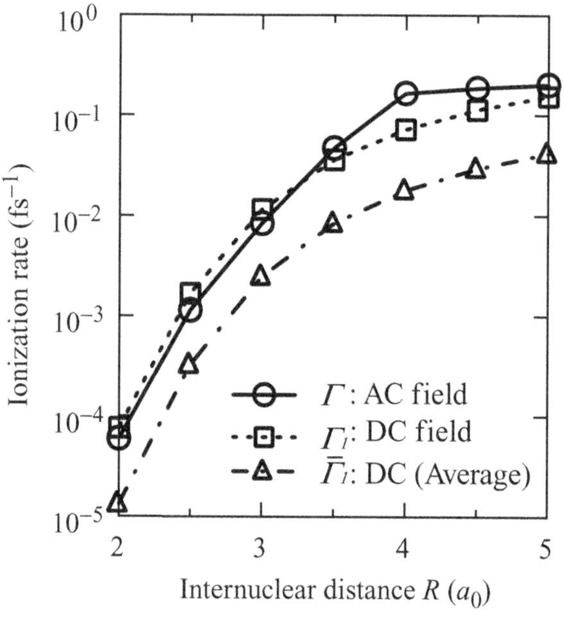

Fig. 3.6. Three types of ionization rates of H_2^+: (Γ denoted by ○) the rate of ionization in a field of $\lambda = 760$ nm and $f(t) = 0.1E_{\rm h}/ea_0$ ($I = 3.5 \times 10^{14}\,\text{W cm}^{-2}$); ($\Gamma_1$ denoted by □) the rate of adiabatic tunnel ionization from the lowest TD adiabatic state in a DC field of $\varepsilon(t) = 0.1E_{\rm h}/ea_0$; ($\bar{\Gamma}_1$ denoted by Δ) the adiabatic tunnel ionization rate averaged over one optical cycle for a field of $f(t) = 0.1E_{\rm h}/ea_0$

indicates that the main ionization mechanism is not the adiabatic tunnel ionization from the lowest TD adiabatic state but the nonadiabatic ionization through higher TD adiabatic states. In the region of $R > 3a_0$, nonadiabatic ionization becomes more dominant. The total population of the higher TD adiabatic states is one-thousandth of the population of the lowest TD adiabatic state, but the ionization rates of higher TD adiabatic states are higher than that of the lowest TD adiabatic state by orders of magnitude [68].

The roles of nonadiabatic response to intense fields in ionization of general polyatomic molecules have been studied experimentally and theoretically [24, 35, 69–71]. Hankin et al. have investigated ionization of 23 organic molecules caused by their interaction with intense 40 fs, 800 nm pulses [71]. All molecules reach saturated ionization at higher intensities than would be expected for atoms of the same ionization potential. Dependence of the ionization rate on the alignment of the molecule with the laser field is ruled out as the cause of the high saturation intensities. For many organic molecules, saturated ionization occurs at higher intensity than expected. Later, Lezius et al. proposed a mechanism that "nonadiabatic multielectron dynamics" is responsible for higher saturated intensities for organic molecules [24]. They emphasized that the adiabatic assumption fails in the case of a large molecule where the size of the molecule (the spatial extent of delocalized electrons) exceeds the path lengths of field-induced electron motion; i.e., the time scale of electron motion inside the binding potential well is longer than the laser period, driving electron dynamics inside the molecule. This is also emphasized in our results of the field-induced intramolecular electron dynamics in H_2^+ or H_2 at *large* internuclear distances. The popular Keldysh parameter, often used as a diagnostic of tunneling behavior, becomes meaningless when electron dynamics inside the molecule become important. They also suggested that collective multielectron behavior should start to dominate the strong-field response as the system increases further in size. The energy poured into a molecule by a laser field can be shared by the dynamic polarization of all electrons.

The existence of various fragmentation pathways in C_{60} [36–40] as well as high saturation intensities for its ionization [38, 41] also support the idea that nonadiabatic multielectron dynamics plays a significant role in ionization, fragmentation, and energetics of a molecule in an intense laser field. Shchatsinin et al. have applied 9-fs pulses to the excitation of C_{60} and observed cations up to C_{60}^{6+} [38]. They found that the yields of high charge parent cations C_{60}^{z+} as a function of the light intensity cannot be fitted by using any sequential ionization model. They ascribed the failure of the sequential model to nonsequential ionization through *nonadiabatic multielectron dynamics*. Even for 9-fs pulses, ionization to C_{60}^{z+} for $z \geq 2$ occurs with a significant amount of large fragments, which is also considered a result of electronic excitation due to nonadiabatic multielectron dynamics.

Applying the dual transformation method to H_2 [22, 23], we have also demonstrated that "doorway states" to ionization of H_2 can be identified by population analysis of TD adiabatic states [23]. We revealed that enhanced ionization of H_2 is due to field-induced formation of a pair of two electrons (charge transfer mechanism).

3.2.5 Coulomb Explosions: Kinetic Energies of Protons

In order to evaluate the kinetic energy release (KER) of protons originating from Coulomb explosions $H^+ + H^+$, we estimate both the critical distance R_c and the critical momentum P_c when ionization to H_2^{2+} occurs. The two humps in Fig. 3.3b correspond to $R_c' \approx 4.2a_0$ and $R_c'' \approx 5.7a_0$. The distribution in conjugate momentum P for molecular vibration can be obtained by using a Fourier transform of $\chi_1(R,t)$ or $\chi_2(R,t)$ with respect to R. The H_2^+ momentum distributions, $|\bar{\chi}_1(P,t)|^2$ and $|\bar{\chi}_2(P,t)|^2$, obtained as a function of time are plotted in Fig. 3.7 as contour maps. As shown in Fig. 3.7b, the momentum distribution obtained from $\chi_2(R,t)$ has two peaks around $P_c' = 7.7\hbar/a_0$ and $P_c'' = 13.0\hbar/a_0$, which correspond to the components around R_c' and R_c'', respectively (This is confirmed by a representation with finite resolution in two-dimensional P–R space). The slower and faster components correspond to Floquet one-photon (1ω) and two-photon (2ω) dissociation channels, respectively [72–74].

The plots in Fig. 3.7 clearly show that the momentum in $|\bar{\chi}_1(P,t)|^2$ increases between two successive avoided crossings as the arrow in Fig. 3.7a indicates, while the momentum in $|\bar{\chi}_2(P,t)|^2$ decreases as the arrow in Fig. 3.7b indicates. This reflects the fact that the motion of the nuclear wavepacket toward dissociation is accelerated between two successive avoided

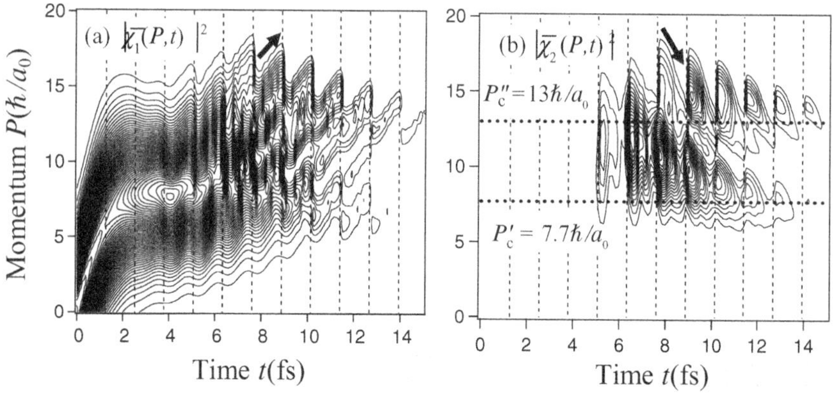

Fig. 3.7. Momentum distributions $|\bar{\chi}_1(P,t)|^2$ and $|\bar{\chi}_2(P,t)|^2$ for molecular vibration of H_2^+ in an intense near-IR laser field. The applied pulse is the same as used for the case in Fig. 3.3. Moments of level crossing for H_2^+, where $\varepsilon(t) = 0$, are indicated by *vertical dotted lines*. The momentum in $|\bar{\chi}_1(P,t)|^2$ increases between two successive avoided crossings, as indicated by an arrow in (**a**); the momentum in $|\bar{\chi}_2(P,t)|^2$ decreases as indicated by an arrow in (**b**)

crossings when it is on $E_1(R,t)$ and the wavepacket is decelerated on $E_2(R,t)$. These features are consistent with the potential shapes of the TD adiabatic states $|1\rangle$ and $|2\rangle$ shown in Fig. 3.2.

The KER of the H^+ fragment for a Coulomb explosion, denoted by K_c, should be given by

$$K_c = \frac{1}{2}\left(\frac{P_c^2}{m_p} + \frac{1}{R_c}\right) \tag{3.16}$$

if the electron is ejected at R_c with the momentum P_c. We have $K_c' = 3.7\,\text{eV}$ for the set of R_c' and P_c'; $K_c'' = 3.6\,\text{eV}$ for R_c'' and P_c''. These values are close to the experimental value. For instance, the peak of the experimental KER distributaion is located at $K_c \approx 3\,\text{eV}$ for a 25-fs, $2 \times 10^{14}\,\text{W cm}^{-2}$ pulse ($\lambda = 795\,\text{nm}$) [75]. A detailed analysis of the KER spectra of proton (deuteron) pairs based on the two-state model is also given in [76].

The internuclear repulsion energy $1/R_c'$ is larger than $1/R_c''$, whereas P_c' is smaller than P_c''. As a result, the values of the KER in both cases are nearly the same. If the kinetic energy at the moment of Coulomb explosion is assumed to be zero, as in the conventional analysis of the experimental results, the critical distance deduced from the KER value of 3.6 or 3.7 eV is $R_c \approx 3.7a_0$. This value is nearly twice as large as the equilibrium internuclear distance R_e of the molecule H_2^+, which feature is commonly observed for other molecules. The KER data obtained by Coulomb explosion imaging gives a rough estimate of the bond length at which Coulomb explosion occurs. Real-time tracking of the nuclear wavepackets evolving on potential curves in H_2^+ has been realized by a fs pump-probe setup of Coulomb explosion imaging [77].

3.3 Dynamics of Polyatomic Molecules in Intense Laser Fields

The description of molecular dynamics in terms of time-dependent adiabatic states developed for H_2^+ and H_2 is applicable to *polyatomic* molecules in intense laser fields. In the near-IR regime, the TD adiabatic state expansion like (3.9) has an advantage over the expansion in terms of field-free states or field-dressed Floquet states [49]. In the expansion using TD adiabatic states, only a small number of *low-lying* adiabatic states are required to describe the electronic and nuclear dynamics of a molecule in intense laser fields. It should be pointed out that many field-free excited states are involved even in the lowest adiabatic state $|1\rangle$. In the next subsection, we present an outline of a TD adiabatic state approach [49 51] for calculating the nuclear wavepacket dynamics of polyatomic molecules.

3.3.1 Description of Dynamics of Polyatomic Molecules in Intense Laser Fields by TD Adiabatic States

In the proposed approach [49–51], the total wavefunction is expanded in terms of "field-following" TD adiabatic states $\{|n(t)\rangle\}$ defined as the eigenfunctions of the "instantaneous" electronic Hamiltonian $H_{\mathrm{el}}(t)$ including the electric dipole interaction with a laser electric field $\varepsilon(t)$: $H_{el}(t)\,|n(t)\rangle = E_n(t)\,|n(t)\rangle$. Nuclear dynamics of a *polyatomic* molecule in a near-IR field can be described by wavepacket propagation on TD adiabatic potentials $E_{\mathrm{n}}(t)$ with *nonadiabatic transitions* between them due to temporal change in $\varepsilon(t)$.

For molecules more complex than H_2^+, we solve the coupled equations for TD adiabatic states as in (3.10) by employing the field-induced terms $\langle i(t)|\,\partial/\partial t\,|j(t)\rangle$ as the dominant nonadiabatic coupling term. Construction of the coupled equations for an arbitrary number of adiabatic states $\{|j(t)\rangle\}$ is thus straightforward. The diagonal elements in the coupled equations are, in addition to the kinetic energy operator for vibrational modes, the energies of TD adiabatic states involved, $\{E_j(t)\}$; the off-diagonal elements are $\{\langle i(t)|\,\partial/\partial t\,|j(t)\rangle\}$. The properties of the adiabatic states of polyatomic molecules can be calculated by ab initio molecular orbital methods. Quantum chemical calculations of the above quantities for TD adiabatic states are required for each charge state. The coupled equations for an arbitrary pulse shape can be solved numerically.

We have already applied the approach to investigate the dynamics of CO_2 in intense laser fields and revealed the origin of the experimentally observed bent structures of Coulomb exploding species CO_2^{2+} [26]; in the CO_2^{2+} stage, symmetric two-bond stretching followed by the occurrence of a large-amplitude bending motion is induced by an intense near-IR field [49,51]. We have also applied the approach to investigate the experimentally observed selective dissociation of C–O and C–C bonds of ethanol in intense laser fields [50, 51]. Itakura et al. experimentally observed the bond selective features that the ratio of $P_{\mathrm{C-O}}/P_{\mathrm{C-C}}$ decreases by a factor of 5 as the linear chirp rate of the pulse decreases to zero (i.e., as the pulse length becomes shorter), where $P_{\mathrm{C-C}}$ is the probability of C–C bond cleavage and $P_{\mathrm{C-O}}$ is that of C–O bond cleavage [78, 79]. We have attributed the origin of the bond selectivity to field-induced avoided crossing in reaction dynamics. Field-induced nonadiabatic transition plays a decisive role in the reaction dynamics of ethanol in intense laser fields [50, 51].

The motion of nuclei can also be treated also classically by evaluting the forces of $E_{\mathrm{n}}(t)$ exerted on the nuclei. We simulated the large-amplitude vibration of C_{60} induced by intense near-IR laser pulses by combining the TD adiabatic state approach with a classical molecular dynamics (MD) method [80]. In the next subsection, as an example of applications of the first-principles TD adiabatic state approach, we present the results of our theoretical investigation of field-induced vibrational dynamics of C_{60}.

3.3.2 Vibrational Excitation of C_{60} by an Intense Near-IR Pulse: Investigation by Combining the TD Adiabatic State Approach with Classical MD Method

Recently, Bhardwaj et al. reported that intact cations of C_{60}^{z+} up to $z \sim 12$ produced by intense laser pulses of wavelength $\lambda \sim 1{,}800\,\text{nm}$($I \sim 10^{15}\,\text{W cm}^{-2}$ and pulse length $T_\text{p} \sim 70\,\text{fs}$ are detected by mass spectrometry almost without fragments [41]. This is the highest charge state of C_{60} ever reported. In contrast, severe fragmentation occurs in the $\lambda = 800\,\text{nm}$ excitation case and the highest charge state of the intact cations produced is C_{60}^{5+} [81].

We first investigated the structural stability of C_{60}^{z+} cations [82, 83]. C_{60}^{z+} cations decay spontaneously to charged fragments if the Coulomb repulsion forces exceed the chemical binding forces. We obtained the optimized structures and vibrational frequencies of C_{60}^{z+} by performing DFT calculations. Normal mode analysis of C_{60}^{z+} has shown that for $z \leq 14$, C_{60}^{z+} has a potential minimum, i.e., a stable structure. We then calculated key energies relevant to dissociation of C_{60}^{z+}, such as the dissociation energies and the excess vibrational energy acquired upon ionization from C_{60}. By comparing the magnitudes of the calculated energies, we found that C_{60}^{z+} cations up to $z \sim 12$ can be produced as a stable or quasistable (µs-order lifetime) intact parent cation, in agreement with the experimental report. The results of Rice–Ramsperger–Kassel–Marcus (RRKM) calculation [84] have suggested that the lifetime of C_{60}^{z+} drastically decreases by ten orders of magnitude as z increases from $z = 11$ to $z = 13$ (the lifetime of C_{60}^{11+} is expected to be on the order of seconds).

The above theoretical conclusion is drawn without taking into account the effects of field-induced polarization on vibrational excitation (impulsive Raman excitation) [85] in the neutral or cation stages of C_{60}. In the following, we present the results of simulation as to how much vibrational energy is acquired by C_{60} or C_{60}^{z+} through the interaction with a near-IR pulse. Here, we assume that the effects of population transfer to higher TD adiabatic states on vibrational dynamics are secondary in the 1,800 nm case. The populations of higher TD adiabatic states in the 1,800 nm case would be very small compared to the 800 nm case, because more photons are required for access to the higher states. For example, for $f(t) = 0.038\,E_\text{h}/ea_0$ ($I = 5 \times 10^{13}\,\text{W cm}^{-2}$), the probability of nonadaibatic transition to the excited electronic state $6^1\text{T}_{1\text{u}}$ per half optical cycle, evaluated by the Landau–Zener formula [58, 59], is as small as 3.2×10^{-5} at $\lambda = 1{,}800\,\text{nm}$ (while it is 0.01 at $\lambda = 800\,\text{nm}$).

Since C_{60} has 174 vibrational degrees of freedom, it is virtually impossible to treat all of the vibrational modes quantum-mechanically. Here, we present the results of simulation obtained by incorporating a classical MD method into the framework of the TD adiabatic state approach. In this combined first-principles approach, we move all the nuclei classically in compliance with the forces of $E_1(\{R_j\}, t)$ of the lowest TD adiabatic state.

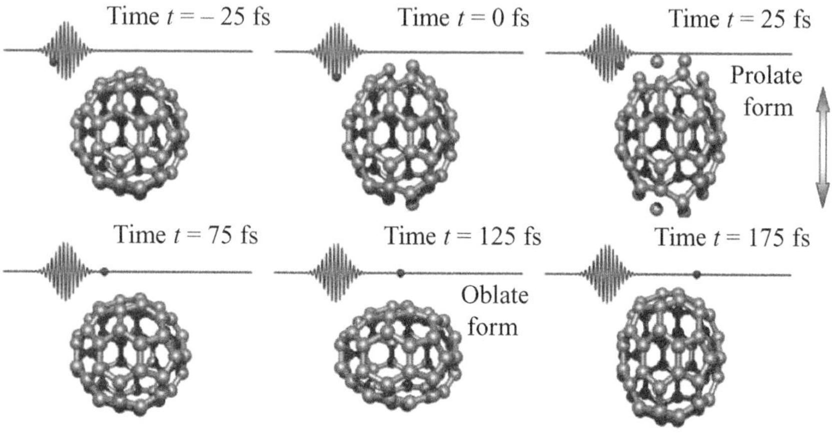

Fig. 3.8. Snapshots of the dynamics of the nuclei of C_{60} obtained by combining the TD adiabatic state approach with a classical MD method. The applied pulse is a Gaussian pulse of $T_p = 70$ fs centered at time $t = 0$ ($\lambda = 1{,}800$ nm and peak intensity $I_{peak} = 7 \times 10^{14}$ W cm^{-2}). All the nuclei are moved classically on the potential surface of the lowest TD adiabatic state, $E_1(\{R_j\}, t)$. The line with *up* and *down arrows* indicates the polarization direction of light. The temporal profile of the applied electric filed is sketched in the upper left corner of each snapshot of C_{60}; a dot in the field profile denotes the field strength at time t

Shown in Fig. 3.8 are snapshots of the nuclei of C_{60} obtained by the first-principles MD. A near-IR Gaussian pulse of $T_p = 70$ fs centered at time $t = 0$ is applied. The energy $E_1(\{R_j\}, t)$ is obtained by using DFT/3–21G implemented in the Gaussian 03 suite of programs [86]. The nuclei keep moving after full decay of the pulse. This indicates that a vibrational energy of ~30 eV is acquired through the interaction with the intense near-IR pulse. The molecule C_{60} is elongated almost in the field polarization direction along the $h_g(1)$ coordinate. About 70% of the energy is localized in the $h_g(1)$ prolate-oblate mode.

The period of one optical cycle is much shorter than the characteristic vibrational periods T_{vib} of C_{60} or C_{60}^{z+}. The vibrational dynamics is thus governed by the effective potential $\bar{E}_1(\{R_j\}, t)$ obtained by averaging $E_1(\{R_j\}, t)$ over one optical cycle; i.e., the single-surface dynamics on $E_1(\{R_j\}, t)$ is independent of ω as long as $2\pi/\omega \ll T_{vib}$ is maintained. The cycle-averaged potential $\bar{E}_1(\{R_j\}, t)$, which is a function of $f(t)$, changes its shape on the time scale of the pulse length T_p. Here, T_p is defined as the full width at half maximum of the light intensity profile $f^2(t)/2$.

The potential minimum of $\bar{E}_1(\{R_j\}, t)$ shifts towards the prolate form of C_{60} as $f(t)$ increases; $\bar{E}_1(\{R_j\}, t)$ is most distorted at $t = 0$ fs, i.e., at the pulse peak. On the other hand, the largest prolate structural deformation occurs with a delay of ~25 fs, as shown in Fig. 3.8. The nuclear motion cannot adiabatically follow the temporal change in $f^2(t)$, which brings rather large excitation (~30 eV) by the end of the incident pulse.

If T_{vib} of a vibrational mode j is much shorter than T_{p}, the motion along the vibrational mode j adiabatically follows the temporal change in $\bar{E}_1(\{R_j\},t)$; the molecule returns to its initial vibrational state at the end of the applied pulse. In the other limit of $T_{\mathrm{vib}} >> T_{\mathrm{p}}$, the vibrational excitation process is nonadiabatic but the energy injected into the vibration mode is small because of its short interaction time. Large vibrational excitation is thus expected for the case of $T_{\mathrm{p}} \sim T_{\mathrm{vib}}/2$ (i.e., if the light intensity profile coincides with the half-cycle vibrational motion) [87]. This is one of the reasons why the $h_g(1)$ mode is selectively excited. The vibrational frequency of the $h_g(1)$ mode is 267 cm^{-1} in the C_{60} case (The $h_g(1)$ mode is the lowest frequency mode of C_{60}). [88]. The corresponding period T_{vib} is 125 fs; the value of $T_{\mathrm{vib}}/2$ is close to $T_{\mathrm{p}} \sim 70$ fs used in the experiment [41]. We also found that the vibrational energy acquired by C_{60} is nearly proportional to $I_{\mathrm{peak}}^{2.5}$, indicating that the order of the nonlinear process in the present intensity regime is slightly higher than in the ordinary second-order Raman process.

The results of our first-principles MD simulation for C_{60} and C_{60}^{z+} implied that the field-induced large-amplitude vibration of the $h_g(1)$ mode persists for a rather long period and the energy initially injected into the $h_g(1)$-like mode does not flow into fission channels of C_{60} at least over the residence time of the cation in the acceleration region of the mass spectrometer (~μs). In conclusion, C_{60} and its cations are extremely robust against field-induced structural deformation.

Acknowledgements

This work was supported in part by a grant-in-aid for scientific research (no. 16350001) and grants-in-aid for scientific research on priority areas, "Control of Molecules in Intense Laser Fields" (Area No. 419) and "Molecular Theory for Real Systems" (Area No. 461), from the Ministry of Education, Culture, Sports, Science and Technology, Japan. M. K. and Y. S. acknowledge financial support from a JSPS Research Fellowship for Young Scientists (No. 18·5252 and 16·3435, respectively).

References

1. K. Yamanouchi, Science **295**, 1659 (2002)
2. S.L. Chin, F. Yergeau, P. Lavigne, J. Phys. B **18**, L213 (1985)
3. N.B. Delone, V.P. Krainov, *Multiphoton Processes in Atoms* (Springer, Berlin, 1994)
4. X.F. Li, A.L. L'Huillier, M. Ferray, L.A. Lompre, G. Mainfray, Phys. Rev. A **39**, 5751 (1989); P. Salières, A. L'Huillier, P. Antoine, M. Lewenstein, Adv. At. Mol. Opt. Phys. **41**, 83 (1999)

5. Y. Nabekawa, T. Shimizu, T. Okino, K. Furusawa, H. Hasegawa, K. Yamanouchi, K. Midorikawa, Phys. Rev. Lett. **96**, 083901 (2006)
6. A. Baltuška, Th. Udem, M. Uiberacker, M. Hentschel, E. Goulielmakis, Ch. Gohle, R. Holzwarth, V.S. Yakovlev, A. Scrinzi, T.W. Hansch, F. Krausz, Nature **421**, 611 (2003)
7. M.F. Kling, C. Siedschlag, A.J. Verhoef, J.I. Khan, M. Schultze, T. Uphues, Y. Ni, M. Uiberacker, M. Drescher, F. Krausz, M.J.J. Vrakking, Science **312**, 246 (2006)
8. G.L. Yudin, A.D. Bandrauk, P.B. Corkum, Phys. Rev. Lett. **96**, 063002 (2006)
9. J.P. Nibarger, S.V. Menon, G.N. Gibson, Phys. Rev. A **63**, 053406 (2001)
10. B. Friedrich, D. Herschbach, Phys. Rev. Lett. **74**, 4623 (1995)
11. T. Seideman, E. Hamilton, Adv. At. Mol. Opt. Phys. **52**, 289 (2005)
12. L. Quaglia, C. Cornaggia, Phys. Rev. Lett. **84**, 4565 (2000)
13. J.H. Posthumus, A.J. Giles, M.R. Thompson, K. Codling, J. Phys. B **29**, 5811 (1996)
14. E. Constant, H. Stapelfelt, P.B. Corkum, Phys. Rev. Lett. **76**, 4140 (1996)
15. T. Zuo, S. Chelkowski, A.D. Bandrauk, Phys. Rev. A **46**, R5342 (1992)
16. L.V. Keldysh, Sov. Phys. JETP **20**, 1307 (1965)
17. M.V. Ammosov, N.B. Delone, V.P. Krainov, Sov. Phys. JETP **64**, 1191 (1986)
18. V.P. Krainov, H.R. Reiss, B.M. Smirnov, *Radiative Processes in Atomic Physics* (Wiley, New York, 1997)
19. S.L. Chin, From Multiphoton to Tunnel Ionization, in *Advances in Multiphoton Processes and Spectroscopy*, vol. 16, ed by S.H. Lin, A.A. Villaeys, Y. Fujimura (World Scientific, Singapore, 2004), p. 249
20. T.D.G. Walsh, F.A. Ilkov, J.E. Decker, S.L. Chin, J. Phys. B **27**, 3767 (1994)
21. I. Kawata, H. Kono, Y. Fujimura, J. Chem. Phys. **110**, 11152 (1998)
22. K. Harumiya, I. Kawata, H. Kono, Y. Fujimura, J. Chem. Phys. **113**, 8953 (2000)
23. K. Harumiya, H. Kono, Y. Fujimura, I. Kawata, A.D. Bandrauk, Phys. Rev. A **66**, 043403 (2002)
24. M. Lezius, V. Blanchet, M.Y. Ivanov, A. Stolow, J. Chem. Phys. **117**, 1575 (2002)
25. I. Kawata, H. Kono, Y. Fujimura, Chem. Phys. Lett. **289**, 546 (1998)
26. A. Hishikawa, A. Iwamae, K. Yamanouchi, Phys. Rev. Lett. **83**, 1127 (1999)
27. A. Assion, T. Baumert, M. Bergt, T. Brixner, B. Kiefer, V. Seyfried, M. Strehle, G. Gerber, Science **282**, 919 (1998)
28. J.R. Levis, G.M. Menkir, H. Rabitz, Science **292**, 709 (2001)
29. S. Shimizu, J. Kou, S. Kawato, K. Shimizu, S. Sakabe, N. Nakashima, Chem. Phys. Lett. **317**, 609 (2000)
30. A.D. Bandrauk, H. Kono, in *Advances in Multi-Photon Processes and Spectroscopy*, vol. 15, ed. by S.H. Lin, A.A. Villaeys, Y. Fujimura (World Scientific, Singapore, 2003), p. 147
31. H. Kono, Y. Sato, Y. Fujimura, I. Kawata, Laser Phys. **13**, 883 (2003)
32. R.S. Mulliken, Phys. Rev. **7**, 20 (1939)
33. H. Harada, M. Tanaka, M. Murakami, S. Shimizu, T. Yatsuhashi, N. Nakashima, S. Sakabe, Y. Izawa, S. Tojo, T. Majima, J. Phys. Chem. A **107**, 6580 (2003)
34. R. Itakura, J. Watanabe, A. Hishikawa, K. Yamanouchi, J. Chem. Phys. **114**, 5598 (2001)

35. L. Robson, K.W.D. Ledingham, A.D. Tasker, P. McKenna, T. McCanny, C. Kosmidis, D.A. Jaroszynski, D.R. Jones, R.C. Issac, S. Jamieson, Chem. Phys. Lett. **360**, 382 (2002)
36. S.C. O'Brien, J.R. Heath, R.F. Curl, R.E. Smalley, J. Chem. Phys. **88**, 220 (1988)
37. M. Boyle, T. Laarmann, I. Shchatsinin, C.P. Schulz, J. Chem. Phys. **122**, 181103 (2005)
38. I. Shchatsinin, T. Laarmann, G. Stibenz, G. Steinmeyer, A. Stalmashonak, N. Zhavoronkov, C.P. Schulz, I.V. Hertel, J. Chem. Phys. **125**, 194320 (2006)
39. I.V. Hertel, T. Laarmann, C.P. Schulz, in *Advances in Atomic, Molecular, and Optical Physics*, vol. 50, ed. by B. Bederson, H. Walther (Academic Press, London, 2005), p. 219
40. T. Laarmann, C.P. Schulz, I.V. Hertel, in: *Progress in Ultrafast Intense Laser Science*, vol. 3, ed. by K. Yamanouchi, S.L. Chin, P. Agostini, G. Ferrante (Springer, Berlin, 2008), to be published
41. V.R. Bhardwaj, P.B. Corkum, D.M. Rayner, Phys. Rev. Lett. **91**, 203004 (2003)
42. H. Kono, A. Kita, Y. Ohtsuki, Y. Fujimura, J. Comput. Phys. **130**, 148 (1997)
43. I. Kawata, H. Kono, J. Chem. Phys. **111**, 9498 (1999)
44. I. Kawata, H. Kono, Y. Fujimura, A.D. Bandrauk, Phys. Rev. A **62**, 031401(R) (2000)
45. H. Kono, I. Kawata, in *Advances in Multi-Photon Processes and Spectroscopy*, vol. 14, ed. by R.J. Gordon, Y. Fujimura (World Scientific, Singapore, 2001), p. 165
46. I. Kawata, A.D. Bandrauk, H. Kono, Y. Fujimura, Laser Phys. **11**, 188 (2001)
47. H. Kono, S. Koseki, *Laser Control and Manipulation of Molecules*, ACS Symposium Series, vol. 821, ed. by A.D. Bandrauk, R.J. Gordon, Y. Fujimura (Oxford, Oxford, 2002), p. 267
48. D.M. Volkov, Z. Phys. **94**, 250 (1935)
49. Y. Sato, H. Kono, S. Koseki, Y. Fujimura, J. Am. Chem. Soc. **125**, 8020 (2003)
50. H. Kono, Y. Sato, N. Tanaka, T. Kato, K. Nakai, S. Koseki, Y. Fujimura, Chem. Phys. **304**, 203 (2004)
51. H. Kono, Y. Sato, M. Kanno, K. Nakai, T. Kato, Bull. Chem. Soc. Jpn **79**, 196 (2006)
52. H. Kono, S. Koseki, M. Shiota, Y. Fujimura, J. Phys. Chem. A **105**, 5627 (2001)
53. J.R. Hiskes, Phys. Rev. **122**, 1207 (1961)
54. M. Thachuk, M.Y. Ivanov, D.M. Wardlaw, J. Chem. Phys. **105**, 4094 (1995)
55. M. Thachuk, M.Y. Ivanov, D.M. Wardlaw, J. Chem. Phys. **109**, 5747 (1998)
56. K. Mullen, E. Ben-Jacob, Y. Gefen, Z. Schuss, Phys. Rev. Lett. **62**, 2543 (1989)
57. Y. Kayanuma, J. Phys. Soc. Jpn **53**, 108 (1984)
58. C. Zener, Proc. Roy. Soc. London **696**, A137 (1932)
59. C. Zhu, Y. Teranishi, H. Nakamura, Adv. Chem. Phys. **117**, 127 (2001) and references therein
60. A. Zavriyev, P.H. Bucksbaum, J. Squier, F. Saline, Phys. Rev. Lett. **70**, 1077 (1993)
61. T.D.G. Walsh, F.A. Ilkov, S.L. Chin, F. Chateauneuf, T.T. Nguyen-Dang, S. Chelkowski, A.D. Bandrauk, O. Atabek, Phys. Rev. A **58**, 3922 (1998)
62. M. Lein, T. Kreibich, E.K.U. Gross, V. Engel, Phys. Rev. A **65**, 033403 (2002)
63. X. Urbain, B. Fabre, E.M. Staicu-Casagrande, N. de Ruette, V.M. Andrianarijaona, J. Jureta, J.H. Posthumus, A. Saenz, E. Baldit, C. Cornaggia, Phys. Rev. Lett. **92**, 163004 (2004)

64. K. Mishima, K. Nagaya, M. Hayashi, S.H. Lin, Phys. Rev. A **70**, 063414 (2004)
65. H. Niikura, P.B. Corkum, D.M. Villeneuve, Phys. Rev. Lett. **90**, 203601 (2003)
66. H. Reiss, Phys. Rev. A **75**, 031404 (R) (2007)
67. L.Y. Peng, D. Dundas, J.F. McCann, K.T. Taylor, I.D. Williams, J. Phys. B **36**, L295 (2003)
68. M. Kanno, T. Kato, H. Kono, Y. Fujimura, F.H.M. Faisal, Phys. Rev. A **72**, 033418 (2005)
69. D. Mathur, F.A. Rajgara, J. Chem. Phys. **124**, 194308 (2006)
70. A.N. Markevitch, D.A. Romanov, S.M. Smith, H.B. Schlegel, M.Yu. Ivanov, R.J. Levis, Phys. Rev. A **69**, 013401 (2004)
71. S.M. Hankin, D.M. Villeneuve, P.B. Corkum, D.M. Rayner, Phys. Rev. Lett. **84**, 5082 (2000)
72. A. Giusti-Suzor, F.H. Mies, L.F. DiMauro, E. Charron, B. Yang, J. Phys. B **28**, 309 (1995)
73. L.J. Frasinski, J.H. Posthumus, J. Plumridge, K. Codling, P.F. Taday, A.J. Langley, Phys. Rev. Lett. **83**, 3625 (1999)
74. I. Maruyama, T. Sako, K. Yamanouchi, J. Phys. B **37**, 3919 (2004)
75. Th. Ergler, A. Rudenko, B. Feuerstein, K. Zrost, C.D. Schröter, R. Moshammer, J. Ullrich, Phys. Rev. Lett. **95**, 093001 (2005)
76. A. Staudte, D. Pavičić, S. Chelkowski, D. Zeidler, M. Meckel, H. Niikura, M. Schöffler, S. Schössler, B. Ulrich, P.P. Rajeev, Th. Weber, T. Jahnke, D.M. Villeneuve, A.D. Bandrauk, C.L. Cocke, P.B. Corkum, R. Dörner, Phys. Rev. Lett. **98**, 073003 (2007)
77. A.S. Alnaser, B. Ulrich, X.M. Tong, I.V. Litvinyuk, C.M. Maharjan, P. Ranitovic, T. Osipov, R. Ali, S. Ghimire, Z. Chang, C.D. Lin, C.L. Cocke, Phys. Rev. A **72**, 030702(R) (2005)
78. R. Itakura, K. Yamanouchi, T. Tanabe, T. Okamoto, F. Kannari, J. Chem. Phys. **119**, 4179 (2003)
79. H. Yazawa, T. Tanabe, T. Okamoto, M. Yamanaka, F. Kannari, R. Itakura, K. Yamanouchi, J. Chem. Phys. **124**, 204314 (2006)
80. H.B. Schlegel, S.S. Iyengar, X. Li, J.M. Millam, G.A. Voth, G.E. Scuseria, M.J. Frisch, J. Chem. Phys. **117**, 8694 (2002)
81. E.E.B. Campbell, K. Hoffmann, H. Rottke, I.V. Hertel, J. Chem. Phys. **114**, 1716 (2001)
82. R. Sahnoun, K. Nakai, Y. Sato, H. Kono, Y. Fujimura, M. Tanaka, Chem. Phys. Lett. **430**, 167 (2006)
83. R. Sahnoun, K. Nakai, Y. Sato, H. Kono, Y. Fujimura, M. Tanaka, J. Chem. Phys. **125**, 184306 (2006)
84. W. Forst, *Unimolecular Reactions* (Cambridge, Cambridge, 2003)
85. Y.-X. Yan, J.E.B. Gamble, K.A. Nelson, J. Chem. Phys. **83**, 5391 (1985)
86. M.J. Frisch, G.W. Trucks, H.B. Schlegel et al., *Gaussian 03* (Gaussian, Inc., Pittsburgh, PA, 2003)
87. K. Nakai, H. Kono, Y. Sato, N. Niitsu, R. Sahnoun, M. Tanaka, Y. Fujimura, Chem. Phys. **338**, 127 (2007)
88. C.H. Choi, M. Kertesz, L. Mihaly, J. Phys. Chem. A **104**, 102 (2000)

4

Molecular Ion Beams Interrogated with Ultrashort Intense Laser Pulses

Itzik Ben-Itzhak

Abstract. Molecules have been interrogated with intense ultrashort laser pulses by many researchers for over a decade. In contrast, molecular-ion studies are scarce, though their number is increasing rapidly in the last few years. In this chapter we will discuss how these studies differ from those conducted on molecular targets, describe the main experimental methods used, and review the findings on the dissociation and ionization of the simplest molecular ion, H_2^+, as well as some of the discoveries concerning more complex molecular ions.

4.1 Introduction

Advances in laser technology yielding intense laser fields which are comparable or even greater than the typical binding in a molecule have resulted in many surprising molecular dissociation mechanisms due to the non-perturbative nature of the interaction in the strong field limit (see, for example, [1–7]). These multi-photon processes are commonly interpreted using the dressed potential energy curves (PEC) of the molecule either in their diabatic or adiabatic Floquet form, shown in Fig. 4.1 for H_2^+. Examples for such discoveries are the bond-softening and vibrational-trapping mechanisms. In the bond-softening (BS) mechanism, the molecule dissociates through the gap formed between the ground state and the first excited state dressed by one photon, thus resulting in a higher dissociation probability with increasing laser intensity [1–3,6,7]. Vibrational trapping (VT), also known as bond hardening (BH), in contrast refers to the reduction in dissociation probability of highly excited vibrational states due to their trapping in the laser-induced well of the upper dressed PEC [3,6,7,9–13] (in this chapter we will use VT to denote this mechanism as we find it more descriptive). Above threshold dissociation (ATD) is the dissociation of the molecule with higher kinetic energy release resulting from the absorption of an excess number of photons than those needed for dissociation [1,2,6,7,14–16]. In addition, molecular alignment effects (e.g., [17–19]) as well as enhanced ionization of "stretched" H_2^+ (e.g., [20–22]) have been the subject of many studies.

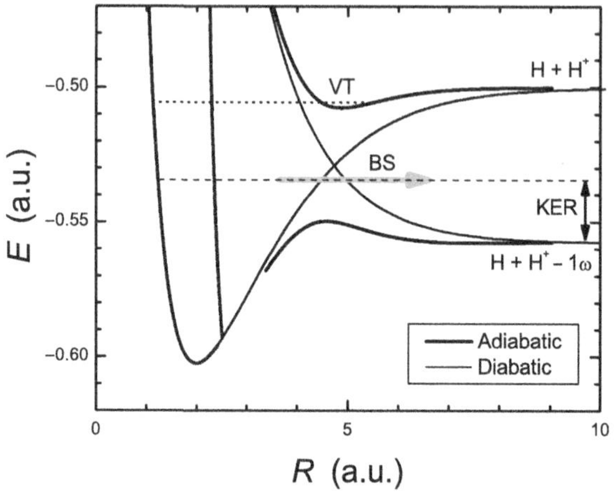

Fig. 4.1. Schematic diabatic (*thin*) and adiabatic (*thick*) Floquet potential energy curves of H_2^+ in an intense laser field (from [8]). The *dashed* and *dotted lines* represent vibrational levels undergoing bond softening (BS) and vibrational trapping (VT), also known as bond hardening (BH). The kinetic energy released (KER) in bond softening is also marked

In addition to the evolution of field strength with time, laser pulses are also becoming shorter than the typical nuclear motion time scale (fs), thus enabling studies of the propagation of nuclear wave functions in real time (e.g., [23–25]).

Most experimental studies of the dissociation and ionization of the simplest molecular ion, namely H_2^+, in an ultrashort intense laser pulse have been conducted by probing a neutral H_2 gas target, as discussed in detail in a few review articles (see, for example, [6,7,26]). In contrast, direct studies of molecular ion targets are few, mainly due to the difficulty introduced by the low target density of an ion beam used as the target. One may wonder why we should attempt such a task instead of probing the H_2^+ ions produced by the same laser pulse that is used to probe the molecular ion. A few answers can be given to this question. The most obvious answer is that theorists would favor studies of the H_2^+ system because it is simpler and because it eliminates the multi-step process involved in their study using an H_2 target. In addition, one would expect dissociation of H_2^+, caused by bond-softening, to start occurring at relatively lower intensities than the intensities required to ionize the H_2. This is more significant for the vibrational states around the curve crossing between the ground state and the dressed first excited state that dissociates at much lower intensities. Pioneering studies of the dissociation of an H_2^+ beam were reported in the beginning of the new millennium by Sändig et al. [27], and by Williams et al. [28], the latter also studying ionization. The experimental resolution of Sändig et al. enabled them to observe the behavior of different vibrational states slightly above and below the curve crossing shown in Fig. 4.1. These highly excited vibrational states ($v = 6$–10) have significant population in an H_2^+ beam produced typically by electron-impact ionization in the ion source, which has approximately a Franck–Condon distribution that is peaked around $v = 2$ and falls off relatively slowly [29] (note that some small deviations from a Franck–Condon distribution have been reported [30]). Experimental evidence showing that the vibrational population of H_2^+ produced by a short intense laser pulse favors the population of lower vibrational states has been recently reported by Urbain et al. [31]. As a result the dissociation of H_2^+ may differ depending on the production mechanism of the H_2^+ that is later dissociated by the laser field.

Ionization studies of H_2^+ created from H_2 by the laser pulse are also expected to differ from those performed on an H_2^+ beam because of the role played by the first ionized electron. This electron can cause the ionization of the H_2^+ by the re-scattering mechanism if a linearly polarized laser beam is used (see, for example, [32–34]). For an H_2^+ beam, in contrast, re-scattering cannot occur and the ionization mechanism is described by charge resonance enhanced ionization (CREI) or by multi-photon ionization (MPI). As the vibrational states slightly below the crossing (see Fig. 4.1), which have significant initial population in an H_2^+ beam, are expected to dissociate at relatively low intensities, one would expect CREI to play a major role for the ionization of these states unless the pulse duration is very short and the intensity high enough to cause significant multi-photon ionization.

Finally, the nuclear wave function of the H_2^+ produced in the laser pulse is significantly different than that of an H_2^+ beam arriving from an ion source. The former is a coherent sum of the populated vibrational states, and it can be expressed by the H_2 ground state wave function projected on the dressed H_2^+ potential energy curve (PEC), later evolving in time under the influence of the laser field. In contrast, an H_2^+ beam should be described by an incoherent sum of the populated vibrational states because of the large time (typically μs) between the creation of the H_2^+ in the ion source and its interaction with the laser pulse.

4.2 Experimental Techniques

Experiments using H_2^+ "targets", though preferred by theorists, are rare due to the additional complexity of using an ion beam as a target. A few such studies have been reported in recent years. These experimental studies of molecular ions in intense laser fields are conducted by focusing an intense laser beam onto an ion beam of a few keV originating from an ion source or low energy accelerator. The molecular fragments are detected downstream from the interaction region as shown in Fig. 4.2, for example, for the main experimental setups used so far.

The laser parameters available for measurements by these groups are somewhat different; however, the main differences between the three experimental

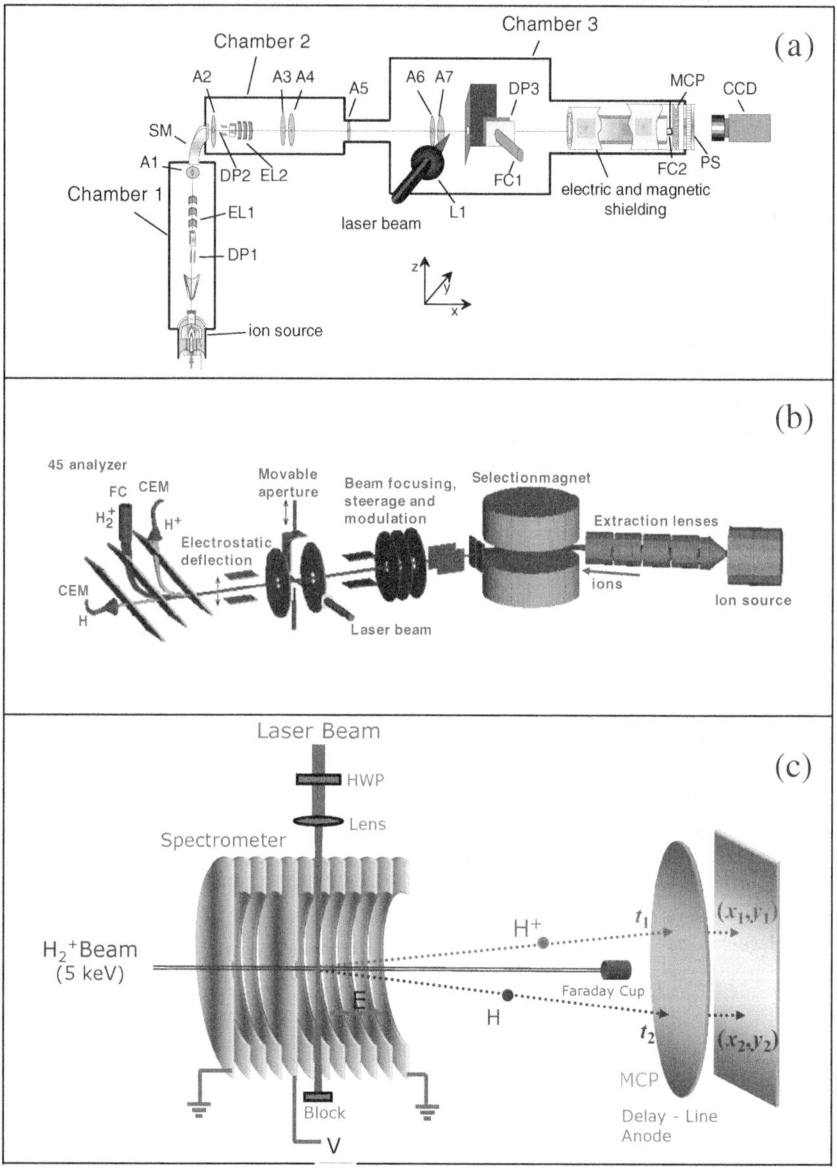

Fig. 4.2. Schematic diagram of the experimental setup used by: (**a**) Figger and Hänsch's group [27,35–39]. DP, deflection plates; EL, Einzel lenses; SM, sector magnet; A, apertures; L1, focusing lens; FC, Faraday cups; MCP, multichannel plate detector; PS, phosphor screen; CCD, (charge-coupled device) camera (from [39]). (**b**) Williams et al. [28, 40–42]. (**c**) Ben-Itzhak's group [8, 43–46]; the molecular ion beam was produced by an electron cyclotron resonance (ECR) ion source and directed to the interaction region by a pair of sector magnets

techniques and setups, shown in Fig. 4.2, are in the different detection schemes of the dissociating fragments. The first group, lead by *Figger and Hänsch* (setup shown in Fig. 4.2a), used a 2D position sensitive detector to image one of the fragments. By using linear polarization aligned parallel to the detector plane they managed to evaluate the 3D momentum information (by an Abel transformation) taking advantage of the cylindrical symmetry around the laser polarization. A transverse electric field was applied by Sändig et al. [27] to eliminate the charged fragments, thus measuring only the dissociation channel. More recently the same group studied ionization, which was identified by the larger kinetic energy release (KER) of the H^+ fragments associated with this breakup channel [37,38]. The second group, working at the *Rutherford Laboratory* (setup shown in Fig. 4.2b), uses an energy analyzer and the time-of flight (TOF) technique to determine the energy spectrum of the charged and neutral fragments, respectively. Ionization is then determined by subtracting the two spectra after matching the low energy peak, which appears in both spectra as shown in Fig. 4.3b [28]. Very recently our group, at the *J.R. Macdonald Laboratory* (JRML) (setup shown in Fig. 4.2c), reported on kinematically complete measurements using coincidence 3D momentum imaging, i.e. the

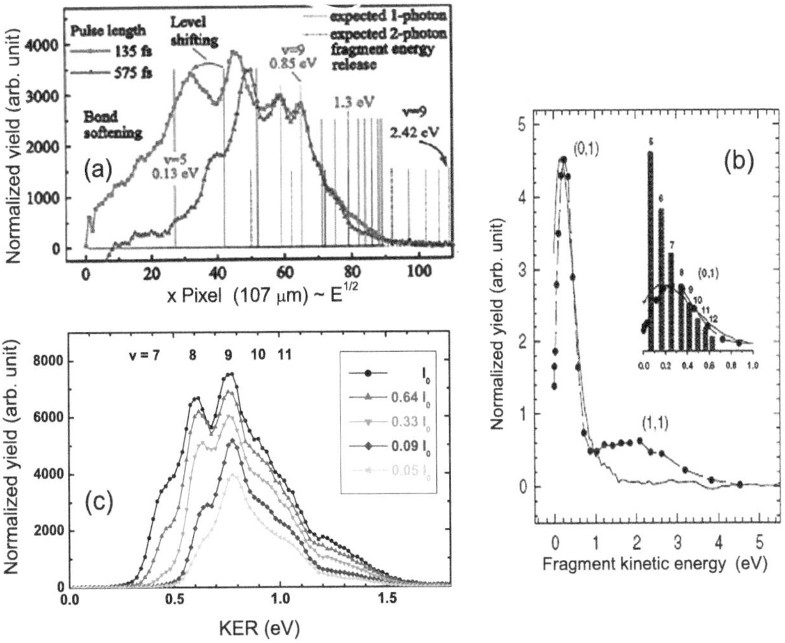

Fig. 4.3. Typical energy spectra of laser induced dissociation of an H_2^+ beam measured with the experimental setups shown in Fig. 4.2, respectively. (**a**) Sändig et al. [27] measurements at 135 and 575 fs with a 1.0 mJ pulse, i.e. $I_0 = 1.5 \times 10^{14}$ W cm^{-2} for the shorter pulse. Note that the horizontal axis is $E^{1/2}$. (**b**) Williams et al. [28] measurements at 65 fs $I_0 = 3 \times 10^{15}$ W cm^{-2}. In this case energy per proton is shown in the figure. (**c**) Wang et al. [45] measurements at 135 fs $I_0 = 2.4 \times 10^{14}$ W cm^{-2}

coincident measurement of the momentum of both fragments. In addition, a weak electric field is introduced in the interaction region, thus separating the H^+ and H fragments by their TOF [43,46]. The energy resolution of both imaging techniques seems superior to that of Williams et al. [28] as can be seen from the typical energy distributions reported by these groups (see Fig. 4.3).

Another important difference between the experimental methods is the interaction region defined by the crossing of the ion and laser beams. The first group collimated their ion beam to 50 by 300 μm^2, i.e. much smaller than their laser waist and Rayleigh length, respectively. This was done in order to minimize the volume effects, i.e. the contributions from lower intensities away from the laser axis. It is important to note that some volume effects will still persist along the ion-beam direction. The other two groups use an ion beam much wider than the laser beam waist, but much smaller than the Rayleigh length. Wang et al. [44] introduced a method for eliminating the contributions from the low-intensity tail altogether for the latter interaction-region configuration assuming a Gaussian laser-beam profile. The conceptual representation of this "intensity-difference spectra (IDS)" method and some typical results are shown in Fig. 4.4.

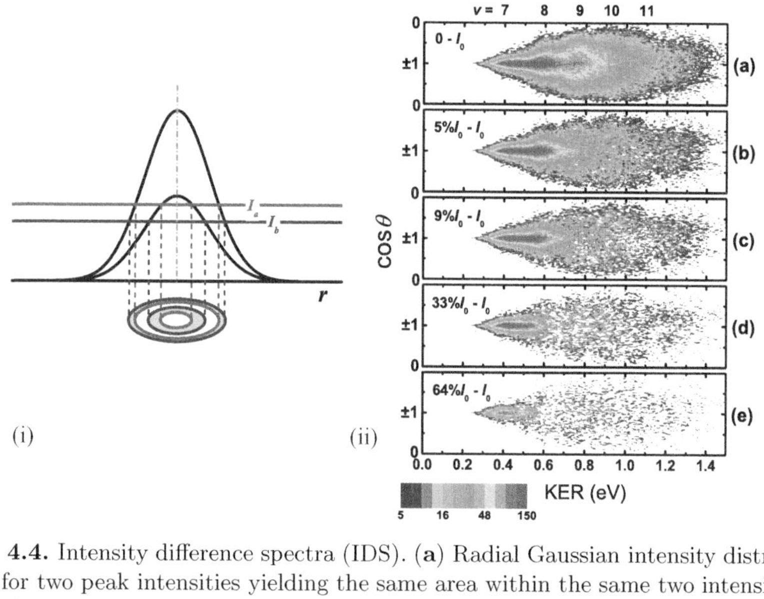

Fig. 4.4. Intensity difference spectra (IDS). (**a**) Radial Gaussian intensity distribution for two peak intensities yielding the same area within the same two intensities, I_a and I_b, as shown by the projected rings (see, [44] for details). (**b**) KER vs. $\cos\theta$ spectra for different intensity ranges evaluated using the IDS method. The top panel includes all intensities starting from the highest intensity measured. The lower panels show a narrow intensity range eventually reaching the range from I_0 to $0.64 \times I_0$. Note that for that intensity range dissociation is dominated by bond-softening of the $v \sim 7$ vibrational state (from [44])

Last but not least, the laser systems used are similar in some aspects but have some significant differences affecting the measurements. The core of all these lasers is a Ti-sapphire laser providing approximately an 800 nm beam. The pulse durations span the range from 100 to 700 fs (all pulse durations in this chapter are FWHM values) in the studies of the first group, was about 65 fs for the second, and 30–200 fs for the third. The laser beam was typically linearly polarized, and it was focused to about a 10–50 μm waist in order to reach the peak intensities of interest, 10^{13}–10^{15} W cm^{-2}. The highest intensity range was measured by Williams et al. [28], however their laser has a lower repetition rate of 10 Hz as compared to the 1 kHz of the other two groups, thus limiting their ability to measure at lower intensities due to low counting rate. In a way, these three experimental approaches are complementary to each other as each has its advantages for some specific conditions.

4.3 One Electron Molecular Ions: H_2^+

This one electron molecular ion is the one favored by theorists, and it serves the same role as atomic hydrogen for more complex atomic systems. As stated before, H_2^+ generated by the laser pulse from H_2 is the most studied molecular system [6, 7, 26]. However, creating the H_2^+ and following its behavior later in the pulse makes it hard to de-convolute the effects of the former from the latter. Below, we discuss experimental studies of the interaction of intense laser fields with H_2^+ beams, where this problem is eliminated. However, one should be aware that such thin targets introduce new problems and challenges, including for the theorists, as such targets typically have a broad population of vibrational states.

4.3.1 Dissociation: $nh\nu + H_2^+ \rightarrow H^+ + H$

The dissociation of an H_2^+ target beam by an ultrashort intense laser pulse was first probed directly by Sändig et al. [27] and by Williams et al. [28]. In both measurements, the dissociation was identified by measuring the neutral fragment from $nh\nu + H_2^+ \rightarrow H^+ + H$. The 2D imaging measurements of Sändig et al. have much higher energy resolution, as can be seen in Fig. 4.3, thus enabling the observation of the vibrational structure of the dissociating H_2^+. The KER of the dissociating fragments is mainly around 0.2–0.9 eV, a range similar to that measured for the dissociation using H_2 targets [7]. The lower KER dissociation is due to bond-softening (BS), i.e. the lowering of the barrier formed by the avoided crossing shown in Fig. 4.1. It can be clearly seen that higher intensities open up this dissociation route for lower vibrational states. The same can be seen from the measurements of Wang et al. [45], shown in Fig. 4.3c for a few intensities and a fixed pulse duration. The $v = 9$ state, which is the closest vibrational state to the curve crossing for $\lambda = 790$ nm (see Fig. 4.1), is the main state dissociating by BS at low laser intensities.

The high KER resolution of Sändig et al. [27] not only enables the observation of the vibrational states but also their shift toward lower energies, as shown in Fig. 4.3a. The shorter and more intense laser pulse causes a downward energy shift of the vibrational energy levels from the expected field-free values marked on the figure. Note that the shift increases for lower vibrational states, which are exposed to higher intensities before they dissociate. In contrast, the $v = 9$ state, which dissociates early in the laser pulse as it requires much lower intensity, is barely shifted. It is also important to note that the $v = 6$ state is shifted down almost to the expected KER for the field-free $v = 5$ state. This level shifting may cause problems in identifying the vibrational states involved in dissociation solely by their field-free energy, especially as the laser intensities are increased.

A similar problem exists when trying to identify the high KER contributions, i.e. higher than the KER expected for the $v = 9$ state. Since this KER range can be due to either the dissociation of the $v > 9$ vibrational states via vibrational trapping (VT) or above threshold dissociation (ATD) of the low vibrational states, both expected KER series are shown in Fig. 4.3a, labelled as 1 and 2 photon, respectively (note that field free vibrational energies are used). The first mechanism (VT) reduces the dissociation rate by trapping these vibrational states in the laser-induced well formed in the upper PEC shown in Fig. 4.1. Above threshold dissociation is somewhat similar to bond-softening, but it is initiated by the avoided crossing between the ground $1s\sigma$ state and the $2p\sigma$ state dressed by three photons (usually denoted as $2p\sigma - 3\omega$). This can be followed by a transition to the dressed $1s\sigma - 2\omega$ state once the molecular ion reaches the avoided crossing between these two states (i.e. the $2p\sigma - 3\omega$ and $1s\sigma - 2\omega$). In this process the molecular ion "absorbs" 3 photons and re-emits 1, thus resulting in a net 2-photon transition. The data shown in Fig. 4.4ii suggests that most of the high KER is due to vibrational trapping, as it vanishes when the intensity is very high for which we expect ATD to become even stronger. It seems that the best way to avoid the interpretation problems caused by the ambiguities about the expected KER values is to work closely with theorists, who can compute the explicit KER values expected for the experimental conditions and are able to trace more easily what contribution leads to each KER. For example, such calculations were conducted by Serov et al. [47] and by Kondorskiy and Nakamura [48] to match Sändig et al. data [27] or by Peng et al. [49] to match Williams et al. data [28].

An analogous problem arises when trying to identify the main features in the $N(\mathrm{KER}, \cos\theta)$ distribution shown in Fig. 4.5, where θ is the angle between the internuclear axis and the laser polarization. Two main features appear in the distribution shown in Fig. 4.5a, a dominant feature aligned along the laser polarization with KER centered around 0.8 eV, and a secondary feature very strongly aligned along the laser polarization with KER <0.4 eV. As discussed above, the $v = 9$ dissociation is expected to occur close to 0.8 eV, and thus one may tend to associate the main feature with BS of the $v = 9$ state in the low intensity regions of the interaction volume, which have a larger "density" of

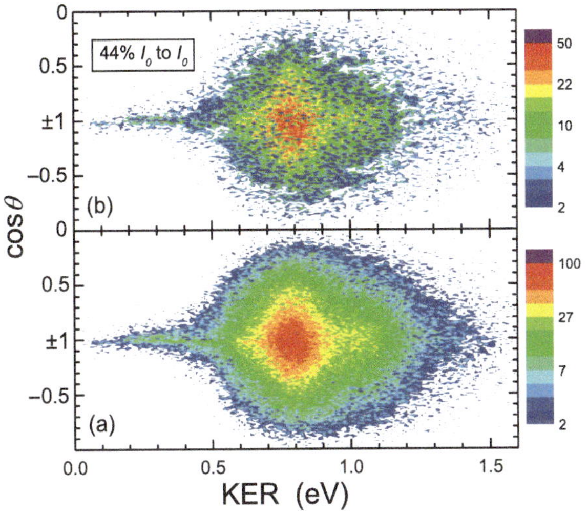

Fig. 4.5. (**a**) The $\cos\theta$-KER distribution for dissociation caused by a 45 fs, $I_0 = 2.5 \times 10^{14}\,\mathrm{W\,cm^{-2}}$ laser pulse, (**b**) the same as (**a**) but including only intensities between 44 and 100% of I_0 (from [46])

target molecules. However, Esry's calculations [8,46] indicate that this feature is mainly due to net 2-photon ATD of the lower vibrational states. Using the IDS method [8,44] the spectrum of the high intensity part of the pulse was created and is shown in Fig. 4.5b. It can be clearly seen that the main feature is truly a high intensity process, as the dissociation of the $v = 9$ state should have disappeared upon subtraction of the low intensity contribution. The weaker low KER feature is due to BS according to the same calculations.

Angular distributions measured by Sändig et al. [27] and Pavičić et al. [50] for H_2^+ and D_2^+, respectively, nicely reveal another feature of the BS mechanism, namely the narrowing of the angular distribution of the lower vibrational states in comparison to the $v = 9$ state, as shown in Fig. 4.6ii. The first fits a $\cos^{12}\theta$ while the latter follows the $\cos^2\theta$ distribution expected at the weak field limit (recall that the effective intensity along the molecular axis is given by $I_{\mathrm{eff}} = I_0 \cos^2\theta$). Williams et al. [28] observed that the angular distribution integrated over KER nicely fits a $\cos^2\theta$ distribution, as shown in Fig. 4.6i.

Recently, Wang et al. [45] suggested a method to highlight the difference in angular distributions of the different vibrational states as well as between the BS and VT mechanisms. Noting that at the weak-field limit one expects a $\cos^2\theta$ angular dependence due to the effective intensity along the internuclear axis of the molecular ion, they suggested plotting the angular distribution as a function of $\cos^3\theta$ instead of the commonly used $\cos\theta$ or θ. Note that a $\cos^2\theta$ angular distribution is flat in the $\cos^3\theta$-coordinate as $\mathrm{d}N \propto \cos^2\theta \mathrm{d}(\cos\theta)$ yields $\mathrm{d}N/\mathrm{d}(\cos^3\theta) = \mathrm{constant}$. Their results for a 135 fs and $\sim 2.4 \times 10^{14}\,\mathrm{W\,cm^{-2}}$ laser pulse are shown in Fig. 4.7. It can be clearly seen that plotting vs. $\cos^3\theta$ rather than $\cos\theta$ highlights the difference between

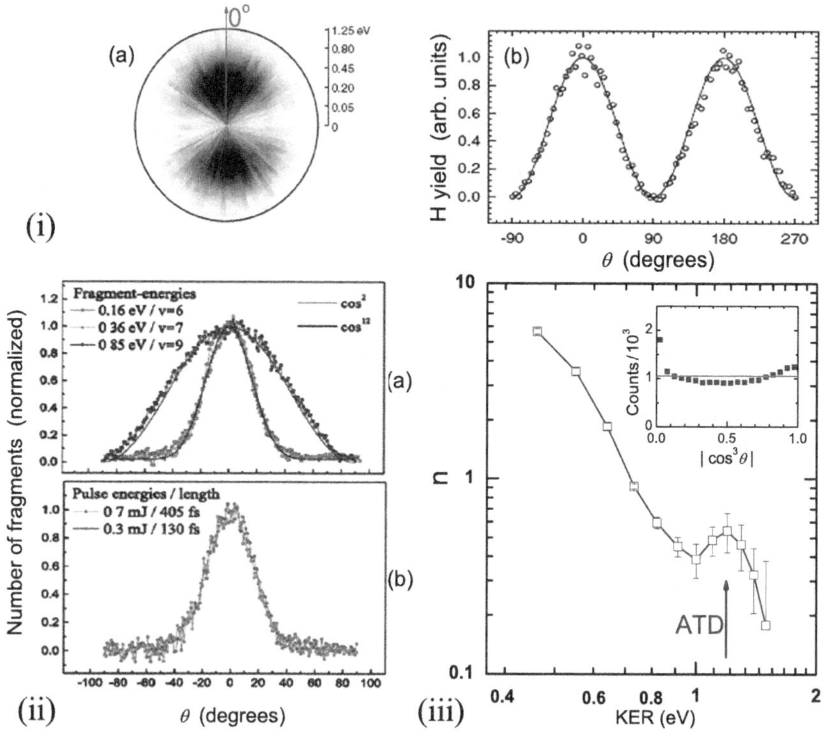

Fig. 4.6. Angular distributions of the photofragments from measurements of (**i**) Williams et al. [28], whose data integrated over KER nicely follows a $\cos^2\theta$ distribution; (**ii**) Sändig et al. [27] (**a**) for various fragment energy releases (measurement in Fig. 4.3a), (**b**) with pulse energies of 0.7 and 0.3 mJ and at pulse lengths of 405 and 130 fs, respectively, from $v = 6$; and (**iii**) n – from a $\cos^{2n}\theta$ fit to the angular distribution – as a function of KER (from [45]). Inset shows angular distribution integrated over all KER

dissociation above and below the curve crossing (shown in Fig. 4.1 and marked by the arrows on Fig. 4.7), i.e. between the bond-softening and vibrational-trapping mechanisms. Furthermore, by fitting the angular distributions for KER slices to a $\cos^{2n}\theta$ distribution it was shown (see Fig. 4.6iii) that the angular distributions are becoming narrower (larger n) as a function of the energy difference from the KER expected for dissociation at the crossing [45]. They also found that the angular distribution after integrating over all KER is well described by a $\cos^2\theta$ distribution, as did Williams et al. (compare Fig. 4.6i and inset in (iii)). This is due to the large contribution from the low intensity dissociation, which has a $\cos^2\theta$ distribution, because of the volume averaging effect (note that the intensity averaging is similar in both these experiments [28,45]).

In spite of the simplicity of H_2^+, it is still challenging to treat all degrees of freedom of this molecule in the strong field produced by the laser pulse.

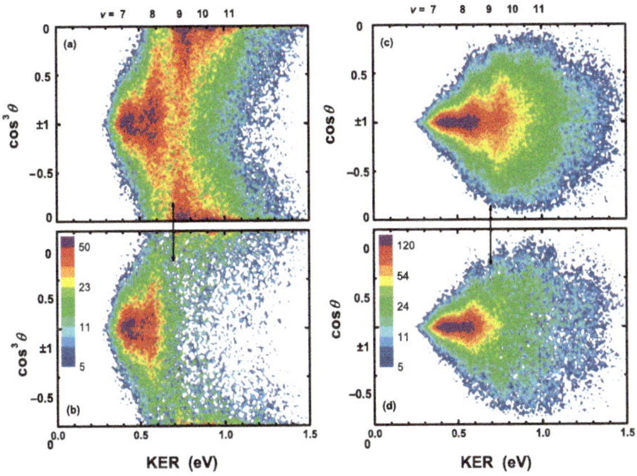

Fig. 4.7. (**a**) The $dN(KER)/d(\cos^3\theta)$ distribution for the dissociation of H_2^+ caused by a 135 fs pulse with a peak intensity of $\sim 2.4\times10^{14}$ W cm^{-2}. (**b**) The same as (**a**) but after subtraction of the contributions from low intensities (see [44]), i.e. intensities lower than 9% of I_0. (**c**) and (**d**) are the same as (**a**) and (**b**), respectively, but as a function of $\cos\theta$. On the top the field-free KER values expected for a few vibrational states are marked. The *arrows* mark the expected KER at the curve crossing shown in Fig. 4.1 (from [45])

Therefore, calculations are usually conducted with reduced dimensionality. For example, Esry's calculations [8] were performed at intensities where ionization is negligible for molecules aligned along the polarization and neglecting nuclear rotation. The calculated dissociation probabilities compare well with the trends observed experimentally for 45 and 135 fs pulses, however some details are not reproduced as well as one would hope (see [8] for details). We suspect that rotation has to be included in order to improve the agreement between measurement and theory.

4.3.2 Ionization: $nh\nu + H_2^+ \rightarrow H^+ + H^+$

The ionization of an H_2^+ target beam by a ultrashort intense laser pulse was first probed directly by Williams et al. [28]. Earlier Gibson et al. [22] tailored their laser pulse intensity to study the enhanced ionization of H_2^+ using an H_2 target. In both studies they attempted to verify the double-peak structure predicted by theory, shown in Fig. 4.8 (see, for example, Zuo and Bandrauk [21]). Gibson et al. [22] reported that ionization is enhanced around 7 and 9 a.u., which compare well with the predicted peaks at 7 and 10 a.u. [21]. In contrast, Williams et al. [28] reported only one peak at about 7 a.u., but claimed that ionization at larger internuclear distances would be hard to separate in their measurements from the dissociation channel as the two processes contribute at the same KER range.

It is important to note that the internuclear distance where ionization occurs, R, is determined from the measured KER by assuming a vertical

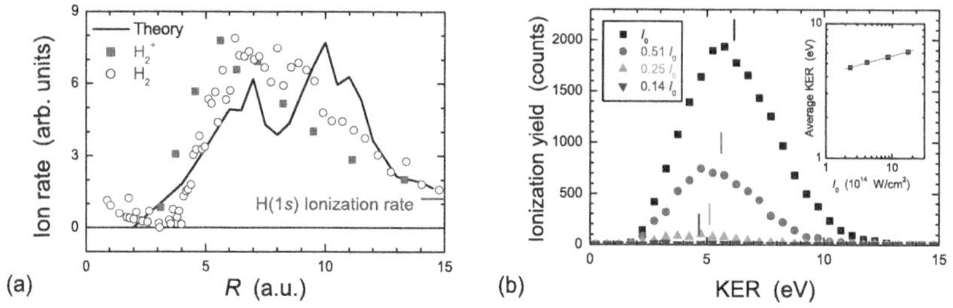

Fig. 4.8. (**a**) Ionization rate of H_2^+ as a function of internuclear distance. H_2^+ measurements of Williams et al. [28] are compared with predictions of Zuo and Bandrauk [21] and the H_2 measurements of Gibson et al. [22]. (**b**) Our H_2^+ ionization KER spectrum for 45 fs laser pulses with peak intensities up to $I_0 = 1.7 \times 10^{15}$ W cm^{-2}. Note the shift toward higher KER with increasing intensity shown explicitly in the insert

transition at that R to the Coulomb repulsion curve. However, there is an ambiguity about the kinetic energy gain before ionization. Gibson et al. [22] first subtracted the average kinetic energy of the main dissociation channel as the wave packet first gains some kinetic energy before ionization occurs, while Williams et al. [28] did not. This uncertainty in converting the measured KER to the internuclear distance can best be removed by theorists providing the expected KER upon enhanced ionization, i.e. by calculations including the vibrational degree of freedom.

Although the question about the predicted double-peak structure in the ionization probability vs. internuclear distance was left unsettled at the time, the experimental data is consistent with the multitude of theoretical calculations suggesting that ionization of H_2^+ is enhanced for some range of internuclear distances of the stretched molecular ion (e.g., [51–53]).

Recently, we attempted to address this question, and our results indicated a single wide peak around the KER value associated with dissociation at about 7 a.u. and no evidence for any ionization peak around the lower KER values expected for the peak at larger R. In these measurements [43] ionization was directly separated from dissociation, as discussed before, therefore removing the uncertainty caused by the subtraction of the dissociation channel in the previous measurements [28, 37, 38]. The measurements discussed above were done at somewhat different pulse duration (65, 40, 100 and 45 fs for Williams et al. [28], Gibson et al. [22], Pavičić et al. [37, 38] and Ben-Itzhak et al. [43], respectively). It is important to note that H_2^+ dissociation is initiated before the 45 fs laser pulse peaks, thus the intensity is high enough to also ionize around the second peak (i.e. $\sim R = 10$ a.u.). One may argue that the second ionization peak is missing because ionization at the first peak depletes the dissociating wave packet completely. However, Ben-Itzhak et al. [43, 54] have measured a significant fraction of dissociation along the laser polarization, thus demonstrating that some of the dissociating wave packet passes through the second CREI peak region without being ionized. These results were further supported by pump-probe measurements of double ionization of D_2 using 10 fs laser pulses [54,55]. These pump-probe results are in contrast to those of Ergler et al. [56, 57] (conducted with 25 fs pulses), in which they reported a structure in double ionization spectra that they attributed to the predicted second CREI peak [21]. Given the clear observation of a single enhanced ionization peak in our measurements, we were wondering *why is this second CREI peak missing*? We suggested that the double peak structure in CREI predicted for "frozen nuclei" (i.e. freezing the nuclear vibration in the calculations) vanishes mainly because of nuclear motion [54]. This claim is supported by calculations by Chelkowski et al. [58], which included nuclear vibration and showed smearing of the structure predicted when this nuclear motion was frozen. The weak remaining structure in the ionization KER is expected to wash out further by intensity averaging as the calculated peaks shift with intensity [54].

The experimental data at peak intensities above about $10^{14}\,\mathrm{W\,cm^{-2}}$ indicate a clear shift of the broad structureless KER distribution toward higher energies with increasing peak intensity of the laser pulse (see [38, 43]), as shown in Figs. 4.8b and 4.9c. This shift indicates that ionization occurs at shorter internuclear distances for stronger laser fields as one might expect qualitatively. Similar shifts in the energy of the fragments from the Coulomb explosion channel were observed in some H_2 measurements (see, for example, [59–61]).

Recently, studies by Pavičić et al. [37, 38] revealed even finer structure in the ionization channel, shown in Fig. 4.9c–d, which they associated with vibrational structure in the Coulomb explosion channel. (Ionization was identified by the higher KER, shown in Fig. 4.9a, and the fact that it does not appear in the equivalent H spectra of dissociation). It can be seen from Fig. 4.9c–d that this fine structure around 1.3–1.5 eV vanishes with increasing intensity. Pavičić et al. suggested a two step process in which dissociation of the different vibrational states is followed by a vertical ionization transition at a critical internuclear distance of $R \sim 13$ a.u. The kinetic energy released in both steps adds up to yield $\mathrm{KER} = \mathrm{KER}(v) + 1/R_{\mathrm{crit}}$ (in a.u.), where the first term is the KER upon dissociation of a specific vibrational state v and $1/R_{\mathrm{crit}}$ is the additional Coulomb explosion energy released after ionization at a narrow R-range around R_{crit}. Explicitly, they used an energy shift of 2.12 and 1.96 eV for H_2^+ and D_2^+, respectively (i.e., $R_{\mathrm{crit}} = 12.8$ and 13.9 a.u.) to match the structure as shown in Fig. 4.9d for H_2^+. Calculations of Peng et al. [62], which followed this report, failed to reproduce this fine vibrational structure of the ionization probability as a function of internuclear distance, but reconfirmed a single wide peak for high laser intensity and a double-peak structure at low intensities as reported previously (see, for example, [7]). Vafaee et al. [63] reported recently more structure in the ionization probability as a function of internuclear distance for laser pulses similar to those used in the experiment of Pavičić et al., however this structure is still much wider than the vibrational structure observed in the ionization KER [37, 38]. Further experimental and theoretical work is needed to shed light on this fine structure in the KER spectrum of ionization.

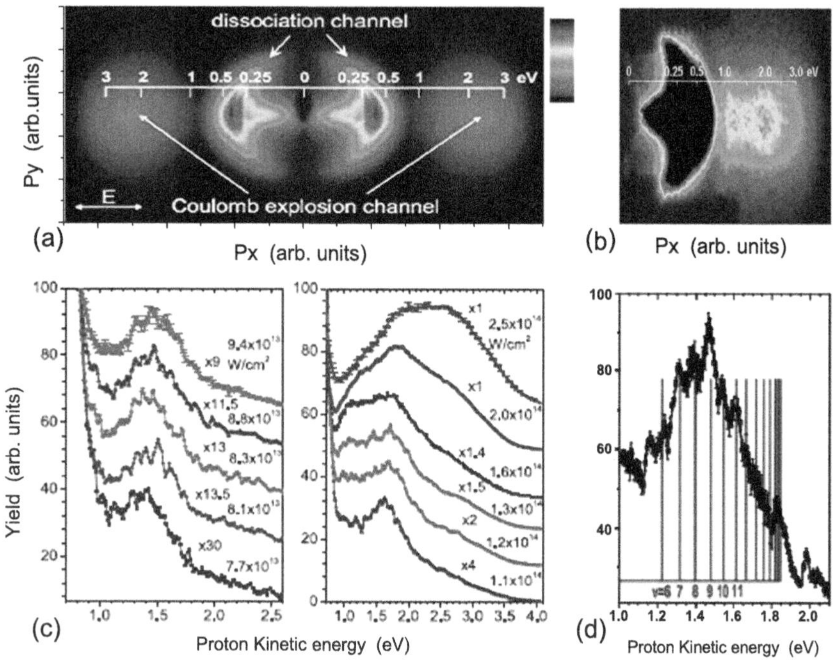

Fig. 4.9. (**a**) Two-dimensional velocity distribution of H_2^+ fragments created by a 10^{15} W cm^{-2}, 100 fs laser pulse. Ionization and dissociation are discerned by their kinetic energies and angular distributions. The kinetic energies of the fragments ejected along the polarization axis are marked by the inserted scale. (**b**) The same distribution for D_2^+ fragments exposed to 10^{14} W cm^{-2}, 350 fs laser pulses. Three E_p peaks around 1.05, 1.4, and 1.85 eV were associated (after subtracting a 0.3 eV average dissociation energy) with enhanced ionization at about 8, 11, and 15 a.u. (**c**) Kinetic energy spectra of protons from the Coulomb explosion of H_2^+, i.e. ionization, integrated over $\pm 2°$ around the laser polarization. (**d**) An extended view of the H_2^+ spectrum measured at 8.8×10^{13} W cm^{-2}. The *vertical lines* mark the fragment energies as expected after one-photon absorption from different vibrational levels, followed by a Coulomb explosion at a critical distance releasing an additional 1.06 eV per fragment (from Pavičić et al. [38])

In addition to the fine KER structure discussed above Pavičić et al. [38] reported their observation of three peaks in the ionization spectrum for 350 fs pulses, which they associated with CREI peaks at $R = 8$, 11 and 15 a.u., as shown in Fig. 4.9b. However, the earlier discussion suggests that structures predicted by "frozen nuclei" calculations are not measurable.

It seems that structure in the KER *spectra of ionization and its origin remain topics of debate even for the simplest* H_2^+ *molecule.*

To add to this discussion, Esry et al. [64] recently discovered structure in the KER spectrum of an H_2^+ beam probed by 45 fs laser pulses near the ionization appearance intensity. The measured peaks and the angular distributions produced by 790 and 395 nm were significantly different from each other, as can be seen in Fig. 4.10a–c. The observed structure was attributed to channel opening of n-photon ionization when the Coulomb PECs, dressed by n photons, cross the H_2^+ PECs as shown in Fig. 4.10e. This model adopts a multi-photon approach and predicts a sequence of peaks in the KER spectra associated with each dissociation pathway (these peak positions are marked by the tics on Fig. 4.10a–c). Inspired by the similar sequences in ATI and ATD, we have named this new mechanism "above threshold Coulomb explosion" (ATCE) [64]. An interesting note is that the 395 nm experiments were conducted due to predictions of this model following its success in interpreting our 790 nm data, thus suggesting predictive power.

Similar KER structures have been observed near the double ionization appearance intensity of H_2 and D_2 by Staudte et al. [65]. Their concurrent findings verify our measurements. Furthermore, their higher statistics data is consistent with our model [66]. Encouraged by this success we applied our model to the 350 fs data shown in Fig. 4.9b provided to us as a KER distribution by Pavičić [38]. Also in this case, our model is in good agreement with the data, thus suggesting that the observed structure is due to ionization channel opening along the dissociation pathways [66].

4.3.3 Ionization Vs. Dissociation

In addition to the interest in the ionization and dissociation of H_2^+ caused by ultrashort intense laser pulses independently from each other, there is also the question of their mutual interdependence, i.e. the relative importance of the two processes as a function of the laser pulse parameters, such as peak intensity, pulse duration, wave length, and pulse shape. This relative importance is given by the branching ratio of the two processes, i.e. $\Gamma_{\text{ion}}/(\Gamma_{\text{ion}} + \Gamma_{\text{dissoc}})$ where Γ_{ion} and Γ_{dissoc} are the rates of ionization and dissociation of H_2^+, respectively. One would expect a higher ionization rate for higher laser intensity or for more energetic photons. Dissociation of H_2^+, in contrast, can occur at lower intensities, especially for vibrational states near the curve crossing between the $1s\sigma$ and $2p\sigma - 1\omega$ potential energy curves as discussed above (see Fig. 4.1). Therefore, ionization is expected to dominate over dissociation when the intensity becomes high enough (see reviews by Giusti-Suzor et al. [6] and Posthumus [7]). It is predicted that this branching ratio depends also on the number of laser cycles in the pulse, i.e. the pulse duration (see, for example, [6, 7, 67]). The first direct measurements of the relative rates of ionization to dissociation for an H_2^+ target were conducted by Williams et al. [28] and are shown in Fig. 4.11 together with the recent measurements by Ben-Itzhak et al. [68]. Both measurements show a more rapid increase in the ionization rate than the dissociation rate, and ionization is much smaller than dissociation over this intensity range in contrast to theoretical predictions. For example, Giusti-Suzor et al. [6] predicted that ionization will be much larger than dissociation for a 1.7×10^{14} W cm^{-2} and 150 fs laser pulse. Similar results

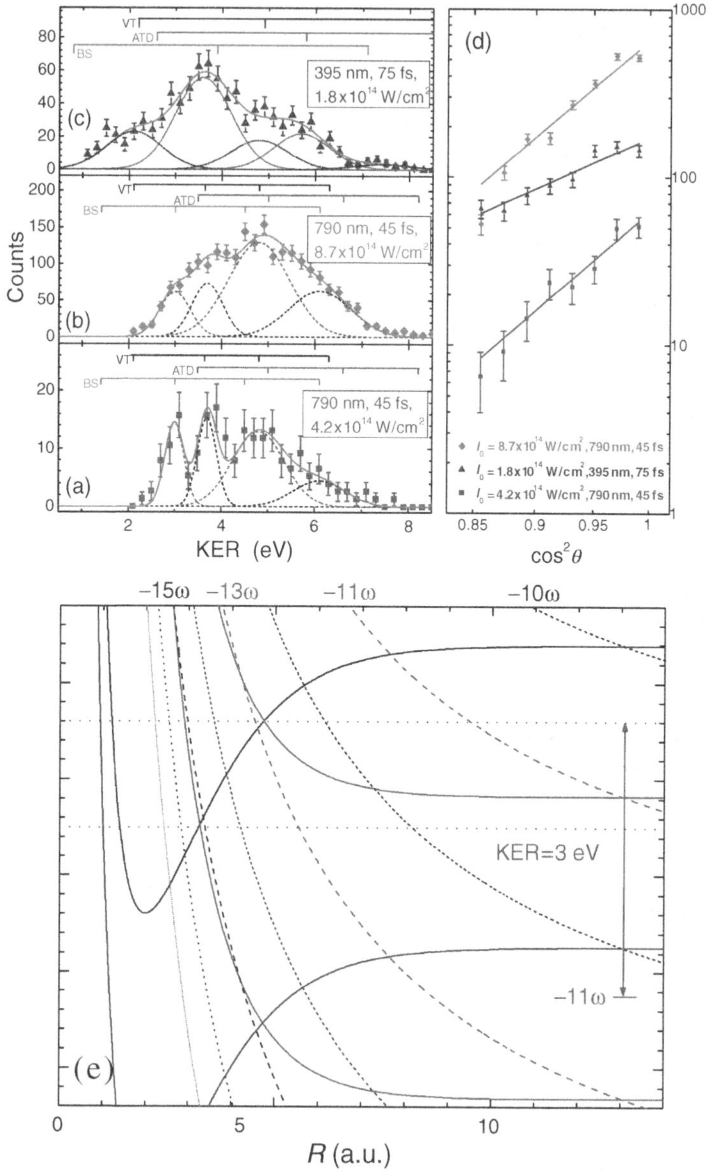

Fig. 4.10. KER distribution of ionization of an H_2^+ beam by (**a**) and (**b**) 790 nm at two peak intensities, and (**c**) 395 nm, 75 fs, 1.8×10^{14} W cm^{-2}. (**d**) A log–log plot of the angular distribution indicating a different number of photons at each wavelength. (**e**) PEC diagram for H_2^+ similar to the one shown in Fig. 4.1, but including dressed Coulomb potentials (*short dash*) to represent ionization channels

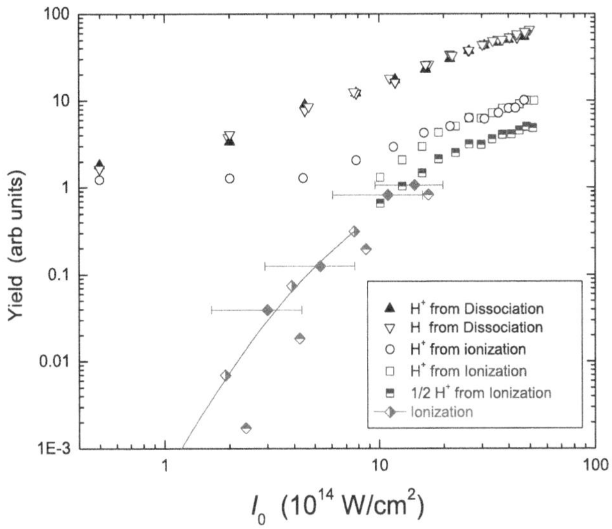

Fig. 4.11. Ionization and dissociation rates of H_2^+ by ultrashort intense laser pulses as a function of laser peak intensity: *triangles* and *circles* – from [28], *squares* – from [41] (a later publication of the same research group, where they also showed that the intensity dependence for the higher intensities is due to the volume effect and corrected their low intensity data), and *diamonds* – our work [68] scaled to the dissociation rate of [28,41] (the half-filled horizontal symbols represent a preliminary measurement in which the beam spot size was not measured, the half-filled vertical symbols connected by the line are the same data scaled to $0.45 \times I_0$ to match the later measurements (*full diamonds*), for which the beam spot was measured). Note that proton or H-atom yields were reported previously [28, 41] while our measurements [68] are dissociation and ionization yields. The *half filled squares* indicate the expected ionization rate from the earlier measurements. Most importantly, ionization is less than 10% of the dissociation even at intensities around 5×10^{15} W cm^{-2}

were reported by Feuerstein and Thumm [69] for a much shorter laser pulse of 25 fs of similar intensity of 2×10^{14} W cm^{-2}. This result seems in contradiction with the data shown in Fig. 4.11, where dissociation clearly dominates at this range of laser intensities. In contrast to the large overestimation of the relative ionization to dissociation rate, the calculated KER distribution [6,69] is in better agreement with the measurement. Explicitly, theory predicts a broad KER distribution centered around about 6 eV, while the experimental distribution, shown in Fig. 4.8b, is also wide and peaks at somewhat lower KER of about 5 eV for this peak intensity.

Part of this disagreement stems from the fact that dissociation occurs at much lower intensities, thus the interaction volume in which dissociation occurs is much larger than that of ionization. Therefore, to properly compare experiment to theory one has to first eliminate the contribution from the low intensity tail of the laser pulse. This can be done, for example by using the intensity difference spectrum (IDS) method discussed before, thus reducing the intensity averaging for the measurement at $1.7 \times 10^{15}\,\mathrm{W\,cm^{-2}}$ to intensities above about $7 \times 10^{14}\,\mathrm{W\,cm^{-2}}$. This simplifies the intensity averaging, as otherwise there is a need to compute both reaction channels over many orders of magnitude of laser intensity. Of course one has to take into account the specific experimental conditions to properly average the intensity. This comparison is further complicated by the fact that ionization is aligned much more than dissociation along the laser polarization [46]. Finally, theoretical treatments of ionization conducted so far are limited in their dimensionality, typically assuming a non-rotating H_2^+ which is aligned along the laser field. As a result of all these effects, it is not surprising that the theoretical values for an aligned H_2^+ exposed to a single peak intensity are in disagreement with the measured relative rates integrated over all molecular alignment and laser intensity within the interaction region, as in the experimental results presented in Fig. 4.11. To circumvent this difficulty, We evaluated the ionization to dissociation branching ratio for H_2^+ aligned within a narrow cone around the laser polarization, i.e. between 1 and the $|\cos\theta|$ value marked on the horizontal axis of Fig. 4.12a. It can clearly be seen that the branching ratio increases from a few percent for large acceptance angles to about 25% for purely aligned H_2^+, i.e. an increase by about a factor of 6. In addition, after applying the IDS method to eliminate low intensities (i.e. intensities lower

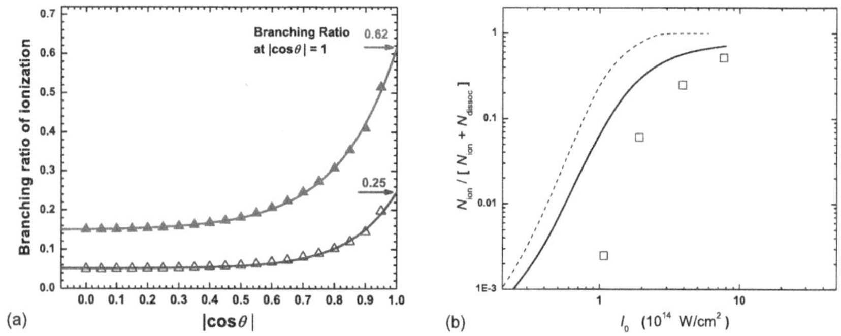

Fig. 4.12. (**a**) The ionization to dissociation branching ratio, i.e. $\Gamma_{\mathrm{ion}}/(\Gamma_{\mathrm{ion}} + \Gamma_{\mathrm{dissoc}})$, for two peak intensities (790 nm, 45 fs pulses) as a function of the acceptance angle around the laser polarization; the open data points include events from all intensities present in the laser pulse (i.e. integrated over the whole interaction region), while the solid data points include only intensities from I_0 to about $I_0/2$, evaluated using the IDS method. (**b**) The ionization to dissociation branching ratio as a function of peak intensity for aligned H_2^+ after elimination of the low intensity contributions [68] (using our scaled I_0 data from Fig. 4.11) Theory: Franck–Condon averaged – *dashed line*, and intensity averaged – *solid line*. Note that agreement with theory improved after selecting aligned molecules and reducing the intensity range experimentally, and integrating the vibrational population and averaging over the relevant peak intensities, although the agreement is still not satisfactory

than about $I_0/2$) the ratio of ionization to dissociation was further increased to about 62% for a peak intensity of about 1.7×10^{15} W cm^{-2}.

The measured branching ratios extrapolated to purely aligned H_2^+, as discussed above (see Fig. 4.12a), for a few peak intensities, are compared to numerical calculations by Roudnev and Esry [68] for H_2^+ aligned along the laser field that include nuclear vibration and the electron motion in 3D (electron dimensionality is reduced by symmetry, see [70]), as shown in Fig. 4.12b. The computed ionization and dissociation rates were averaged over the narrow intensity range of the measurement. Though the agreement between theory and experiment is greatly improved (by more than an order of magnitude at about 10^{15} W cm^{-2}), there is no doubt that further experimental and theoretical work is needed.

4.4 Many-Electron Molecular Ions

A few other molecular ions have been studied; however, these reports are even scarcer than those of H_2^+. Surprisingly, the first measurements of the interaction of a molecular ion target with an intense laser field was done on a many-electron molecular ion, namely Ar_2^+ by Figger and Hänsch's group [71, 72]. These measurements were done using a 532 nm Nd:Yag laser beam with about 5×10^{12} W cm^{-2} peak intensity, and more importantly, much longer laser pulses of about 30 ns. In spite of the long laser pulse the interaction time between the Ar_2^+ and the laser field was much shorter, about 30 ps, which is the time the ions traversed the laser beam. Wunderlich et al. [71, 72] nicely demonstrated that the dissociation of Ar_2^+ is well described by the laser induced molecular potentials (i.e. the dressed potential energy curves) taking advantage of the simple electronic structure of this molecular ion [71, 72].

Assion et al. [73] studied the dissociation of Na_2^+ by ultrashort intense laser pulser over a wide range of laser intensities (10^{11}–10^{14} W cm^{-2}), and their data indicate that the dissociation of Na_2^+ produces four well distinguished KER distributions shown in Fig. 4.13a as forward and backward TOF peaks. Note that the highest two KER distributions (C and D) appear only at high laser intensities, and they were associated with CREI. The higher energy of peak D is due to CREI at a shorter internuclear distance than peak C, however, direct ionization can not be excluded as a source of fragments with the KER of peak D. The lower KER distributions (denoted by A and B) were associated with dissociation due to the coupling of the ground state with two different excited states. Explicitly, one-photon absorption couples the $X^2\Sigma_g^+$ ground state to the $1^2\Sigma_u^+$ state that dissociates to Na^+ and Na(3s) fragments (channel B), while two-photon absorption couples the $X^2\Sigma_g^+$ ground state to the $2^2\Sigma_g^+$ state that dissociates to Na^+ and Na($3p$) fragments (channel A) (see Assion et al. [73] for further details). Recent studies of O_2^+ dissociation and ionization were conducted by Sayler et al. [74] using coincidence 3D momentum imaging (see earlier discussion of Fig. 4.2c). The KER distribution of the O_2^+

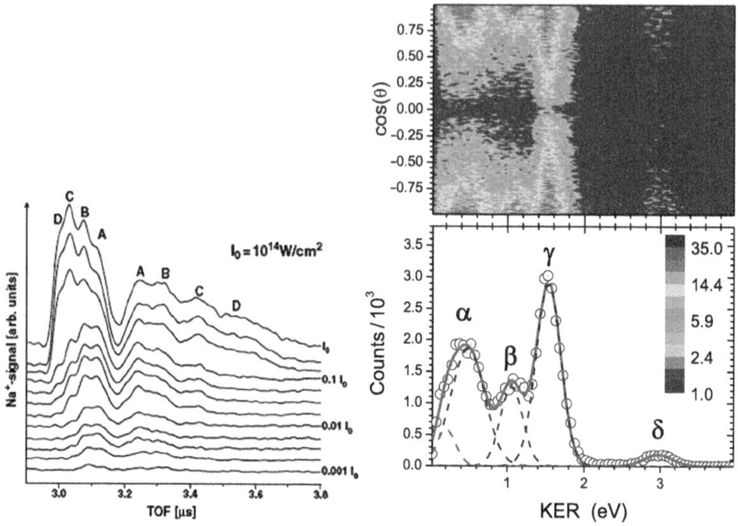

(a) (b)

Fig. 4.13. (**a**) Time-of-flight spectra of Na^+ fragments obtained from fragmentation of Na_2^+ $X^2\Sigma_g^+(v)$ in intense 790 nm, 80 fs laser fields at laser intensities from 10^{11} to 10^{14} W cm^{-2}. The laser polarization was set parallel to the TOF axis. Four fragmentation channels with a released kinetic energy of 0.5 eV (A), 1.3 eV (B), 2.9 eV (C), and 4.7 eV (D) are observed (from Assion et al. [73]). (**b**) KER distribution of the dissociation of O_2^+ by a 790 nm, 50 fs and 8×10^{14} W cm^{-2} laser pulse (from Sayler et al. [74])

dissociation also shows a rich structure. Moreover, the different KER peaks exhibit different angular distributions as shown in the top panel of Fig. 4.13b. Sayler et al. [74] have recently shown that the different angular distributions are caused by the different dissociation pathways through the number of parallel and perpendicular transitions (parallel and perpendicular to the linear polarization) and the number of photons involved in each of these transitions. Note that parallel and perpendicular transitions lead to $\cos^2\theta$ and $\sin^2\theta$ angular distributions, respectively, for transitions involving a single photon. When the number of photons n is higher, these distributions are raised to the power of n. (see Sayler et al. [74] for further details).

Atomic Ions

Studies of the interaction of ultrashort intense laser fields with atomic ion targets are even fewer than those with molecular ions (see, Greenwood et al. [75] and Newell et al. [76]). Greenwood et al. [75] studied the suppression of multiple ionization of Ar^+ ions caused by intense laser fields. They used the same setup as Williams et al. [28], described earlier, and a 790 nm, 60 fs laser beam with peak intensity up to about 10^{16} W cm^{-2}. More recently the same group studied the ionization of C^+ ions using the same technique [76].

These pioneering measurements are mentioned here for completeness of our review of ion–intense laser interactions.

4.5 Summary and Future Directions

Interrogating molecular-ion beams, H_2^+ in particular, with intense ultrashort laser pulses has demonstrated the behavior of the vibrational states during the laser pulse, some expected and some surprising. Dissociation phenomena like level shifting or variation of alignment with vibrational states are examples. Ionization of H_2^+, thought to be well understood, is still a source of surprises, such as the recently reported KER structures. It is encouraging to know that even more surprises await us in the near future, and that exploring molecular ions in intense laser fields will remain a topic of exciting debate.

In spite of the fact that ion beams are much more complex targets than gas phase molecules, which makes the experimental work challenging, it is encouraging to see that more groups around the world are joining this field of research.

One may wonder in which direction the attention in this field will drift. Few cycle laser pulses, which are becoming more readily available, are also making their way to studies of molecular ion beams. These few fs pulses used in a pump-probe scheme have the right time scale to enable investigations of the time evolution of the dissociation and ionization processes. Furthermore, for such short pulses the carrier envelope (CE) phase, i.e. the phase between the peaks of the carrier and the envelope, is expected to affect the outcome of the measurement. For example, Roudnev et al. [77] predicted that the ratio of HD^+ dissociation into $H^+ + D(1s)$ or $H(1s) + D^+$ can be controlled by the CE phase of a 10 fs laser pulse at intensities above about 5×10^{14}. Using a CE phase locked laser Kling et al. [78] have recently reported a somewhat similar effect in which the direction of the D^+ fragment from dissociative ionization of D_2 relative to the laser polarization was a function of the CE phase. This effect was attributed to interference of the wave packets dissociating in the $1s\sigma$ and $2p\sigma$ electronic states, where the latter was excited by recollision of the ionized electron [78]. Probing molecular ions with few-cycle pulses seems to be one of the most promising research directions in this field.

Acknowledgements

Thanks are due to present and former members of my research group, A.M. Sayler, M.A. Smith, M. Leonard, Dr. J.F. Xia, Dr. P.Q. Wang, and Professor K.D. Carnes, for their contributions to the research presented here. Our experimental work benefits greatly from the theoretical support of Professor B.D. Esry and his former research assistant V. Roudnev. The laser work presented here was made possible by Dr. B. Shan and Professor Z. Chang who provide the intense laser beams for our experiments, and Dr. C. Fehrenbach for his help with the ion beams. Last but not least, I wish to thank many members of the JRML group who have contributed to our understanding of the laser-induced processes described above. This work was supported by the Chemical Sciences, Geosciences and Biosciences Division, Office of Basic Energy Sciences, Office of Science, U.S. Department of Energy.

References

1. P.H. Bucksbaum, A. Zavriyev, H.G. Muller, D.W. Schumacher, Phys. Rev. Lett. **64**, 1883 (1990)
2. A. Zavriyev, P.H. Bucksbaum, H.G. Muller, D.W. Schumacher, Phys. Rev. A **42**, 5500 (1990)
3. A. Zavriyev, P.H. Bucksbaum, J. Squier, F. Saline, Phys. Rev. Lett. **70**, 1077 (1993)
4. L.J. Frasinski, J.H. Posthumus, J. Plumridge, K. Codling, P.F. Taday, A.J. Langley, Phys. Rev. Lett. **83**, 3625 (1999)
5. J.H. Posthumus, J. Plumridge, L.J. Frasinski, K. Codling, E.J. Divall, A.J. Langley, P.F. Taday, J. Phys. B **33**, L563 (2000)
6. A. Giusti-Suzor, F.H. Mies, L.F. DiMauro, E. Charron, B. Yang, J. Phys. B **28**, 309 (1995)
7. J.H. Posthumus, Rep. Prog. Phys. **67**, 623 (2004)
8. P.Q. Wang, A.M. Sayler, K.D. Carnes, J.F. Xia, M.A. Smith, B.D. Esry, I. Ben-Itzhak, Phys. Rev. A **74**, 043411 (2006)
9. A. Giusti-Suzor, F.H. Mies, Phys. Rev. Lett. **68**, 3869 (1992)
10. G. Yao, S.I. Chu, Chem. Phys. Lett. **197**, 413 (1992)
11. G. Yao, S.I. Chu, Phys. Rev. A **48**, 485 (1993)
12. E.E. Aubanel, G.M. Gauthier, A.D. Bandrauk, Phys. Rev. **48**, 2145 (1993)
13. E.E. Aubanel, A. Conjusteau, A.D. Bandrauk, Phys. Rev. **48**, R4011 (1993)
14. A. Giusti-Suzor, X. He, O. Atabek, F.H. Mies, Phys. Rev. Lett. **64**, 515 (1990)
15. B. Yang, M. Saeed, L.F. DiMauro, A. Zavriyev, P.H. Bucksbaum, Phys. Rev. A **44**, R1458 (1991)
16. G. Jolicard, O. Atabek, Phys. Rev. A **46**, 5845 (1992)
17. J.H. Posthumus, J. Plumridge, L.J. Frasinski, K. Codling, A.J. Langley, P.F. Taday, J. Phys. B **31**, L985 (1998)
18. L.J. Frasinski, J. Plumridge, J.H. Posthumus, K. Codling, P.F. Taday, E.J. Divall, A.J. Langley, Phys. Rev. Lett. **86**, 2541 (2001)
19. F. Rosca-Pruna, M.J.J. Vrakking, Phys. Rev. Lett. **87**, 153902 (2001)
20. K. Codling, L.J. Frasinski, P.A. Hatherly, J. Phys. B **22**, L321 (1989)
21. T. Zuo, A.D. Bandrauk, Phys. Rev. A **52**, R2511 (1995)
22. G.N. Gibson, M. Li, C. Guo, J. Neira, Phys. Rev. Lett. **79**, 2022 (1997)
23. S. Chelkowski, P.B. Corkum, A.D. Bandrauk, Phys. Rev. Lett. **82**, 3416 (1999)
24. B. Feuerstein, U. Thumm, Phys. Rev. A **67**, 063408 (2003)
25. A.S. Alnaser, X.M. Tong, T. Osipov, S. Voss, C.M. Maharjan, P. Ranitovic, B. Ulrich, B. Shan, Z. Chang, C.D. Lin, C.L. Cocke, Phys. Rev. Lett. **93**, 183202 (2004)
26. K. Codling, L.J. Frasinski, J. Phys. B **26**, 783(1993)
27. K. Sändig, H. Figger, T.W. Hänsch, Phys. Rev. Lett. **85**, 4876 (2000)

28. I.D. Williams, P. McKenna, B. Srigengan, I.M.G. Johnston, W.A. Bryan, J.H. Sanderson, A. El-Zein, T.R.J. Goodworth, W.R. Newell, P.F. Taday, A.J. Langley, J. Phys. B 33, 2743 (2000)
29. Z. Amitay, A. Baer, M. Dahan, J. Levin, Z. Vager, D. Zajfman, L. Knoll, M. Lange, D. Schwalm, R. Wester, A. Wolf, I.F. Schneider, A. Suzor-Weiner, Phys. Rev. A **60**, 3769 (1999)
30. F. von Busch, G.H. Dunn, Phys. Rev. A **5**, 1726 (1972)
31. X. Urbain, B. Fabre, E.M. Staicu-Casagrande, N. de Ruette, V.M. Andrianarijaona, J. Jureta, J.H. Posthumus, Phys. Rev. Lett. **92**, 163004 (2004)
32. P.B. Corkum, N.H. Burnett, F. Brunel, Phys. Rev. Lett. **62**, 1259 (1989)
33. P.B. Corkum, Phys. Rev. Lett. **71**, 1994 (1993)
34. A.S. Alnaser, T. Osipov, E.P. Benis, A.Wech, B. Shan, C.L. Cocke, X.M. Tong, C.D. Lin, Phys. Rev. Lett. **91**, 163002 (2003)
35. K. Sändig, H. Figger, T.W. Hänsch, *Multiphoton Processes: ICOMP VIII, 8th International Conference*, vol. CP 525, ed. by L.F. DiMauro, R. Freeman, K.C. Kulander (American Institute of Physics, 2000), p. 502
36. K. Sändig, H_2^+ in intensiven Lichtfeldern. PhD Thesis, Ludwig-Maximilians-Universität München (2000)
37. D. Pavičić, A. Kiess, T.W. Hänsch, H. Figger, Euro. Phys. J. D **26**, 39 (2003)
38. D. Pavičić, A. Kiess, T.W. Hänsch, H. Figger, Phys. Rev. Lett. **94**, 163002 (2005); and private communication
39. D. Pavičić, Coulomb Explosion and Intense-Field Photodissociation of Ion-Beam H_2^+ and D_2^+. PhD Thesis, Max-Planck-Institut für Quantenoptik (2004)
40. I.D. Williams, B. Srigengan, P. McKenna, W.R. Newell, J. H. Sanderson, W.A. Bryan, A. El-Zein, P.F. Taday, A.J. Langley, Phys. Scr. **T 80**, 534 (1999)
41. J.H. Sanderson, W.A. Bryan, A. El-Zein, T.R.J. Goodworth, I.D. Williams, P. McKenna, I.M.G. Johnston, P.F. Taday, E.J. Divall, A.J. Langley, *CLF Annual Report* (2000), p. 65
42. I.D. Williams, P. McKenna, B. Srigengan, I.M.G. Johnston, W.A. Bryan, J.H. Sanderson, A. El-Zein, T.R.J. Goodworth, W.R. Newell, P.F. Taday, A.J. Langley, *Application of Accelerators in Research and Industry – Sixteenth Int'l Conf.*, vol. CP 576, ed. by J.L. Duggan, I.L. Morgan (American Institute of Physics, 2001), p. 44
43. I. Ben-Itzhak, P.Q. Wang, J.F. Xia, A.M. Sayler, M.A. Smith, K.D. Carnes, B.D. Esry, Nucl. Instrum. Meth. B **233**, 56 (2005)
44. P.Q. Wang, A.M. Sayler, K.D. Carnes, B.D. Esry, I. Ben-Itzhak, Opt. Lett. **30**, 664 (2005)
45. P.Q. Wang, A.M. Sayler, K.D. Carnes, J.F. Xia, M.A. Smith, B.D. Esry, I. Ben-Itzhak, J. Phys. B **38**, L251 (2005)
46. I. Ben-Itzhak, P.Q. Wang, J.F. Xia, A.M. Sayler, M.A. Smith, K.D. Carnes, B.D. Esry, Phys. Rev. Lett. **95**, 073002 (2005)
47. V.N. Serov, A. Keller, O. Atabek, N. Billy, Phys. Rev. A **68**, 053401 (2003)
48. A. Kondorskiy, H. Nakamura, Phys. Rev. A **66**, 053412 (2002)
49. L.Y. Peng, I.D. Williams, J.F. McCann, J. Phys. B **38**, 1727 (2005)
50. D. Pavičić, T.W. Hänsch, H. Figger, Phys. Rev. A. **72**, 053413 (2005)
51. S. Chelkowski, T. Zuo, O. Atabek, A.D. Bandrauk, Phys. Rev. A **52**, 2977 (1995)
52. J.H. Posthumus, L.F. Frasinski, A.J. Giles, K. Codling, J. Phys. B **28**, L349 (1995)
53. T. Seideman, M.Yu. Ivanov, P.B. Corkum, Phys. Rev. Lett. **75**, 2819 (1995)

54. A.M. Sayler, Measurements of ultrashort intense laser-induced fragmentation of simple molecular ions. PhD Thesis, Kansar State University (2008)
55. A.S. Alnaser, I. Litvinyuk, T. Osipov, B. Ulrich, A. Landers, E. Wells, C.M. Maharjan, P. Ranitovic, I. Bochareva, D. Ray, C.L. Cocke, J. Phys. B **39**, S485 (2006)
56. Th. Ergler, A. Rudenko, B. Feuerstein, K. Zrost, C.D. Schroẗer, R. Moshammer, J. Ullrich, Phys. Rev. Lett. **95**, 093001 (2005)
57. Th. Ergler, A. Rudenko, B. Feuerstein, K. Zrost, C.D. Schroẗer, R. Moshammer, J. Ullrich, J. Phys. B **39**, S493 (2006)
58. S. Chelkowski, C. Foisy, A.D. Bandrauk, Phys. Rev. A **57**, 1176 (1998)
59. J. Ludwig, H. Rottke, W. Sandner, Phys. Rev. A **56**, 2168 (1997)
60. M.R. Thompson, M.K. Thomas, P.F. Taday, J.H. Posthumus, A.J. Langley, L.J. Frasinski, K. Codling, J. Phys. B **30**, 5755 (1997)
61. C. Trump, H. Rottke, M. Wittmann, G. Korn, W. Sandner, M. Lein, V. Engel, Phys. Rev. A **62**, 063402 (2000)
62. L.Y. Peng, D. Dundas, J.F. McCann, K.T. Taylor, I.D. Williams, J. Phys. B **36** L295 (2003)
63. M. Vafaee, H. Sabzyan, J. Phys. B **37**, 4143 (2004)
64. B.D. Esry, A.M. Sayler, P.Q. Wang, K.D. Carnes, I. Ben-Itzhak, Phys. Rev. Lett. **97**, 013003 (2006)
65. A. Staudte, D. Pavičić, S. Chelkowski, D. Zeidler, M. Meckel, H. Niikura, M. Schöffler, S. Schössler, B. Ulrich, P.P. Rajeev, Th. Weber, T. Jahnke, D.M. Villeneuve, A.D. Bandrauk, C.L. Cocke, P.B. Corkum, R. Dörner, Phys. Rev. Lett. **98**, 073003 (2007)
66. B.D. Esry, I. Ben-Itzhak, Phys. Rev. A (submitted for publication)
67. K.C. Kulander, F.H. Mies, K.J. Schafer, Phys. Rev. A **53**, 2562 (1996)
68. I. Ben-Itzhak I, P.Q. Wang, A.M. Sayler, K.D. Carnes, V. Roudnev, B.D. Esry (in preparation)
69. B. Feuerstein, U. Thumm, Phys. Rev. A **67**, 043405 (2003)
70. V. Roudnev, B.D. Esry, Phys. Rev. A **71**, 013411 (2005)
71. C. Wunderlich, E. Kobler, H. Figger, T.W. Hänsch, Phys. Rev. Lett. **78**, 2333 (1997)
72. C. Wunderlich, H. Figger, T.W. Hänsch, Phys. Rev. A **62**, 023401 (2000)
73. A. Assion, T. Baumert, U. Weichmann, G. Gerber, Phys. Rev. Lett. **86**, 5695 (2001)
74. A.M. Sayler, P.Q. Wang, K.D. Carnes, B.D. Esry, I. Ben-Itzhak, Phys. Rev. A **75**, 063420 (2007)
75. J.B. Greenwood, I.M.G. Johnston, P. McKenna, I.D. Williams, T.R.J. Goodworth, J.H. Sanderson, W.A. Bryan, A. El-Zein, W.R. Newell, A.J. Langley, E.J. Divall, Phys. Rev. Lett. **88**, 233001 (2002)
76. W.R. Newell, I.D. Williams, W.A. Bryan, Eur. Phys. J. D **26**, 99 (2003)
77. V. Roudnev, B.D. Esry, I. Ben-Itzhak, Phys. Rev. Lett. **93**, 163601 (2004)
78. M.F. Kling, Ch. Siedschlag, A.J. Verhoef, J.I. Khan, M. Schultze, Th. Uphues, Y. Ni, M. Uiberacker, M. Drescher, F. Krausz, M.J.J. Vrakking, Science **312**, 246 (2006)

5

Dynamical Reaction Theory for Vibrationally Highly Excited Molecules

Mikito Toda

Abstract. We review the present status of the dynamical reaction theory, which is based on the concepts of the Arnold webs and normally hyperbolic invariant manifolds (NHIMs). First, we discuss reaction processes under laser fields where the Arnold web in the potential well is nonuniform. In particular, we present the possibility of controlling reaction processes utilizing cooperative effects of laser fields and the Arnold web. Second, we present existence of fractional behavior for processes in nonuniform Arnold webs. Based on the fractional behavior, we cast doubt on the very existence of the concept of the reaction rate *constant*. Third, we show that the concept of transition states (TSs) itself has limitations because of chaos on the NHIM. These limitations become manifest for those reaction processes when bath modes are highly excited above the saddle by strong laser fields. We further discuss the possibility of extending the dynamical reaction theory to processes of going over multiple saddles and to those in quantum systems.

5.1 Introduction

Reaction processes in ultrafast intense laser fields provide a challenging research arena where the basic assumptions of the statistical reaction theory should be reexamined. In the statistical reaction theory, separation of time scales between relaxation and the reaction is crucial. In other words, relaxation within the potential well is assumed to take place much faster than the reaction. Then, thermal equilibrium within the well is supposed to be maintained during the whole processes of the reaction. To the contrary, reaction processes in intense laser fields occur in situations which are far from equilibrium. This is because laser fields put large amount of energy to specific degrees of freedom so that reactions can take place *before* the energy is redistributed statistically to other degrees of freedom. Moreover, ultrafast laser fields enable us to observe reaction processes within a time scale which is much shorter than that for the redistribution. Thus, it is possible to see the dynamical processes where the energy flows within the molecule. Therefore, we need the dynamical theory of reactions for understanding processes under ultrafast intense laser fields.

The purpose of this review is to provide an overview of the recent development of the dynamical theory of reactions. Roughly speaking, dynamical processes of reactions consist of the following three, i.e., redistribution of energy among vibrational modes in the well, going over the potential saddles, and their dynamical connection. For each of the former two processes, perturbation approach is possible to construct the normal form. These normal forms offer basic understanding of the processes. For the distribution, the Dunham expansion is the normal form, which describes how the vibrational modes exchange energy in the well. They are, in general, nonlinear modes with their frequencies dependent on their amplitudes. When nonlinear resonance takes place, the perturbation theory breaks down, and energy exchange among vibrational modes is enhanced dramatically there. In the action space, resonant regions constitute the network called the Arnold web. Therefore, properties of the Arnold web play an important role in our topics. For the processes of going over the saddle, the normal form theory is developed recently, which provides a mathematically sound foundation of the concept of transition states (TSs). The theory is based on the phase space structures called normally hyperbolic invariant manifolds (NHIMs). It enables us to define the boundary between the reactant and the product, and to single out the reaction coordinate, at least locally in the phase space near the saddles of index one. Thus, the dynamical theory of reactions is based on the studies of the above two, and on understanding how these processes are connected in the phase space.

In this review, we focus our attention to the processes where limitations of the conventional statistical theory become manifest. First, we discuss reaction processes under laser fields where resonances exist nonuniformly in the potential well. In particular, we point out that understanding of the Arnold web is crucial for the strategy of controlling reaction processes. This is because cooperative effects of external fields and resonances play an important role in processes where the web is nonuniform. Second, we point out that doubt should be cast on the very existence of the reaction rate *constant* for processes when the Arnold web is nonuniform. The doubt results from fractional behavior of the reaction processes. Moreover, we discuss a possibility of the Maxwell's demon system in molecules. This opens a new horizon to the problem of information processing in molecular levels. Third, we show that the concept of TSs itself has limitations because of chaos on the NHIM. The conventional statistical theory assumes that the reaction coordinate is one-dimensional and is separable from the other degrees of freedom, i.e., the bath modes. Moreover, the bath modes are supposed to be harmonic near the saddle. However, the dynamical theory of reactions indicates that these assumptions do not necessarily hold when the bath modes start to exhibit chaos near the saddle. Such processes can take place for reactions under strong laser fields since the laser field excite the bath modes so strongly that nonlinear coupling among them cannot be neglected even near the saddle.

The contents of this review is as follows. In Sect. 5.2, we briefly discuss the assumptions which underlie the conventional statistical reaction theory. After that, we present in Sect. 5.3 the basic concepts of the dynamical reaction theory. In particular, the following concepts are explained such as nonlinear resonance, the Arnold web, NHIMs, and their stable/unstable manifolds. In Sect. 5.4, we discuss intramolecular vibrational-energy redistribution (IVR) from the standpoint of the Arnold web. Vibrationally highly excited acetylene is studied as an example where the Arnold web is nonuniform. It is shown that, due to the symmetry of the molecule, most of the resonant interactions are prohibited leading to a sparse Arnold web. Moreover, we investigate how cooperative effects of external laser fields and resonances in the web lead the system toward isomerization from acetylene to vinyliden. This indicates that, in designing laser fields for control of the reaction, distribution of resonances in the web plays a crucial role. In Sect. 5.5, definability of the rate *constant* is discussed. Using a prototype model for isomerization, we show that sparse distribution of resonances results in fractional behavior in the reaction. Existence of the fractional behavior casts a doubt on definability of the rate *constant*. Moreover, we discuss that existence of fractional behavior provides resources for information processing in molecular levels. In other words, this presents a possibility to construct the Maxwell's demon system in molecules. In Sect. 5.6, limitations of the concept of TSs are pointed out based on the dynamical theory of reactions. We discuss that, for vibrationally highly excited molecules, the bath modes can exhibit chaos even near the saddle. Then, we show that the conventional idea of the reaction coordinate breaks down. In general, this problem lies outside of the applicability of the perturbation analysis. Here, we use a simple model Hamiltonian to obtain an intuitive idea of how breakdown of the perturbation is related to the limitations of the concept of TSs. In Sect. 5.7, we briefly discuss those issues which are not the main topics of this review.

5.2 Statistical Reaction Theory in a Nutshell

Here, we briefly explain the assumptions underlying the statistical reaction theory. It is assumed that relaxation in the potential well is much faster compared to the reaction. Even when reaction processes cause a slight deviation from equilibrium, such deviations disappear quickly because of fast relaxation in the well so that the next process takes place in equilibrium. This assumption leads us to the phenomenological treatment of the reaction, where the rate equation with the rate *constant* k,

$$\frac{\mathrm{d}}{\mathrm{d}t}P(t) = -kP(t), \tag{5.1}$$

describes how the density $P(t)$ of the reactant changes. Then, $P(t)$ exhibits the exponential decay, and the inverse of the rate *constant* gives the characteristic time scale of the processes.

Thus, the most important task of the statistical reaction theory is to estimate the rate *constant* k. Here, the concept of TSs comes into play. In the conventional theory, a TS is a saddle of the potential in the configuration space, where the Hessian matrix there has one negative eigenvalue (It is called a saddle of index one.). Then, the eigenvector corresponding to the negative eigenvalue directs along the way the reaction takes place locally around the saddle. By extending somehow the path along the eigenvector further into the well, the reaction coordinate is supposed to be defined.

Assuming further that orbits never come back once they cross the saddle, we can derive the formula for the reaction rate *constant* $k(E)$,

$$k(E) = \frac{G^{+}(E - E_0)}{h\rho(E)}, \tag{5.2}$$

where E is the total energy of the system, $\rho(E)$ is the density of states in the reactant side, $G^{+}(E - E_0)$ is the sum of the number of the states from 0 to $E - E_0$ at the saddle, and h is the Planck constant [1]. The formula indicates that the rate constant $k(E)$ exhibits a staircase as the energy $E - E_0$ above the saddle is increased, when a new state at the saddle begins to contribute to the reaction. Such a staircase is actually observed in reaction processes where the energy above the saddle is not too large [2, 3]. This observation is considered to indicate that the concept of TSs is correct.

However, the above ideas suffer from the recrossing problem, that is, the TS thus defined in the configuration space is not a real boundary between the reactant and the product. It is known that some orbits return to the reactant well just after they cross the saddle in the configuration space. Existence of such orbits leads to overestimation of the reaction rate. Moreover, the formula given by (5.2) cannot be used for reaction processes where the assumption for the separation of the time scales is questionable. This limits applicability of the statistical reaction theory.

5.3 Dynamical Reaction Theory

5.3.1 Normally Hyperbolic Invariant Manifolds (NHIMs)

The recrossing problem is completely solved by the dynamical reaction theory, which is recently developed based on the analysis of phase space structures. The key concept here is NHIMs and their stable/unstable manifolds.

We start our explanation on NHIMs by the simplest case, i.e., the NHIM around a saddle of index one. Suppose that a Hamiltonian of N degrees of freedom is expanded around a saddle located at $(q_1, \cdots, q_N) = (0, \cdots, 0)$ as follows:

$$H = \left(\frac{p_1^2}{2} - \frac{\mu_1^2 q_1^2}{2}\right) + \sum_{j=2}^{N}\left(\frac{p_j^2}{2} + \frac{\omega_j^2 q_j^2}{2}\right) + \sum_{n=3}^{\infty} H_n, \tag{5.3}$$

where μ_1 and ω_j for $j \geq 2$ are real, and H_n consists of the terms of nth order in $(\mathbf{q}, \mathbf{p}) = (q_1, \cdots, q_N, p_1, \cdots, p_N)$. Then, the coordinate q_1 lies, locally near the saddle, along the direction of the eigenvector corresponding to the negative eigenvalue of the Hessian matrix at the saddle. This means that, in the conventional theory, (q_1, p_1) are the reaction coordinate and its canonically conjugate momentum, respectively. Thus, the boundary between the reactant and the product is supposed to be given by the hyperplane $q_1 = 0$ in the configuration space. The rest of the degrees of freedom $(\mathbf{q}_b, \mathbf{p}_b) = (q_2, \cdots, q_N, p_2, \cdots, p_N)$ are supposed to be the bath modes.

However, the recrossing problem ruins this idea. The problem is caused by the nonlinear coupling terms in the expansion of the Hamiltonian (5.3). Thus, in order to define the true boundary, we need to eliminate all of the coupling terms between the reaction coordinate and the bath degrees of freedom by transforming the original coordinates and momenta to new ones. Existence of such a transformation is guaranteed by the fact that no "resonance" takes place between the mode with the negative eigenvalue of the Hessian matrix and those with the positive ones. This is the essence of the theory of NHIMs for saddles of index one.

In general, a NHIM is a manifold where instability (either in a forward or a backward direction of time) along its normal directions is much stronger than that along its tangential directions [4–6]. Using the Lyapunov exponents, which quantitatively measure instability, we can define a NHIM as a manifold where the absolute values of the Lyapunov exponents along its normal directions are much larger than those along its tangential directions. Its stable/unstable manifolds consist of those orbits which approach or leave the NHIM, respectively. For saddles of index one, these geometric structures enable us to identify the one-dimensional reaction coordinate by the normal directions to the NHIM, and thereby to define the TS as the the dividing hypersurface of co-dimension one locally near the saddle. The TS thus defined is free from the problem of recrossing orbits, and can decompose the phase space into the distinct regions of the reactants and the products [7–10] (see also the recent reviews [11–14] and the book [15]). Moreover, the stable/unstable manifolds of the NHIM provide us with the reaction conduit through which all the reactive trajectories pass from the reactant to the product or vice versa. Thus, these manifolds offer a crucial clue to understand controllability of the reaction, and to investigate dynamical correlation in reaction processes taking place over multiple saddles.

NHIMs are structurally stable under perturbations. The wider the gap of instability is between the normal directions and the tangential ones, the more stable it is. However, as we raise the energy above the saddle, chaos can emerge on the NHIM [16], which is caused by nonlinear resonances among the bath degrees of freedom on the NHIM. It can lead to breakdown of normal hyperbolicity, because the Lyapunov exponents tangent to the NHIM have a possibility of being comparable to those normal to the NHIM [12]. Then, we can no longer define a dividing hypersurface even locally near the saddle.

Thus, breakdown of normal hyperbolicity raises serious questions concerning applicability of the concept of TSs for reactions taking place high above the saddle. To the contrary, the dynamical theory of reactions provides us with a strategy for investigating such cases beyond the conventional concept. We will discuss this problem in Sect. 5.6.

5.3.2 Arnold Web

Processes taking place in the well can be understood using the Arnold web. Suppose that a Hamiltonian of N degrees of freedom is expanded near a minimum located at $(q_1, \cdots, q_N) = (0, \cdots, 0)$ as follows:

$$H = \sum_{j=1}^{N} \left(\frac{p_j^2}{2} + \frac{\omega_j^2 q_j^2}{2} \right) + \sum_{n=3}^{\infty} H_n, \tag{5.4}$$

where ω_j for $j \geq 1$ are real, and H_n consists of the terms of nth degree in $(\mathbf{q}, \mathbf{p}) = (q_1, \cdots, q_N, p_1, \cdots, p_N)$. Note the difference from the expansion (5.3) around the saddle of index one. This difference plays a crucial role in the following argument.

Given the Hamiltonian (5.4), we try to eliminate as many coupling terms as possible by transforming the original coordinates and momenta to new ones. If we succeed in eliminating all of the coupling terms, those modes represented by the new coordinates and momenta describe independent motions in the well. In other words, the system is integrable then, and there exist no processes of energy transfer among the modes thus represented. However, such transformations do not exist in general because of nonlinear resonances [17]. Nonlinear resonances generally constitute a network in the action space (The network is called the Arnold web.), and resonance overlap results in globally chaotic motions [18]. Thus, characteristics of the Arnold web play an important role in intramolecular vibrational-energy redistribution (IVR) [19, 20]. In particular, whether the Arnold web is uniform or not and how regions of resonance overlap are distributed are crucial in understanding limitations of the statistical reaction theory. We will discuss this point in Sects. 5.4 and 5.5.

5.3.3 Dynamical Connection Between Arnold Web and NHIM Around Saddle

In studying reaction processes beyond the statistical reaction theory, global aspects of the dynamics should be taken into account, that is, dynamics near the saddles, that around the minima, and how they are connected [12].

These aspects are displayed by the schematic pictures in Fig. 5.1, where the hills indicate the potential barriers and the rectangles on the top of the hills show the NHIMs around the saddles. Their stable and unstable manifolds are shown by the arrows, respectively. In Fig. 5.1a, the stable and unstable

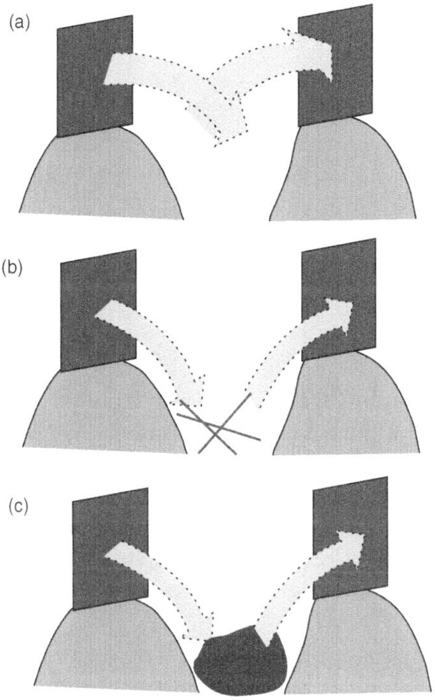

Fig. 5.1. Schematic pictures showing how NHIMs are connected by their stable and unstable manifolds. (**a**) Their stable and unstable manifolds have direct intersections without going deeply in the reactant well. (**b**) They land on the Arnold web in the well, where the web is sparse. (**c**) They land on the Arnold web in the well, where the web is dense. Reprinted figure with permission from Shojiguchi, Li, Komatsuzaki and Toda, Communications in Nonlinear Science and Numerical Simulation **13**, 857 (2008), Copyright 2008 by Elsevier

manifolds have intersections without going deeply into the well. On the other hand, in Fig. 5.1b, c, they land on the Arnold web in the well. While, in Fig. 5.1b, the Arnold web is sparse, the web is dense in Fig. 5.1c.

These characteristics of the web will manifest themselves in the distribution of the residential times, thereby leading to the difference in the statistical features of the reaction processes. In Sect. 5.5, we pay attention to this problem. In particular, we will show that there exist fractional behavior for reaction processes where the Arnold web is sparse and nonuniform. Existence of fractional behavior casts a doubt on definability of the reaction rate *constant*.

5.4 Intramolecular Vibrational-Energy Redistribution (IVR)

In the traditional theory of reactions, IVR in the potential well is supposed to be statistical. However, this idea is not applicable when ergodicity of the processes in the well is questionable. This is the point where the dynamical reaction theory comes into play. As explained in Sect. 5.3.2, the processes of energy exchange result from nonlinear resonances among vibrational modes. Thus, the properties of the Arnold web is crucial in studying IVR. Here, we discuss vibrationally highly excited acetylene as a typical example where the Arnold web is sparse leading to nonergodicity of the IVR.

Acetylene is among those molecules where detailed analysis of the IVR is going on both experimentally and theoretically [21]. One of its interesting features is a hierarchical structure in spectra for vibrationally highly excited states [22]. Such a hierarchical structure in spectra implies a hierarchy in bottlenecks for the IVR, where each bottleneck would hinder relaxation within the potential well for a finite duration. This idea can be contrasted with the one behind the traditional theory, where relaxation is supposed to take place exponentially fast with a single time constant.

Nonstatistical features of the IVR in acetylene is clearly revealed by the recent experiments of the group of Field [23]. They showed that, for certain highly excited states, the IVR is unexpectedly regular. In other words, the equi-energy surface of acetylene at this energy is nonergodic. This result contradicts the naive expectation that IVR for highly excited molecules would be strongly chaotic. In order to understand these features, we show the Arnold web of acetylene using the effective Hamiltonian which is constructed by the Dunham expansion [24]. Here, we also point out that symmetry of the molecule is important in considering the Arnold web.

In the electronic ground state, acetylene is a linear molecule. Using the harmonic approximation, vibrational energy levels for small amplitude oscillations can be labeled as $(v_1, v_2, v_3, v_4^{l_4}, v_5^{l_5})^l$ where v_1 is the quantum number of the symmetric CH stretch, v_2 the CC stretch, v_3 the antisymmetric CH stretch, v_4 the *trans*-bend with the vibrational angular momentum l_4, v_5 the *cis*-bend with the vibrational angular momentum l_5.

As the molecule is vibrationally excited, couplings among these harmonic modes become important. These couplings are taken into account by the Dunham expansion. The Dunham expansion for the vibration–rotation energy of a linear polyatomic molecule above the zero point level is given by

$$H_0 = \sum_i \omega_i v_i + \sum_{i \leq j} x_{ij} v_i v_j + \sum_{t \leq s} g_{ts} l_t l_s - B_v l^2, \tag{5.5}$$

and

$$B_v = B_0 - \sum_i \alpha_i v_i + \sum_{i \leq j} \gamma_{ij} v_i v_j + \sum_{t \leq s} \gamma^{ts} l_t l_s. \tag{5.6}$$

In this expression, we put the total angular momentum $J = 0$ for simplicity. The values of the constants appearing in the above expression are taken from Table 1 of [22]. In (5.5), the effective Hamiltonian H_0 is expressed solely by the quantum numbers $v_i (i = 1 \cdots 5)$ and $l_i (i = 4, 5)$. This means that these quantities are its conserved quantities. The eigenstates of H_0, which are in general anharmonic, are specified by the conserved quantities. In other words, (5.5) represents the diagonal part of the true Hamiltonian H.

IVR results from nondiagonal couplings among the anharmonic modes of the effective Hamiltonian H_0. The effects of these couplings are most significant when resonances among the modes take place. The locations where nonlinear resonances can occur are given by

$$\boldsymbol{n} \cdot \frac{\partial H_0}{\partial \boldsymbol{v}} = 0, \tag{5.7}$$

where $\boldsymbol{n} = (n_1, n_2, n_3, n_4, n_5)$ and $\boldsymbol{v} = (v_1, v_2, v_3, v_4, v_5)$, and $n_i (i = 1 \cdots 5)$ are integers. Here, we limit our argument to those resonances with $l_4 = 0$ and $l_5 = 0$ for simplicity. We will call $\sum_{i=1}^{5} |n_i|$ the order of a resonances $\boldsymbol{n}$. In these locations, the perturbation series are doomed to break down.

For acetylene, we calculate these location where resonances can take place. This calculation does not mean that those nondiagonal terms actually exist. Its purpose is to indicate the locations where their effects are most magnificent if they exist. In other words, when experiments are planned to look for their effects, these calculations show where to search.

In Fig. 5.2, a section of the Arnold web of acetylene is displayed where nonlinear resonances up to seventh order are estimated. The value of the energy is chosen to be 18,797 cm^{-1}. Since, for systems of more than three degrees of freedom, we cannot directly display the whole web, we have to take a section. Here, the section is defined by $(v_1 = 0, v_2, v_3 = 0, v_4, v_5)$ and $(l_4 = 0, l_5 = 0)$. Although these sections offer only a partial information on how the web is connected, we can derive some useful consequences as will be shown in the following.

The first thing we should notice in Fig. 5.2 is that the distribution of the resonances is very sparse. There, only three major resonances exist: $\boldsymbol{n}_{44/55} = (0, 0, 0, 2, -2)$, $\boldsymbol{n}_{1/244} = (1, -1, 0, -2, 0)$ and $\boldsymbol{n}_{1/255} = (1, -1, 0, 0, -2)$. This is because the selection rules of the symmetry prohibit a large part of coupling terms. Thus, the symmetry of the molecule plays a crucial role in studying the Arnold web.

The second point to notice is that these calculations correspond to the experiments. All of the three major terms in Fig. 5.2 are actually observed in the IVR processes of acetylene. Thus, these calculations play a role of guidelines for experiments. Furthermore, our results explains the experimental results by the group of Field [25]. They showed that IVR from pure bending dynamics becomes regular as the energy increases from 10,000 cm^{-1}. It corresponds to our results showing that the initial states are located far apart from the resonance $\boldsymbol{n}_{44/55} = (0, 0, 0, 2, -2)$ as their energies increase.

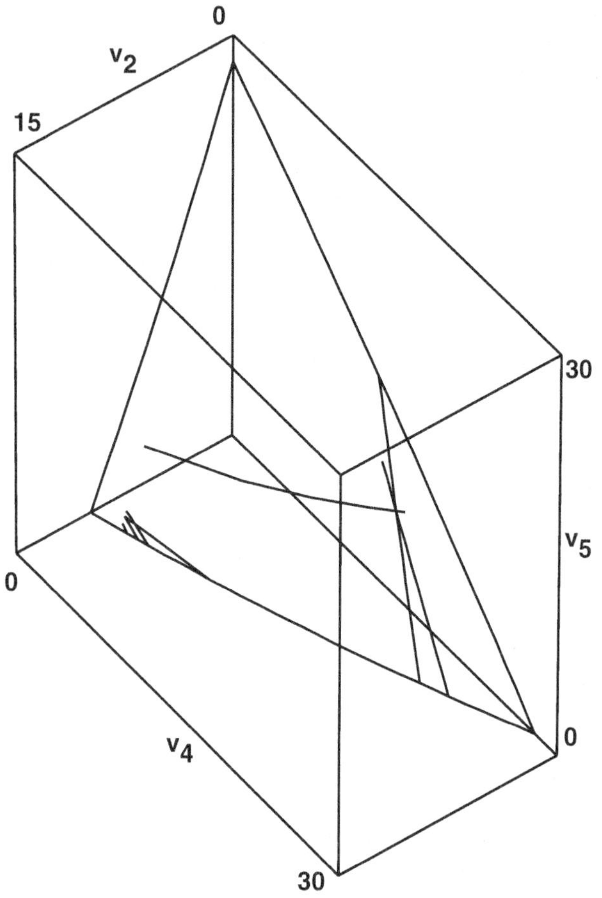

Fig. 5.2. A section of the Arnold web for acetylene. Reprinted figure with permission from Toda, Advances in Chemical Physics **123**, 153 (2002), Copyright 2002 by John Wiley & Sons

Up to now, the following coupling terms are already identified in the experiments: $\mathbf{n} = (-2,0,2,0,0)$, $(0,0,0,-2,2)$, $(0,-1,1,-1,-1)$, $(-1,1,0,0,2)$, $(-1,1,0,2,0)$, $(-1,0,1-1,1)$, and $(-1,-1,2,-2,0)$ [24]. In these experiments, only the SEP bright states $(0,v_2,0,v_4,0)$ are accessible as the initial conditions. One remarkable aspect about these terms is that the following quantities

$$P = 5v_1 + 3v_2 + 5v_3 + v_4 + v_5, \tag{5.8}$$

$$R = v_1 + v_2 + v_3, \tag{5.9}$$

$$L_z = l_4 + l_5, \tag{5.10}$$

are conserved [26]. We call these quantities Kellman's constants.

By enumerating up to seventh order all the resonances which satisfy the selection rules, we confirm that almost all of them conserve the Kellman's constants. Existence of conserved quantities generally means that the IVR does not take place on the whole equi-energy surface, but is limited within a lower dimensional region specified by these quantities. For acetylene, the existence of the Kellman's constants indicates that the following limitation exists for the IVR starting from the SEP bright states.

By expressing (5.8) and (5.9) in the following way

$$v_1 + v_3 = R - v_2 \tag{5.11}$$

$$v_4 + v_5 = P - 5R + 2v_2, \tag{5.12}$$

we see that the sum $v_4 + v_5$ cannot increase in the IVR from the SEP initial states. In order for the value of the sum to increase, the value of v_2 should increase because of (5.12). In order for the value of v_2 to increase, the value of the sum $v_1 + v_3$ should decrease because of (5.11). For the SEP initial states, however, the value of the sum $v_1 + v_3$ cannot decrease, because it already attains the lowest possible value, i.e., 0.

Thus, the existence of Kellman's constants means that no IVR takes place from the SEP bright states which increases the value of the sum $v_4 + v_5$. This limitation has the following implications for a vibrationally highly excited acetylene. For a vibrationally highly excited acetylene, the isomerization process from acetylene to vynilidene is of interest. In this process, both of the bending modes v_4 and v_5 should be excited as one of the hydrogen atoms goes half way round the two carbon atoms. Therefore, the reaction coordinate of the isomerization is chosen to be the sum $v_4 + v_5$. However, the limitation imposed by the Kellman's constants means that no IVR occur from the SEP bright states which is directed toward the isomerization to vinylidene.

Thus, in order for the isomerization to take place from the SEP bright states, we have to seek for the following two possibilities. The first is to find coupling terms which violate the conservation of Kellman's constants. The second is to put the molecule under external fields which do not conserve the constants. In the following, we pursue the second possibility.

Coupling with external electric fields is possible only for v_3 and v_5 because of the selection rules. Here, we put the molecules under the electric field which is resonant with the mode v_5, thereby violating the conservation of Kellman's constants. However, coupling the mode v_5 with the electric field is not sufficient, because both v_4 and v_5 need to be excited for the present purpose. Then, in order to make the energy flow from v_5 to v_4, we utilize the resonance $n_{44/55}$ by locating the SEP initial conditions near its resonant region. A schematic explanation of our ideas is shown in Fig. 5.3.

Figure 5.3a displays the processes of the energy flow which consists of the two steps, that is, excitation of the mode v_5 by the external field and that of the mode v_4 by the resonance $n_{44/55}$. In Fig. 5.3b, the mechanism to lead the molecule toward isomerization is shown. There, the two arrows with solid lines

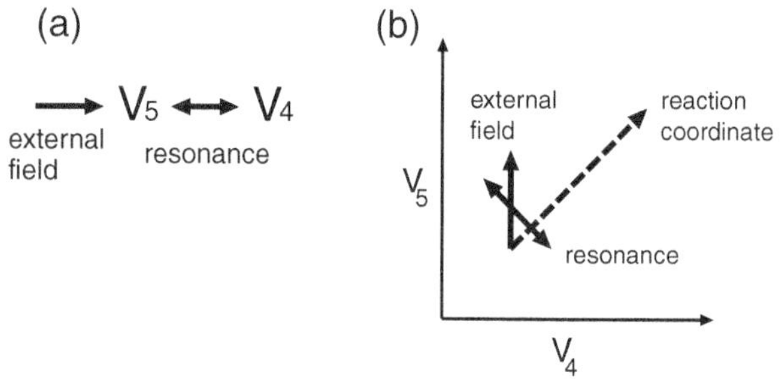

Fig. 5.3. A schematic explanation for leading the molecule toward isomerization. Reprinted figure with permission from Toda, Advances in Chemical Physics, **123**, 153 (2002), Copyright 2002 by John Wiley & Sons

show the movements caused by the resonance $n_{44/55}$ and the external field, respectively. The arrow with broken lines indicate the behavior leading toward the isomerization. Our aim is to induce the behavior indicated by the broken arrow by combining the two movements shown by the solid arrows. Thus, for designing external fields to promote specific reactions in a sparse Arnold web, we need cooperative effects of external fields and nonlinear resonances. In numerical simulation, we have confirmed that the above idea works in the well [27]. In order to excite further the molecule into the isomerization, we need information concerning the potential surface near the saddle. At present, such information is not available either experimentally or theoretically.

5.5 Fractional Behavior in Molecular Systems

In the conventional reaction theory, existence of the rate *constant* is presumed. This is based on the assumption that the processes in the well take place statistically. Then, the distribution $P(t)$ of the reactant obeys the exponential decay following the phenomenological rate equation. However, as we have discussed in the previous section, this assumption does not necessarily hold in general. In particular, in processes under strong laser fields, the reaction can occur *before* vibrational energy is distributed statistically to other modes. Thus, we should ask what are the characteristic features of the reaction processes when ergodicity is not achieved.

In this section, we discuss that fractional behavior is a universal feature in nonergodic reaction processes. Here, the term "fractional behavior" means the following; Relaxation processes exhibit power-law decays and diffusion processes do not follow the normal Brownian type. Moreover, power spectra show $1/f$ dependence on frequencies. Existence of such behavior implies that time developments deviate from what we expect based on the conventional statistical laws [28]. Fractional behavior is ubiquitous in complex systems such as amorphous semiconductors [29], fluid mechanics [30], Hamiltonian systems

[31,32], and biophysics where long-time memories and anomalous diffusion are observed for a variety of biomolecules [33–35]. Fractional behavior can be seen even in econophysics [36] where prices in stock markets, for example, show anomalous fluctuation. The origins of such fractional behavior are attributed to existence of hierarchical structures and/or impossibility of separating characteristic time scales. They lead to breakdown of the underlying assumptions of the central limit theorem, giving rise to deviation from the normal diffusive behavior. As is discussed in the previous section, hierarchical structures exist in spectra of vibrationally highly excited molecules. This leads us to speculate that fractional behavior can be also observed in reaction processes.

Here, we present a typical model Hamiltonian which indicates a mechanism to give rise to fractional behavior in reactions. The model is of three degrees of freedom with a double-well potential where the Arnold web there is sparse and nonuniform. It is regarded as a prototype of isomerization reactions with nonergodic dynamics in the well. The Hamiltonian is as follows:

$$\begin{aligned} H &= H_0 + H_1, \\ H_0 &= \frac{p_1^2}{2} - \frac{\lambda^2 q_1^2}{2} + \frac{\lambda^2}{4} q_1^4 + \sum_{i=2}^{3} \left(\frac{p_i^2}{2} + \frac{\omega_i^2 q_i^2}{2} + b q_i^4 \right), \\ H_1 &= \exp\left[-(q_1 - 1)^2/\sigma^2\right] \left[a_1 q_2^2 q_3^2 + a_2 (q_1 - 1)^2 (q_2^2 + q_3^2)\right], \end{aligned} \tag{5.13}$$

Here q_1 is a coordinate which is the lowest order approximation of the reaction coordinate, and q_i $(i = 2, 3)$ are coordinates which are the lowest order approximation of the bath modes. The frequencies are $\omega_1 = 1.02$, $\omega_2 = 0.94$, and $\omega_3 = 1.04$, where $\omega_1 = \sqrt{2}\lambda$ is the unperturbed frequency at $q_1 = 1$, i.e., the bottom of the well. The parameters $\sigma = 0.5$, $a_1 = 0.1$, $a_2 = 0.9$ and $b = 0.5$ are chosen to ensure that the system is trapped in the well by energy exchange between the bath and the reaction coordinates. The coupling function is selected so that the interactions take place in the well more than near the saddle. This is validated for an energy slightly above the saddle [37].

The unperturbed Hamiltonian of the bath coordinates $h_i = \frac{p_i^2}{2} + \frac{\omega_i^2 q_i^2}{2} + b q_i^4$ $(i = 2, 3)$ are nonlinear oscillators, respectively. They are integrable with the constants of motion J_i $(i = 2, 3)$, which are represented using the elliptic functions. Expanding the Hamiltonian (5.13) using Lie canonical perturbation theory (LCPT) at the bottom of the well, we obtain the nonlinear frequencies for the bath coordinates, up to the first order of the action variables, as follows:

$$\bar{\omega}_i(J_i) = \frac{\partial H_0}{\partial J_i} = \omega_i + \frac{3b}{\omega_i^2} J_i + O(J_i^2), \quad (i = 2, 3). \tag{5.14}$$

The reaction coordinate has the action variable J_1 in the well, where the unperturbed Hamiltonian of the reaction coordinate is expressed as $h_1 = \frac{p_1^2}{2} - \frac{\lambda^2 q_1^2}{2} + \frac{\lambda^2}{4} q_1^4$. The nonlinear frequency of the reaction coordinate in the well is given, up to the first order of the action variable by

$$\bar{\omega}_1 = \frac{\partial H_0}{\partial J_1} = \omega_1 - \frac{3}{4} J_1 + O(J_1^2). \tag{5.15}$$

Then, the system has the primary resonances given by

$$\bar{\omega}_1 = \bar{\omega}_2, \quad \bar{\omega}_1 = \bar{\omega}_3, \quad \bar{\omega}_2 = \bar{\omega}_3. \tag{5.16}$$

We will see that the Arnold web of the primary resonances is sparse. On the other hand, there exist regions where multiple resonances meet, i.e., resonance junctions. Thus, the Arnold web is nonuniform. These features will play an important role in the following results.

Initial conditions are chosen to be uniformly distributed on the unstable manifold of the NHIM near the saddle. Then, the residential time is estimated for each trajectory by counting the total number n of crossing points with the plane $q_1 = 1$ under the condition $p_1 > 0$, until it leaves the well. We call n the crossing number. By adding the number of trajectories with their crossing numbers from n to infinity, we obtain the number of trajectories which remain in the well up to the crossing number n. The ratio of this quantity to the total number of the trajectories is the survival probability $P(n)$.

In Fig. 5.4, we show the survival probability $P(n)$ as a function of the crossing number n. There, two ranges of n exist where $P(n)$ behaves differently. One is the range where $P(n)$ varies as the power-law decay $n^{-\gamma}$, and the other where $P(n)$ changes exponentially as $\exp(-\alpha n)$. The range of the power-law decay extends from about $n = 10$ up to about $n = 100$ with $\gamma = 0.82$, and that of the exponential decay does from about $n = 100$ up to about $n = 5{,}000$ with $\alpha = 0.0015$. Moreover, the Fourier spectra of the action variables correspondingly exhibit different characteristics [38]. For trajectories in the range of the power-law decay, the Fourier spectrum exhibits $1/f$ dependence. On the other hand, it shows the Lorentzian feature for those in the range of the exponential decay.

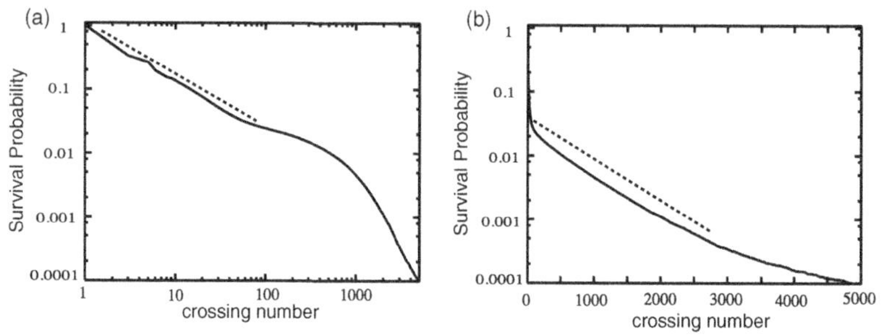

Fig. 5.4. Survival probability $P(n)$ as a function of the crossing number n is shown. (**a**) Log vs. log plot. The line indicates fitting with the function $n^{-\gamma}$. (**b**) Log vs. linear plot. The line indicates fitting with the function $\exp(-\alpha n)$. Reprinted figure with permission from Shojiguchi, Li, Komatsuzaki and Toda, Communications in Nonlinear Science and Numerical Simulation **13**, 857 (2008), Copyright 2008 by Elsevier

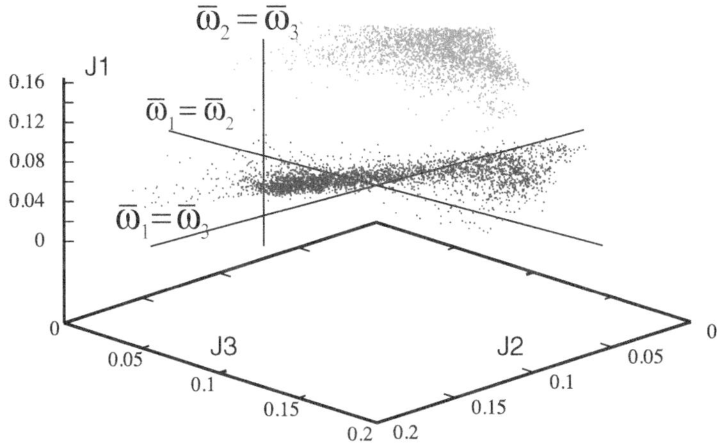

Fig. 5.5. The average locations of trajectories in the action space are plotted for initial conditions which are in the range of power-law (*gray dots*), and those in the range of exponential (*black dots*). Reprinted figure with permission from Shojiguchi, Li, Komatsuzaki and Toda, Communications in Nonlinear Science and Numerical Simulation **13**, 857 (2008), Copyright 2008 by Elsevier

We study the relation between those features of $P(n)$ and the characteristics of the Arnold web. In Fig. 5.5, we plot the average location of a trajectory in the action space (J_2, J_3, J_1). A location is recorded each time when the trajectory crosses the plane $q_1 = 1$, and the average of the locations is taken over its residential time. The gray (black) points are shown for those trajectories in the range where $P(n)$ changes as the power-law decay (exponentially). There, we also indicate the locations of the primary resonances for comparison. Thus, the figure reveals how the variance of $P(n)$ is related to the processes of energy exchange between the bath modes and the reaction coordinate.

We can see in Fig. 5.5 that the Arnold web of the primary resonances is sparse. On the other hand, there exist resonance junctions, where resonance overlap leads to fully chaotic chaos. These features indicate that the Arnold web is nonuniform. In Fig. 5.5, while the gray points are distributed in a region far from the resonance junctions, the black points are spread over the junctions wandering fully chaotic regions. Thus, the two types of the trajectories experience separate regions in the phase space. Therefore, dynamical barriers exist which prevent them from exploring the whole phase space. Moreover, the difference in the decays of the survival probability corresponds to whether the trajectories experience fully developed chaos or not. Recently, we have also found fractional behavior in isomerization of HCN [39]. Based on the analysis using wavelet (see, for example, [40]), we are confirming that the behavior results from a sparse distribution of nonlinear resonances. These findings suggest that fractional behavior is a universal feature for reaction processes taking place in the Arnold web where resonances are sparse.

The results we have presented indicate that hierarchical structures leading to fractional behavior exist in phase space for systems of more than two degrees of freedom. For Hamiltonian systems of two degrees of freedom, fractional behavior is well established [32]. It is shown that a hierarchy of resonant tori exist trapping nearby trajectories for a finite time [17]. Based on this property, they are described as "sticky". Then, the hierarchy of time scales leads to a power-law distribution of residential times. However, when the degrees of freedom is larger than two, it is not obvious that the same argument holds because the dimension of tori is not large enough to work as dynamical barriers. Therefore, fractional behavior in systems of more than two degrees of freedom is a new phenomenon in the study of Hamiltonian chaos.

Moreover, the dynamical connection between the NHIM around the saddle and the Arnold web is crucial for the survival probability. In our model system, it is shown that the unstable manifold of the NHIM lands on the Arnold web, where the resonances are sparse [41]. This results in the fact that the survival probability exhibits a power-law decay for shorter residential times and an exponential one for longer times. Depending on the way the NHIM is connected to the web, other features of the survival probability are possible (see [41] for a detailed discussion). Thus, not only the properties of the Arnold web but also how the web is connected to the NHIM is important for understanding reaction processes.

We now discus whether our results are relevant for quantum dynamics. It is known in general that "quantum chaos" is less statistical compared to classical chaos although the discrepancy is shown to be smaller as the number of the degrees of freedom is larger [42]. Therefore, we expect that fractional behavior of classical chaos manifest itself in the corresponding quantum system. In fact, a recent study shows, for a system of two degrees of freedom, that fractional behavior of classical chaos has corresponding effects in the quantum one [43]. Thus, we think that the fractional behavior we have shown is relevant for reaction processes, which are quantum in nature.

We should point out importance of fractional behavior in reaction processes from the standpoint of information processing by molecules. Fractional behavior in reaction processes provides a clue to understand possibility of Maxwell's demon in molecules. This idea results from recent studies where Maxwell's demon is discussed in terms of fractional behavior in Hamiltonian dynamics [44]. An important aspect of these studies is that the problem is treated from a purely dynamical point of view without assuming existence of heat bath. To the contrary, most of the studies on Maxwell's demon so far have supposed that the demon is surrounded by heat bath thereby being under thermal fluctuations [45, 46]. Then, the only possibility for the demon is to wait for rare thermal fluctuations [47]. However, if fractional behavior exists in molecular levels, this provides new resources for Maxwell's demon to utilize. It is because fractional behavior in dynamics implies that long-lasting memories exist there. Their existence offers conditions under which Maxwell's demon can work [48]. On the other hand, decays of the exponential type lead to fast recovery of equilibrium, where Maxwell's demon cannot operate. Thus, studies of fractional behavior in reaction processes open a new research arena where possibility of information processing by molecules is considered.

5.6 Limitations of the Concept of Transition States

In this section, we discuss limitations of the concept of TSs from the standpoint of the dynamical reaction theory. As we have discussed, the concept obtains a sound mathematical foundation based on the geometric structures of phase space, i.e., NHIMs. However, the recent study of NHIMs also reveals that the concept has limitations as the energy above the saddle increases [49]. The origin of the limitations is chaos on the NHIM and breakdown of normal hyperbolicity resulting from it [12]. Suppose that we raise the energy above the saddle, i.e., the energy of the bath modes there. Then, nonlinear coupling terms among the bath modes in general create chaos on the NHIM. Therefore, the Lyapunov exponents along the tangential directions to the NHIM are no longer zero. The stronger the chaos on the NHIM is, the larger their absolute values are. Thus, some of the Lyapunov exponents along the tangential directions can become comparable to that along the normal one. Then, the condition for the existence of NHIMs is lost. The problem resulting from this is the following; What would happen to the NHIM after the condition for its existence is lost? What are the implications of this loss for reaction processes?

In order to acquire an intuitive idea for the problem, we first study a simpler and tractable case. It is a Hamiltonian with a saddle of index two, i.e., a saddle where the Hessian matrix there has two negative eigenvalues. Suppose that the Hamiltonian is expanded around the saddle located at $(q_1, \cdots, q_N) = (0, \cdots, 0)$ as follows:

$$H = \left(\frac{p_1^2}{2} - \frac{\mu_1^2 q_1^2}{2}\right) + \left(\frac{p_2^2}{2} - \frac{\mu_2^2 q_2^2}{2}\right) + \sum_{j=3}^{N}\left(\frac{p_j^2}{2} + \frac{\omega_j^2 q_j^2}{2}\right) + \sum_{n=3}^{\infty} H_n, \quad (5.17)$$

where μ_1, μ_2 and ω_j for $j \geq 3$ are real. Furthermore, we assume that $\mu_1 = k\mu_2$ holds with an integer k larger than 1, that is, a "resonance" condition is satisfied between the instability along (q_1, p_1) and that along (q_2, p_2) (This is a Hamiltonian version of the example discussed in [50–54].). Even for this case, the NHIM exists with its normal directions along the reaction coordinate and momentum, where (q_1, p_1) give their lowest order terms, respectively. However, in the transformation to obtain the reaction coordinate and momentum, we cannot eliminate those coupling terms of the form $q_1^{m_1} q_2^{km_2}$ because of the "resonance" condition. This means that the reaction coordinate is not separable when coupling terms of the form $q_1^{m_1} q_2^{km_2}$ exist in the Hamiltonian. Then, its dynamics of approaching/leaving the NHIM is affected by the movement of (q_2, p_2).

Note that, in this case, another NHIM exists with its normal directions where *both* (q_1, p_1) and (q_2, p_2) are their lowest order terms. In other words, two-dimensional reaction coordinates can be constructed from *both* (q_1, p_1) and (q_2, p_2). These coordinates would be useful for describing how the reaction proceeds, i.e., as a skeleton of the reaction paths. However, the NHIM thus constructed does not determine a dividing hypersurface because its dimension is not large enough. Thus, in this case, we cannot define a TS in the sense of the traditional reaction theory.

Based on the intuition obtained by the above example, we will discuss those cases when chaos on the NHIM exists. At present, we do not have analytical methods to investigate them. However, we can proceed in the way which is similar to the previous one. Suppose that the Hamiltonian is expanded as in (5.3). For the energy slightly above the saddle, the NHIM exists, which is structurally stable. However, as we raise the energy further, chaos emerges on the NHIM. Then, take a periodic orbit in the chaotic region on the NHIM, and estimate its linear stability using the Poincaré section locally near the orbit. It has at least two positive Lyapunov exponents; one along the normal direction and the other along the tangential one to the NHIM. When a "resonance" condition does not exist between these two exponents, the reaction coordinate can be constructed from (q_1, p_1), which is separable locally near the orbit. However, when a "resonance" condition is satisfied, the reaction coordinate constructed from (q_1, p_1) is not separable when the corresponding coupling terms exist. In other words, the dynamics of approaching/leaving the NHIM is affected, locally near the orbit, by the movement of the tangential direction involved in the "resonance". Note that, in the resonant cases, another NHIM would exist constructed from *both* the normal and the tangential ones involved in the "resonance". The NHIM thus constructed would work as a skeleton of the reaction paths locally near the orbit. Thus, the analysis can be performed based on the linear stability of periodic orbits in the chaotic region on the NHIM.

In general, periodic orbits are densely distributed in the chaotic regions. Therefore, contrary to the saddle of index two, their possibility of satisfying "resonance" conditions is large. In particular, when chaos on the NHIM becomes stronger, "resonance" conditions can be satisfied by simpler rational numbers. This means that the corresponding coupling terms are of lower orders. Thus, the dynamics of approaching/leaving the NHIM become more sensitive to the movement of the tangential directions.

Such effects can be clearly seen in Fig. 5.6, which is reprinted from Fig. 3 of [49]. The picture displays representative reactive trajectories projected on the space of the reaction coordinate and momentum (q'_1, p'_1). Here, the reaction coordinate and momentum are obtained using the partial LCPT and applying Padé approximation. In Fig. 5.6, the more energy is distributed into the bath modes, i.e., the less energy the reaction coordinate and momentum have, thereby passing nearer to the saddle, the more pronounced singularities of the

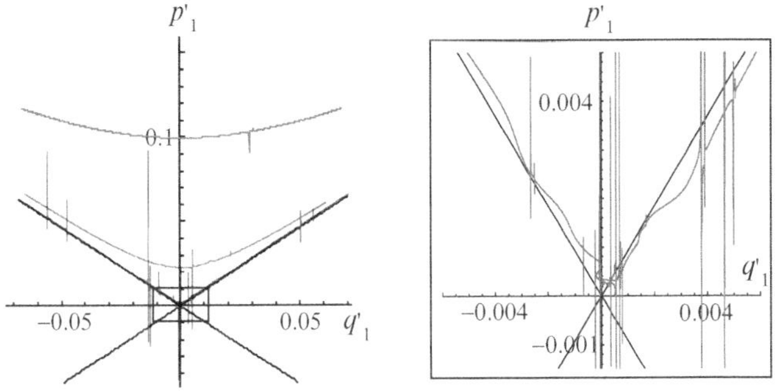

Fig. 5.6. Projection of representative reactive trajectories on the space of the reaction coordinate and momentum. As the trajectories pass closer to the saddle, they exhibit more pronounced singularities. Reprinted figure with permission from Li, Shojiguchi, Toda and Komatsuzaki, Phys. Rev. Lett., **97**, 128301 (2006), Copyright 2006 by the American Physical Society

Padé approximation are exhibited. This is exactly what is expected from the argument above.

We should note that the linear stability analysis of unstable periodic orbits is applicable even after the condition for the existence of NHIMs ceases to hold. It is because unstable periodic orbits themselves are structurally stable. Therefore, we can develop the dynamical reaction theory further beyond the breakdown of the conventional concept of TSs.

We now present possible implications of the above discussion for experiments. The above idea implies that reaction processes do depend on "resonance" conditions between the reaction coordinate and the bath modes. In other words, the reaction rate can exhibit a pronounced increase when "resonance" results in enhanced energy flow from the bath modes to the reaction coordinate. In fact, such an increase is observed in the dissociation reaction of keten when the molecule is highly excited above the saddle by the laser field [2]. Then, the reaction rate exhibits some peaks which look like resonant effects. Therefore, we need to confirm these speculations concerning experimental implications of the "resonance" conditions on the NHIM.

5.7 Summary and Future Problems

In this review, we have briefly explained the dynamical reaction theory, which has developed recently based on the phase space structures called NHIMs and their stable/unstable manifolds. The theory not only gives a sound mathematical foundation for the concept of TSs, but also reveals the limitations of the conventional concept. In addition, the theory is applicable to those reaction processes where the assumption of the conventional statistical theory does not hold. These aspects of the new theory indicates that it is suitable to analyze reaction processes under ultrafast intense laser fields.

In particular, we have pointed out the methodology of controlling reaction processes by utilizing cooperative effects of laser fields and nonuniform Arnold webs. This indicates that, in designing laser fields, it is crucial to take into account nonuniform features of Arnold webs. Furthermore, there exists the possibility that, under strong laser fields, we can change the Arnold web to optimize the reaction processes by deforming the molecular structures. We have also discussed the limitation of the traditional concept of TSs based on chaos on the NHIM. The limitation becomes manifest when the processes of going over the saddle accompany chaos in the bath modes. Then, we need to develop a dynamical reaction theory beyond the conventional concept of TSs. The theory explains new phenomena such as "resonant reactions" and will further develop to involve chaotic itinerancy (see [55] for more details.). These phenomena will become possible for reactions where the bath modes are vibrationally highly excited by strong laser fields.

We can further develop the dynamical theory of reactions to reaction processes taking place over multiple saddles. Processes of going over multiple saddles become important in reactions involving biomolecules such as protein folding. In these reactions, we can set initial conditions precisely using laser fields, and observe how processes of reactions depend on the choice of the initial conditions. Then, dynamical correlation manifests itself, when reactions proceed *before* equilibration is attained in potential wells. Such dynamical correlation can be crucial in considering efficiency of biological reaction processes.

Here, we have limited our attention to reaction processes described by classical dynamics. However, the recent development shows that it is possible to extend the theory to quantum dynamics. The extension is achieved by using the idea of complex NHIMs and quantum normal forms. It is applied to analyze, for example, tunneling phenomena and resonances in multi-dimensional systems [56–59]. Thus, the dynamical reaction theory is developing further to cover a wider range of reaction processes.

Acknowledgments

The author would like to acknowledge Prof. Komatsuzaki, Dr. Shojiguchi and Dr. ChunBiu Li for their collaboration. The results presented in Sects. 5.5 and 5.6 are done as joint studies with them. The author is supported by the Japan Society for the Promotion of Science, Grant-in-Aid for Research on Priority Area "Control of Molecules in Intense Laser Fields" by Japan Ministry of Education Culture Sports Science and Technology, and the discretionary research grant of University President for researchers of Nara Women's University.

References

1. R.D. Levine, *Molecular Reaction Dynamics.* (Cambridge, Cambridge, 2005)
2. E.R. Lovejoy, S.K. Kim, C.B. Moore, Science **256**, 1541, 1992
3. E.R. Lovejoy, C.B. Moore, J. Chem. Phys. **98**, 7846, 1993
4. N. Fenichel, Indiana Univ. Math. J. **21**, 193, 1971
5. N. Fenichel, Indiana Univ. Math. J. **23**, 1109, 1974
6. N. Fenichel, Indiana Univ. Math. J. **26**, 81, 1977
7. A.M. Ozorio De Almeida, N. De Leon, M.A. Mehta, C.C. Marston, Physica D **46**, 265, 1990
8. N. De Leon, M.A. Mehta, R.Q. Topper, J. Chem. Phys. **94**,8310, 1991
9. T. Uzer, C. Jaffé, J. Palacián, P. Yanguas, S. Wiggins, Nonlinearity **15**, 957, 2002
10. S. Wiggins, Physica D **44**, 471, 1990
11. M. Zhao, J. Gong, S.A. Rice, Adv. Chem. Phys. **130A**, 1 (2005)
12. M. Toda, Adv. Chem. Phys. **130A**, 337 (2005)
13. C. Jaffe, S. Kawai, J. Palacian, P. Yanguas, T. Uzer, Adv. Chem. Phys. **130A**, 171 (2005)
14. T. Komatsuzaki, R.S. Berry, Adv. Chem. Phys. **130A**, 143 (2005)
15. M. Toda, T. Komatsuzaki, T. Konishi, R.S. Berry, S.A. Rice, eds, in *Geometric Structures of Phase Space in Multidimensional Chaos: Applications to Chemical Reaction Dynamics in Complex Systems*, Adv. Chem. Phys. **130A, 130B** (2005) and references therein
16. C.B. Li, Y. Matsunaga, M. Toda, T. Komatsuzaki, J. Chem. Phys. **125**, 184301, 2005
17. A.J. Lichtenberg, M.A. Lieberman, *Regular and Chaotic Dynamics*, 2nd edn. (Springer, Berlin, 1992)
18. B.V. Chirikov, Phys. Rep. **52**, 263, 1979
19. C.C. Martens, M.J. Davis, G.S. Ezra, Chem. Phys. Lett. **142**, 519, 1987
20. S.A. Schofield, P.G. Wolynes, Chem. Phys. Lett. **217**, 497, 1994
21. M. Herman, J. Lievin, J.V. Auwera, A. Campargue, Global and accurate vibration Hamiltonian from high-resolution molecular spectroscopy, Adv. Chem. Phys. **108** (1999)
22. K. Yamanouchi, N. Ikeda, S. Tsuchiya, D.M. Joans, J.K. Lundberg, G.W. Adamson, R.W. Field, J. Chem. Phys. **95**, 6330, 1991
23. M.P. Jacobson, R.W. Field, J. Phys. Chem. **A104**, 3037, 2000
24. M.A. Temsamani, M. Herman, J. Chem. Phys. **102**, 6371, 1995
25. M.P. Jacobson, J.P. O'Brien, R.J. Silbey, R.W. Field, J. Chem. Phys. **109**, 121, 1998
26. M.E. Kellman, J. Chem. Phys. **93**, 6630, 1990
27. M. Toda, Adv. Chem. Phys. **123**, 153, 2002
28. R. Metzler, J. Klafter, Phys. Rep. **339**, 1, 2000
29. H. Scher, E.W. Montroll, Phys. Rev. B **12**, 2455, 1975
30. W. Young, A. Pumir, Y. Pomeau, Phys. Fluids A **1**, 462, 1989
31. Y.Y. Yamaguchi, Int. J. Bifurcat. Chaos **7**, 839, 1997
32. G.M. Zaslavsky, *Hamiltonian Chaos and Fractional Dynamics.* (Oxford, Oxford, 2004)
33. M. Takano, T. Takahashi, K. Nagayama, Phys. Rev. Lett. **80**, 5691, 1998
34. X. Yu, D.M. Leitner, J. Chem. Phys. **119**, 12673, 2003

35. W. Min, G. Luo, B.J. Cherayil, S.C. Kou, X.S. Xie, Phys. Rev. Lett. **94,** 198302, 2005
36. R.N. Mantegna, H.E. Stanley, *An Introduction to Econophysics.* (Cambridge, Cambridge, 2000)
37. T. Komatsuzaki, R.S. Berry, J. Chem. Phys. **110**, 9160, 1999
38. A. Shojiguchi, C.B. Li, T. Komatsuzaki, M. Toda, Phys. Rev. E **75**, 0352046 2007
39. A. Shojiguchi, C.B. Li, T. Komatsuzaki, M. Toda, (to be published)
40. C. Chandre, S. Wiggins, and T. Uzer, Physica D **181**, 171, 2003
41. A. Shojiguchi, C.B. Li, T. Komatsuzaki, M. Toda, Phys. Rev. E **76**, 056205, 2007; Phys. Rev. E 77, 019902 (E), 2008
42. M. Toda, S. Adachi, K. Ikeda, Prog. Theor. Phys. Suppl. **98**, 323, 1989
43. G. Casati, I. Guarneri, G. Maspero, Phys. Rev. Lett. **84**, 63, 2000
44. G.M. Zaslavsky, M. Edelman, Phys. Rev. E **56**, 5310, 1997
45. L. Brillouin. *Science and Information Theory.* (Academic Press, New York, 1956)
46. H.S. Leff, A.F. Rex, ed., *Maxwell's Demon 2.* (IoP, Bristol, 2003)
47. C.M. Caves, Phys. Rev. Lett. **64**, 2111, 1990
48. M.M. Millonas, Phys. Rev. Lett. **74**, 10 (1995). In this study, he discussed possibility of Maxwell's demon by *assuming* non-thermal fluctuations. However, he did not present the origin of such fluctuations in molecular levels
49. C.B. Li, A. Shojiguchi, M. Toda, T. Komatsuzaki, Phys. Rev. Lett. **97**, 028302, 2006; Phys. Rev. Lett. **97**, 129901, 2006
50. K.R. Meyer, SIAM Rev. **28**, 41, 1986
51. S. Sternberg, Duke Math. J. **24**, 97, 1957
52. S. Sternberg, Am. J. Math. **80**, 623, 1958
53. S. Wiggins. *Introduction to Applied Nonlinear Dynamical Systems and Chaos.* (Springer, New York, 2000)
54. S. Sternberg, Am. J. Math. **79**, 809, 1957
55. M. Toda, Adv. Chem. Phys. **130A**, 337, 2005
56. S.C. Creagh, Nonlinearity **17**, 1261, 2004
57. S.C. Creagh, Nonlinearity **18**, 2089, 2005
58. R. Schubert, H. Waalkens, S. Wiggins, Phys. Rev. Lett. **96**, 218302, 2006
59. H. Waalkens, R. Schubert, S. Wiggins, Nonlinearity **21**, R1, 2008

6

Relativistic Quantum Dynamics in Intense Laser Fields

Guido R. Mocken, Yousef I. Salamin, and Christoph H. Keitel

Abstract. We investigate the relativistic dynamics of electrons in intense laser fields. Examples of both free and bound electron dynamics are discussed using the approach appropriate for each particular case, i.e., either classical relativistic mechanics or relativistic quantum mechanics. The algorithm for numerically solving the Dirac equation is explained in detail before showing results that were obtained for both free and bound electronic wave packets in interaction with laser fields. In the case of the former, we discuss Volkov wave packets and point out features such as Lorentz contraction, spin and non-dipole effects. A sevenfold charged oxygen ion in a counterpropagating beam illustrates the latter and demonstrates a method for generating high-energy electron–nucleus collisions. Furthermore, we briefly outline the procedure for solving the classical equations of motion in arbitrary electromagnetic fields. A special field configuration (that of a radially polarized laser beam) is considered as an example. We discuss the fields that result from solving Maxwell's equations and calculate the energy that may be gained by a single electron in interaction with this particular configuration.

6.1 Introduction

Electron dynamics in the presence of a radiation field may be discussed at a number of different levels dictated indirectly by the intensity of the radiation field. Weak fields cause small-amplitude and slow on-average oscillatory electron motion. It suffices for many practical purposes to discuss the motion in this regime semiclassically and taking only the electric component of the radiation field into consideration. A semiclassical treatment involves representing the electron by a wave-packet built from solutions to the appropriate Schrödinger equation and treating the radiation field classically as a set of oscillating electric and magnetic fields. With m and $-e$ standing for the electron's mass and charge, respectively, and assuming a sinusoidal electric field dependence upon the time, the amplitude of the resulting oscillatory motion of the electron may be written as

$$\bar{v} = \frac{eE_0}{m\omega}, \tag{6.1}$$

where E_0 is the electric field amplitude and ω is its frequency. However, in the presence of fields whose intensity and frequency are of such magnitudes as to render $\bar{v}$ comparable to c, the speed of light in vacuum, not only does the magnetic component of the radiation field become important but the relativistic effects associated with the dynamics may not be ignored anymore. For some applications, like particle acceleration, the center-of-mass motion of the particle only is important and it suffices to treat the electron classically, but relativistically. On the other hand, a full account of the quantum effects must be based on a fully quantum representation. This calls then for the construction of electron wave-packets from solutions to the appropriate Dirac equation.

In this chapter the dynamics of a single electron, free as well as bound, will be discussed in the regime of relativistic field intensities. In Sect. 6.2 we review progress in the quantum theoretical treatment of relativistic (free and bound) electron wave-packets in intense laser fields. As a recent example in which the classical description is sufficient, the dynamics of a free electron in a radially polarized Gaussian laser beam will be discussed in Sect. 6.3.

6.2 Quantum Dynamics

Interaction of light and matter is responsible for a plethora of interesting phenomena which are essentially caused by the extremely high laser intensities that are available today. Depending on the nuclear charge and laser intensity, the electric field strength due to the light can compensate (e.g., $2.7 \times 10^{10}\,\mathrm{V\,cm^{-1}}$ at an intensity of $10^{18}\,\mathrm{W\,cm^{-2}}$) the one due to the nuclear charge (e.g., averaged $1.0 \times 10^{10}\,\mathrm{V\,cm^{-1}}$ for the ground state of hydrogen) or even exceed it ($8.7 \times 10^{11}\,\mathrm{V\,cm^{-1}}$ at $10^{21}\,\mathrm{W\,cm^{-2}}$) substantially.[1]

For the theoretical treatment of phenomena such as MPI[2], ATI[3], HHG[4] and OTBI[5], this means that, with increasing laser intensity, one eventually has to discard perturbative treatments which consider the incoming laser as a source of merely a small perturbation compared to the binding energy of the electrons to the ions and atoms. Just as well one also has to abandon the so-called dipole approximation, which neglects the magnetic component

[1] For extremely highly charged ions, the field strengths (e.g., averaged $8 \times 10^{15}\,\mathrm{V\,cm^{-1}}$ for the ground state of hydrogen-like Uranium, according to a non-relativistic calculation) are still out of reach.

[2] Multi Photon Ionization [1].

[3] Above Threshold Ionization [2].

[4] High Harmonic Generation [3,4].

[5] Over The Barrier Ionization [5,6].

of the light. Highly charged ions, such as the ones produced, for example, at CERN[6] in Geneva [7] or at GSI[7] in Darmstadt [8], and the ever increasing laser intensities [9], demand a relativistic treatment. Whereas the magnetically induced forward drift, and even wave-packet spreading, of electrons in a laser beam can be described classically, a detailed examination of the wave packet dynamics and the interaction with an ionic nucleus demand a true quantum mechanical treatment, i.e., solution of the Dirac equation. This is especially the case under conditions in which intrinsic quantum features such as interferences, tunneling, or spin effects become important.

The relativistically correct, i.e., Lorentz-invariant, description of the quantum mechanical behaviour of a single particle in the presence of arbitrary time- and position-dependent classical (non-quantized) electromagnetic fields is given by the famous Dirac equation [10]. Its solution has always been both a difficult task on the one hand and a fascinating challenge on the other [11,12]. Analytical solutions are known for just a few simple cases such as the hydrogen atom (bound dynamics) [13,14] and a charged particle in the field of an electromagnetic plane wave (free dynamics) [15]. For more complicated systems, numerical solutions are inevitable.

6.2.1 Numerical Solution of the Dirac Equation

In atomic physics, the so-called split-operator technique [16,17] has been successfully employed in numerous Schrödinger calculations [18–20], most of which, with just a few exceptions [21,22], have been non-relativistic. However, this method can be employed to solve the Dirac equation just as well. The size of the numerical grid, the inherently required high temporal resolution, and the number of spatial dimensions that is achievable, represent the main numerical challenges. The high computational demand is the main reason why only a small number of attempts to address the Dirac problem numerically have been undertaken. Maquet et al. have recently solved the Klein–Gordon equation numerically [23]. There have already been applications of the two-spinor Pauli equation [18] and the four-spinor Dirac equation [24–29] in intense laser physics, too. Besides the aforementioned full quantum mechanical treatments, strong-field interactions have been investigated extensively by semiclassical and classical methods, such as the semiclassical relativistic Monte-Carlo method [30] and particle-in-a-cell (PIC) simulations [31].

There are other algorithms for solving the Dirac equation that rely on finite-difference techniques, rather than Fourier transforms. Although these are typically limited in their success by the so-called fermion doubling problem [32], they are more suitable for distributed parallel processing than any codes based on Fast Fourier Transforms (FFT). The latter can at least benefit from symmetric multiprocessing environments (SMP), where all grids can

[6] CERN: Conseil Européen pour la Recherche Nucléaire.
[7] GSI: Gesellschaft für Schwerionenforschung.

be kept in a single contiguous memory space. Employing a multi-threaded FFT implementation such as the one provided by the FFTW library [33], can also be of great benefit in this case. For all *local* grid operations, such as the special ones given below in (6.6) (leaving out the Fourier transforms) and (6.7), there also exists a rather obvious parallel implementation: one can simply split the grid into slices and feed each slice to one particular CPU. If one makes sure that, for any total grid size, the difference in size of the individual grid parts is always less than or equal to a single row, then the execution times of the separate threads are guaranteed to be almost equal, and the waiting times for thread synchronization are minimized. When using size-adaptive grids and a large number of CPUs, it is particularly important to pay attention to this, because here it becomes difficult (or at least inefficient) to maintain grid sizes that are divisible by the number of CPUs without a remainder for all times [34].

Series expansions of the Dirac Hamiltonian [35] can, to some extent, be used to extend the regime of validity of the intrinsically nonrelativistic Schrödinger codes into the relativistic regime, while avoiding the immense computational effort. Apart from concerns about convergence [36], adding higher and higher-order terms, with increasing particle velocity, also makes the code more and more demanding. In fact, the full Dirac equation is formally much simpler than even the Schrödinger equation (especially if one includes the spin and relativistic corrections), because the Hamiltonian is only linear in the momentum, not quadratic. This fact simplifies the algorithm significantly. The typical running times however, are still about two orders of magnitude longer, because of the omnipresent rest mass energy mc^2 and the resulting need for a much higher temporal resolution, even at fairly low kinetic energies.

In addition to its advantage of formal simplicity, the Dirac equation also takes into account the particle's spin without the need for additional terms. It accurately describes both particles and antiparticles and, to some extent, even transitions between both states, i.e., it can be employed to model pair creation processes. A matrix formalism is the usual choice for mathematically representing both the particle's spin and its sign of energy. As a consequence, the Hamiltonian itself is a four by four matrix, and the wave function is a complex four-spinor.

We write the Dirac equation in the same form as the Schrödinger equation, namely

$$\mathrm{i}\hbar\frac{\partial\psi}{\partial t} = \mathbb{H}\psi. \tag{6.2}$$

The Dirac Hamiltonian for a fermionic particle with charge q and rest mass m then reads as follows:

$$\begin{aligned}\mathbb{H} &= -\mathrm{i}\hbar c\alpha^i\frac{\partial}{\partial x^i} + \beta mc^2 + q\left(A_0(\boldsymbol{x}) + \alpha^i A_i(\boldsymbol{x})\right),\\ &= \qquad \mathbb{H}_{\boldsymbol{\partial}} \qquad + \qquad \mathbb{H}_{\boldsymbol{x}},\end{aligned} \tag{6.3}$$

where the second line indicates the split of the whole operator into a derivative-dependent part and a position-dependent part. Here, α^i, β represent the usual Dirac matrices with $i \in \{1,2,3\}$ and A_0 and $\boldsymbol{A} = (A_1, A_2, A_3)$ stand for the scalar and vector potentials of the ions and radiation field, respectively. As in the Schrödinger case, we consider the short-time propagation of a wave function at time t

$$\psi_{(t+\Delta t)} = \mathrm{e}^{-\frac{\mathrm{i}\Delta t \mathbb{H}}{\hbar}} \psi_t. \tag{6.4}$$

The split-operator approximation then reads as follows: [26]

$$\psi_{(t+\Delta t)}(\boldsymbol{x}) \approx \exp\left(-\frac{\mathrm{i}}{\hbar}\frac{\Delta t}{2}\,\mathbb{H}_{\boldsymbol{\partial}}\right)\exp\left(-\frac{\mathrm{i}}{\hbar}\Delta t\,\mathbb{H}_{(t+\frac{\Delta t}{2},\boldsymbol{x})}\right)\exp\left(-\frac{\mathrm{i}}{\hbar}\frac{\Delta t}{2}\,\mathbb{H}_{\boldsymbol{\partial}}\right)\psi_t(\boldsymbol{x}) + O(\Delta t^3). \tag{6.5}$$

Again, from well-established Schrödinger treatments, it is known that fast Fourier transforms are an appropriate means to evaluate the action of the exponential momentum space operator. In the Dirac case, however, the matrix character of $\mathbb{H}_{\boldsymbol{\partial}}$ also has to be taken into account. This is most conveniently accomplished with the help of a series expansion. The result can be written as follows:

$$\exp\left(-\frac{\mathrm{i}}{\hbar}\frac{\Delta t}{2}\,\mathbb{H}_{\boldsymbol{\partial}}\right) = \mathbb{F}^{-1}\left(\cos(\phi) - \mathrm{i}\sin(\phi)\frac{\beta\frac{mc}{\hbar} - \alpha^i k_i}{\sqrt{\left(\frac{mc}{\hbar}\right)^2 + \boldsymbol{k}^2}}\right)\mathbb{F},$$
$$\phi = \frac{c\Delta t}{2}\sqrt{\left(\frac{mc}{\hbar}\right)^2 + \boldsymbol{k}^2}, \tag{6.6}$$

where $\mathbb{F}$ indicates the Fourier transform and $\mathbb{F}^{-1}$ its inverse. Similarly, one obtains

$$\exp\left(-\frac{\mathrm{i}}{\hbar}\Delta t\,\mathbb{H}_{(t+\frac{\Delta t}{2},\boldsymbol{x})}\right) = \left[\cos\left(\frac{qA_0}{\hbar}\Delta t\right) - \mathrm{i}\sin\left(\frac{qA_0}{\hbar}\Delta t\right)\right] \times \left[\cos\left(\frac{qA}{\hbar}\Delta t\right) - \mathrm{i}\sin\left(\frac{qA}{\hbar}\Delta t\right)\frac{\alpha^i A_i}{A}\right], \tag{6.7}$$

with $A = |\boldsymbol{A}|$. The main problem of the Dirac treatment as compared to the Schrödinger one is the temporal step size $\Delta t \lesssim \hbar/E$ required, which, for the algorithm to converge properly, has to be significantly smaller. The reason for this effect is the large rest mass energy mc^2 that is contained in the particle's total energy E.

Adaptive-Grid

Highly intense laser fields typically result in very extended particle trajectories in position space, especially at low frequencies. For rather compact wave packets, however, for most of the time that the simulation has to cover, the largest fraction of the numerical grid would only hold zero values of the wave function. In such a case it is advisable to restrict the actual grid to a box only slightly larger than the support of the wave function, and to dynamically adapt it in both position and size to the actual requirements as the simulation time evolves [34]. Since one of the most demanding operations of the method that was discussed before is the two-dimensional Fast Fourier Transform, which scales as $O(N^2 \log N)$, where $N \times N$ equals the grid size, it is possible to save a lot of CPU time by means of such adaptive numerical grids. This size-adaptive approach is also a pragmatic solution to the well-known boundary problem [12, 37] in Dirac calculations: the wave function is multiplied by a suitable damping function in order to simulate an absorbing boundary around the edge of the numerical grid. This admittedly insufficient solution however effectively only comes into play after the grid has reached its maximum size, which is merely given by memory and runtime constraints. Before this happens, the boundary problem simply does not exist, because the wave packet is located far away from the boundary. A rather elaborate algorithm that will not be covered here is needed to automatically maintain this favorable condition for as long as possible [38].

6.2.2 Free Dynamics

The simplest system that one can think of is probably that of a single, free electron in a laser field. An analytical solution for this problem exists, namely the Volkov states [15], which read as follows in a popular notation [39–42]

$$\psi_p^r(x) = \frac{1}{(2\pi)^{\frac{3}{2}}} \left(1 + \epsilon^r \frac{q \not{k} \not{A}_{(k\cdot x)}}{2c(k\cdot p)}\right) \frac{1}{\sqrt{\gamma(p)}} u^r_{(p)} \mathrm{e}^{\mathrm{i}S^r_p(x)},$$

$$\text{where} \quad S_p^r(x) = -\epsilon^r \frac{\mathrm{i}}{\hbar} \left(p\cdot x + \int_{-\infty}^{k\cdot x} \mathrm{d}\eta \, \frac{2\frac{q}{c}(p\cdot A_{(\eta)}) - \epsilon^r \frac{q^2}{c^2} A^2_{(\eta)}}{2(k\cdot p)} \right),$$

$$\text{and} \quad \gamma(p) = \sqrt{1 + \left(\frac{\boldsymbol{p}}{mc}\right)^2}. \tag{6.8}$$

In the above, $A^\mu = (0, \boldsymbol{A})$ represents the potential four-vector of the laser field, which only depends on position and time via the scalar phase $\eta = k \cdot x = \omega t - \boldsymbol{k} \cdot \boldsymbol{x}$. Here, k^μ is the laser's wave four-vector, and p^μ is the particle's energy-momentum four-vector. For the definition of ϵ^r and the associated four free spinor solutions u^r ($r = 1, \ldots, 4$) for spin up and down, positive and negative energy, see a textbook on relativistic quantum theory [43]. Although assumed to be normalized and complete for decades, only recently were some of the Volkov states' mathematical properties actually proven [44]. Since the Volkov states themselves are not square integrable, it is reasonable to build wave packets from them by integrating their product with a weight function

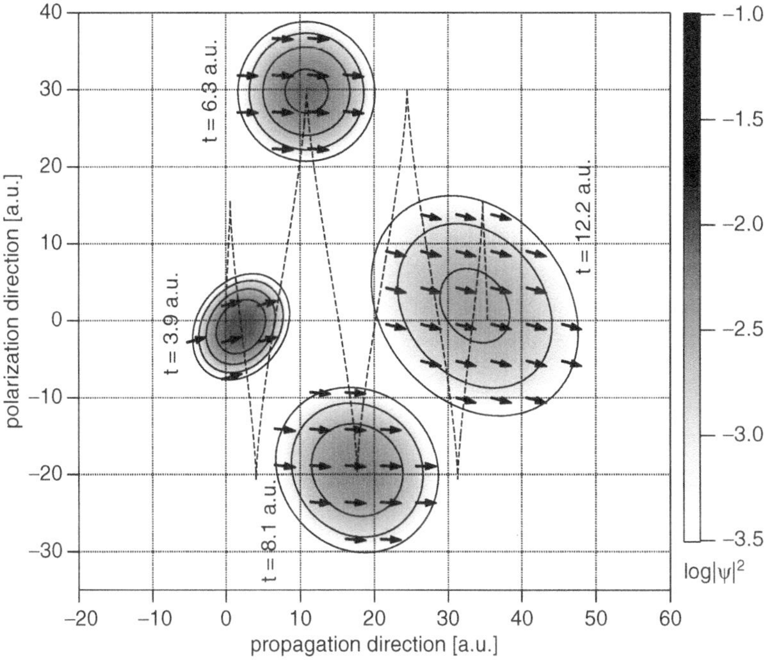

Fig. 6.1. Dynamics of a free electronic wave packet when exposed to the field of a laser pulse. *Arrows* indicate the local orientation of spin (compare [46, 47]). The remaining parameters are $E_0 = 100\,\text{a.u.}$, $\omega = 2\,\text{a.u.}$, and a five-cycle $\sin^2$ pulse (a 2-cycle turn-on, a one-cycle flat top and a 2-cycle turn-off). The dashed line represents the center of mass trajectory

over all momentum space. This procedure, however, is difficult to carry out entirely analytically, and one therefore has to resort to numerics one way or the other. The free motion in a laser field has first been extensively investigated by Roman et al. [45].

For the two-dimensional case, the resulting dynamics are shown in Fig. 6.1. One can easily see the deformation and apparent rotation of the initially round Gaussian wave packet due to relativistic contractions and phase differences along the propagation direction (no dipole approximation has been used in the calculation). The center-of-mass trajectory coincides, within the limitations of numerical accuracy, with the solution of the classical equations of motion, and the spin shows an oscillatory behavior as predicted by Bargmann et al. [46].

The relativistic contraction is only due to the particle velocity, as can be understood from Fig. 6.2 which essentially shows the same Gaussian wave packet (with rather high constant velocity in field-free space) four times: once from the point of view of an observer located in the particle's rest frame, once in the laboratory frame, and in both position and momentum space for both cases.

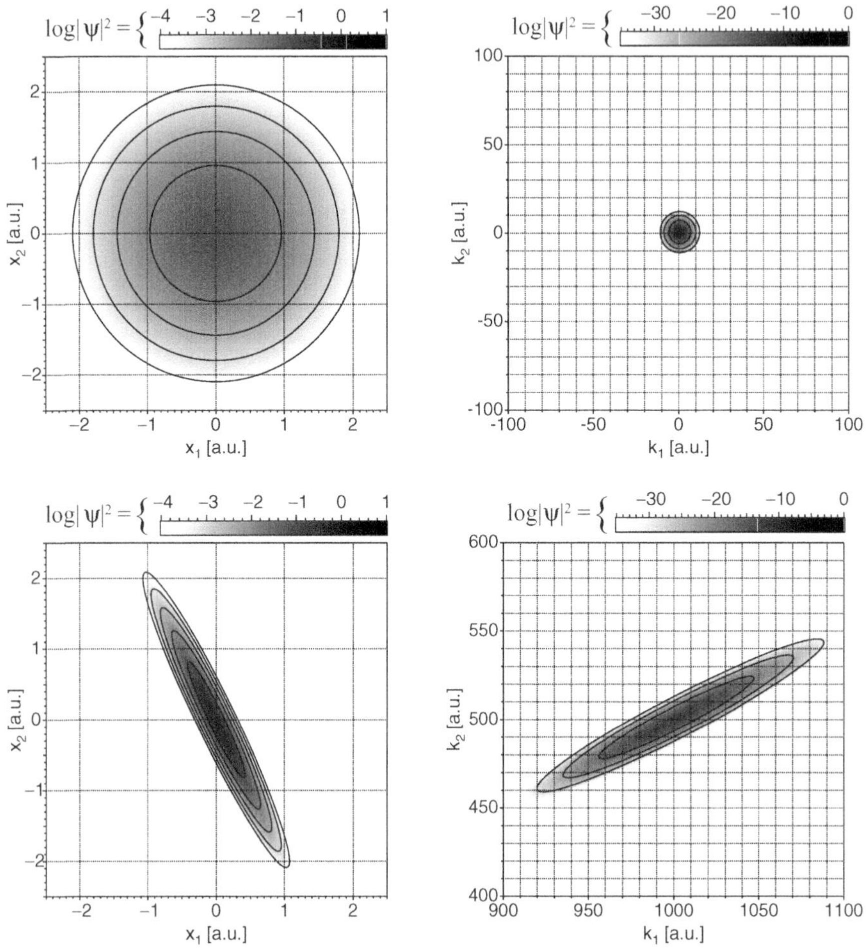

Fig. 6.2. Relativistic electrons with constant velocity in field-free space. All four plots show an electronic wave packet resulting from a 2D Gaussian distribution in momentum space (*right column*) with a $1/e$ half width of 2 a.u. centered around (1,000; 500) a.u. Whereas the wave packet appears circular in its rest frame (*top row*), it shows significant contraction along (perpendicular to) the propagation direction in position (momentum) space in the laboratory frame (*bottom row*)

Whereas both a 2D wave packet (by definition) and a spinless classical charged particle are restricted in their resulting motion to a plane, Walser et al. [47] were the first to point out an additional drift in the magnetic field direction for particles with spin in 3D. Their results from a weakly relativistic Schrödinger treatment could also be reproduced later for the fully relativistic Dirac case [48].

6.2.3 Bound Dynamics

The previous section was devoted to the more simple dynamics of free electrons in a laser field in order to introduce into the topic and to illustrate the general behavior of charged particles in such a field. We now move to the more complicated case of bound electrons in the combined field of both their parent ion and the laser.

Depending on the laser parameters (field strength E_0 and frequency ω) and the nuclear charge (Z) employed, one distinguishes three different regimes. The first is the so-called multi-photon regime, in which the binding potential still dominates over the averaged free-electron kinetic energy (the so-called ponderomotive potential) and where one can observe multi-photon processes, provided the laser intensity is high enough. Then there is a second regime where this relation is reversed, the so-called tunneling-regime, which draws its name from the fact that, at high enough electric field strength, the bound electron can leave the ionic core by tunneling through a momentary potential barrier, provided that the frequency is not too high. Finally there is the over-the-barrier regime, in which the barrier is suppressed below the (formerly) bound state's energy level, thereby releasing the electron.

The tunneling-regime is perhaps the most interesting to deal with due to its non-perturbative character as far as the laser field is concerned while, at the same time, the influence of the nucleus is important. Usually the motion is described in terms of the well-known three-step model for single-atom ionization-recollision dynamics [4, 49]. In broad outline, the description follows along the following lines. If the field is high and its frequency is low enough, a wave packet can tunnel through the barrier. Once outside, this wave packet can gain energy from the laser field during an otherwise "free" motion. This energy plus the ionization potential is released on a subsequent re-collision with the core. The maximum re-collision energy, according to a simple classical estimate, also proven to be correct by more elaborate quantum calculations, turns out to be $3.17\,U_\mathrm{p} + I_\mathrm{p}$, where $U_\mathrm{p} = q^2E_0^2/(4m\omega^2)$ is the so-called ponderomotive potential, and I_p is the ionization potential of the atom.

With the advent of high energy lasers, there is now hope to create electrons with keV or even MeV re-collision energies this way. Those energies can be released in terms of electromagnetic radiation, for example. Because it is a periodic process, it is able to generate coherent high harmonics of the laser frequency. However, at large laser intensities, of the order $10^{19}\,\mathrm{W\,cm^{-2}}$, magnetic field effects become important, and tend to prevent the re-collision of the electron with its parent ionic core by inducing a considerable drift in the laser propagation direction.

In [22], the weakly relativistic regime has been explored, which showed both tunneling and multi-photon behavior (depending on the parameters chosen), high harmonics up to the keV regime, and very energetic ATI electrons. Non-relativistic, non-dipole SFA[8] calculations [50] have been carried out to

[8] SFA: Strong Field Approximation.

show that mid-range keV high harmonics are not severely suppressed in a special scenario where the considered ion is moving at a high speed against the laser propagation direction. The same scenario, but in the MeV regime, has been explored using the above-mentioned fully relativistic Dirac integration techniques in [51], in an attempt to find a more convenient tool to initiate nuclear reactions than standard particle accelerators. In the following, we explain this scheme while restricting the discussion to a simpler, but less realistic, parameter set that is more easily understood than the one shown in the original work [51].

We consider multiply charged hydrogen-like ions at very high velocities under the influence of a very intense counter-propagating laser pulse. The starting point of our calculation is the electronic ground-state wave function of the considered ion. We choose an oxygen ($Z = 8$) core in order to have the nuclear charge lie well inside the regime where relativistic effects are important [52]. As was shown in [34], the numerical generation of the Dirac ground-state wave function for such an ion is a computationally demanding task by itself. The radial extension r_0 of the ground-state wave function, which we measure from the center of the core to a point where the probability density has gone down to 10^{-4}, is about 0.93 a.u.

A charged particle that is placed in a laser field does not only oscillate with an amplitude $\frac{1}{2}\Delta x_{\text{pol}}$ in the polarization direction ($\hat{\boldsymbol{e}}_2$), but it also experiences a drift per cycle Δx_{prop} in the propagation direction ($\hat{\boldsymbol{e}}_1$). As shown in [53], a classical relativistic trajectory calculation in a harmonic vector potential, say

$$\boldsymbol{A}(\eta) = -\hat{\boldsymbol{e}}_2 \frac{cE_0'}{\omega'}[\sin\eta - \sin\eta_0], \tag{6.9}$$

that only depends upon the Lorenz invariant phase $\eta = \omega t - \boldsymbol{k}\cdot\boldsymbol{x}$, and fulfills the initial conditions $\boldsymbol{A}(\eta_0) = \boldsymbol{0}$, $\eta_0 = 0$, and $\boldsymbol{\beta}(\eta_0) = \boldsymbol{0}$, has

$$\Delta x_{\text{prop}} = \frac{\pi q^2 E_0'^2}{2m^2 c\omega'^3}, \qquad \Delta x_{\text{pol}} = \frac{2|q|E_0'}{m\omega'^2}, \tag{6.10}$$

for a particle with mass m and charge q. For a successful re-collision with the ionic core after just one oscillation period, the forward drift per cycle Δx_{prop} has to be smaller than half the width of the returning wave packet, i.e., $\Delta x_{\text{prop}} \lesssim r_0$ if we assume negligible wave packet spread. Thus our first assumption is $\Delta x_{\text{prop}} \lesssim 1$ a.u. A second assumption is, since we are interested in nuclear reactions, to have an average kinetic energy U_{p} in the MeV regime, so we select $U_{\text{p}} = 1.5\,\text{MeV} \gtrsim 1.022\,\text{MeV} = 2mc^2$, where $m = 1$ a.u. is the mass of the electron and $c = 137.036$ a.u. is the speed of light. Assuming free motion, the corresponding maximum momentum in the polarization direction is $p_{\text{pol,max}} = |q|E_0'/\omega' = \sqrt{4mU_{\text{p}}} = 469.8$ a.u. The combination of these constraints with (6.10) leads to a laser frequency of

$$\omega' = \frac{\pi p_{\text{pol,max}}^2}{2q^2 m^2 c \Delta x_{\text{prop}}} = 2{,}529 \text{ a.u.}, \tag{6.11}$$

and an electric field amplitude of

$$E'_0 = \frac{\omega'}{|q|} p_{\text{pol,max}} = 1.188 \times 10^6 \text{ a.u.} \tag{6.12}$$

In these calculations the electron's charge is $q = -e = -1\,\text{a.u.}$

The Doppler effect is used to provide the high frequency that is needed. For experimental realization, one could think of using an advanced Ti:Sapphire laser system, which means $\lambda = 800$ nm corresponding to $\omega = 0.0569\,\text{a.u.}$ and then, to reach our desired frequency of $\omega' = 2{,}529\,\text{a.u.}$, a large relativistic Doppler shift is needed. For the special case of exact counter-propagation and an electromagnetic field only perpendicular to the propagation direction at a scaled velocity $\beta = v/c$, the Doppler effect is easily described by the following formulas

$$\boldsymbol{E}' = \sqrt{\frac{1+\beta}{1-\beta}}\boldsymbol{E}, \qquad \boldsymbol{B}' = \sqrt{\frac{1+\beta}{1-\beta}}\boldsymbol{B}, \qquad \omega' = \sqrt{\frac{1+\beta}{1-\beta}}\omega. \tag{6.13}$$

Here the primed quantities refer to the moving coordinate frame, whereas the plain ones are given in the laboratory frame. Using $\gamma = \left(1-\beta^2\right)^{-1/2} = 22{,}223$ leads to the desired result in terms of frequency. It also converts $E'_0 = 1.188 \times 10^6\,\text{a.u.}$ back to $E_0 = 26.7\,\text{a.u.}$ in the laboratory system, which is feasible [54] and corresponds to an intensity of $I = 2.5 \times 10^{19}\,\text{W cm}^{-2}$. The corresponding high value of γ is not likely to be reached soon in an experiment. It must be noted, however, that our results will show that we have estimated the requirements to reach the MeV regime generously. The actual threshold for the accelerator's kinetic energy, or the laser's intensity, is lower than the values just indicated. Also see [51] for another set of more realistic parameters.

If one keeps in mind that the maximum kinetic energy of a classical particle during such an oscillatory motion is given by $E_{\text{kin,max}} = q^2 E_0'^2/(2m\omega'^2)$ and the wavelength $\lambda' = 2\pi c/\omega'$, then using (6.10) one easily arrives at

$$\Delta x_{\text{prop}} = \frac{E_{\text{kin,max}}}{2mc^2}\lambda'. \tag{6.14}$$

Thus, if we further seek the maximum kinetic energy to be at least 1 MeV, we automatically arrive at $r_0 \gtrsim \Delta x_{\text{prop}} > \lambda'$. Then $r_0 > \lambda'$, which means that the pulse's wavelength should not be larger than the extent of the state. This estimate is valid for the case $\eta_0 = 0$, but even for the optimized case featuring an initial phase $\eta_0 \simeq 0.3$ (resulting in a drift of $0.55\Delta x_{\text{prop}}$ at return time) and a requirement of $3.17\,U_{\text{p}} \gtrsim 1$ MeV, one only gains a factor $3.17/(2 \times 0.55) = 2.88$, i.e., one obtains $2.88\,r_0 > \lambda'$, which is still the same order of magnitude. Thus the classical expressions, and even the more so the expectation values derived from wave packets, are quite restricted in their implications. Depending on their position in propagation direction, different parts of the ground-state wave function are now subjected to significantly different phases of the pulse. Therefore, there is no compact wave packet leaving the core and re-colliding with it later on. The general behavior is entirely different. The wave function stays *on-average* almost at rest, albeit is considerably distorted by the laser pulse. After the pulse has gone by, the wave function regains its original shape almost exactly, however with its origin being displaced by Δx_{prop}. The time evolution of the expectation values of the position and momentum operators is similar to the classical point particle estimate. However, the sharp edges of the position space trajectory are somewhat smoothed by comparison to the classical counterpart.

While the above discussion shows that "re-collisions" in the original meaning of the word are inherently impossible in the MeV regime, it also paves a new way to electron–nucleus interactions with such energies. A laser wavelength that is smaller than the ground state's radius permits to accelerate the electron to and above MeV kinetic energies within a very short time corresponding to $\lambda'/4$ in spatial extent. More specifically, at least fractions of the extended wave packet should be capable of acquiring MeV kinetic energies *before* passing by closely at the nucleus and leaving the ion. For considerably larger wavelengths, the ejected electron will gain the required high energies only *after* it has left the vicinity of the ionic core. After all, a sufficiently small wavelength is required.

We have numerically solved the Dirac equation in the frame of the fast-moving oxygen core using the Lorentz-transformed laser parameters E_0' and ω' mentioned above. In a series of contour plots Fig. 6.3 depicts the time evolution of the $1s^1\uparrow$ state (spin with respect to the axis coming out of the coordinate plane) under the influence of a single-cycle laser pulse modeled by the above-mentioned vector potential and a soft-core potential given by

$$A_0 = \frac{Ze}{\sqrt{a + x_1^2 + x_2^2}}, \tag{6.15}$$

where $e = 1$ a.u. is the elementary charge, $Z = 8$ is the atomic number, and the parameter $a = 0.01$ gives rise to a ground-state energy $E_{\text{g}} = 18{,}746.8$ a.u., sufficiently close to the analytical ground-state energy of a true Coulomb potential in three dimensions $E_{\text{g}} = mc^2\sqrt{1-(Z\alpha)^2} = 18{,}746.84$ a.u. The four snapshots illustrate the dynamics. At such a high frequency, the wavelength is only $\lambda = 2\pi c/\omega' = 0.34$ a.u., which is clearly smaller than the width of the electronic state (2×0.93 a.u.). As a result, not every part of the wave function feels the laser at any given time. Only in those regions of space where the laser electric field is non-zero, the corresponding fraction of the wave function responds to the field with an appropriate distortion. Therefore, one obtains an image of the laser pulse, and it is even possible to verify the laser wave length by reading off the horizontal extension of the perturbation of the wave function in Fig. 6.3c. The laser-induced distortion of the ground state wave function resembles the cosine-shaped displacement of charges following the laser pulse's electric field. It is further modified to a small degree by the influence of the ionic core, which also causes reflections and interferences.

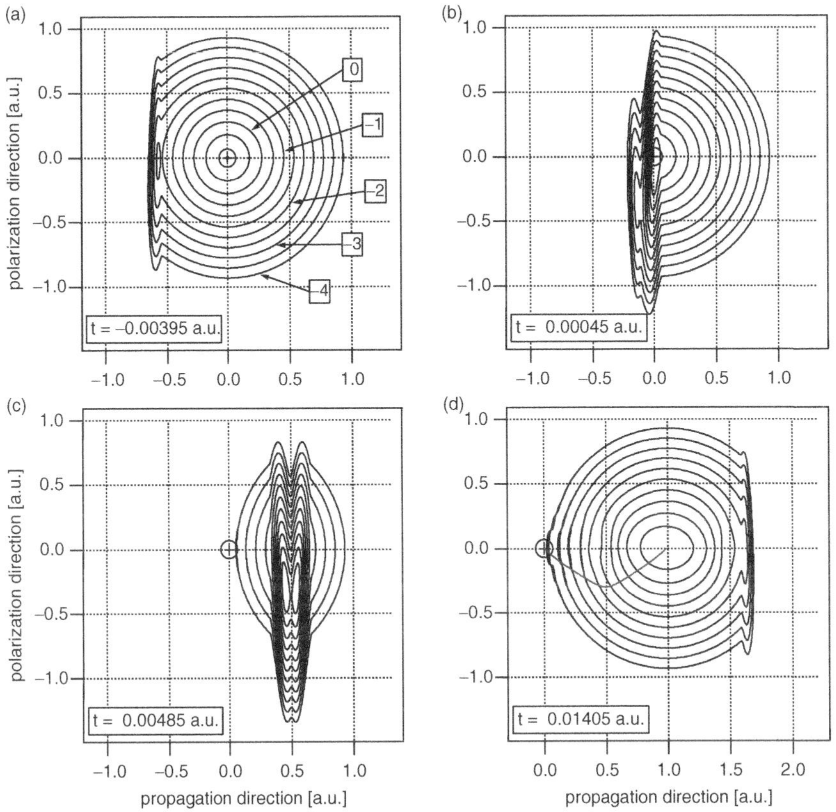

Fig. 6.3. Snapshots of $\rho = \psi^\dagger\psi$ as a function of the two spatial coordinates taken at (**a**) $t = -0.00395$ a.u., (**b**) $t = 0.00045$ a.u., (**c**) $t = 0.00485$ a.u., and (**d**) $t = 0.01405$ a.u. In all four plots the contour lines indicate the logarithm of the probability density ρ ranging from -4 to $+1$ with spacing 0.5 as shown by the five labels in (**a**). The marker "+" indicates the ionic core's position right at the origin. In addition to that, in (**d**) the *thick gray line* that starts at the origin marks the trajectory of the position operator's expectation value. Also note the change of origin in this subfigure. In this simulation we are considering a single-cycle cosine pulse with a Lorentz-transformed field amplitude $E_0' = 1.188 \times 10^6$ a.u. and a frequency $\omega' = 2{,}529$ a.u. The ionic core's charge is $Z = 8$ times the elementary charge. Taken from [55] with permission

After the pulse has passed, the ground state is shifted by about 1.0 a.u. in the forward direction, while the much heavier ionic core ($m_O = 29{,}168$ a.u.) does not noticeably change its state of motion (compare (6.10)). This "final" state is non-stationary, but its further evolution is not discussed here.

Relativistic quantum theory defines the probability current density as [43]

$$j^\mu = c\psi^\dagger\gamma^0\gamma^\mu\psi, \quad \text{where} \quad \gamma^0 = \beta, \quad \text{and} \quad \boldsymbol{\gamma} = \beta\boldsymbol{\alpha}, \tag{6.16}$$

which can be easily decomposed into the vectorial current density

$$\boldsymbol{\jmath} = c\psi^\dagger \boldsymbol{\alpha}\psi, \tag{6.17}$$

and the scalar density

$$\rho = \psi^\dagger \psi. \tag{6.18}$$

Furthermore, the local momentum density $\boldsymbol{P}$ is defined by

$$\boldsymbol{P} = -\mathrm{i}\hbar\psi^\dagger \boldsymbol{\nabla}\psi - \frac{q}{c}\psi^\dagger \boldsymbol{A}\psi. \tag{6.19}$$

By dividing $\boldsymbol{P}$ by ρ we arrive at a local momentum $\boldsymbol{p}$ which, when averaged over all space, is consistent with the overall expectation value.

$$\boldsymbol{p} = \frac{-\mathrm{i}\hbar\psi^\dagger \boldsymbol{\nabla}\psi - \frac{q}{c}\psi^\dagger \boldsymbol{A}\psi}{\psi^\dagger \psi}. \tag{6.20}$$

This quantity is represented by the arrows in Fig. 6.4 (left). On the right of Fig. 6.4 the current density $\boldsymbol{\jmath}$ from (6.17) is shown in a similar fashion. As one can see, with $|\boldsymbol{p}| \simeq 618.1$ a.u., the momentum magnitude close to the ionic

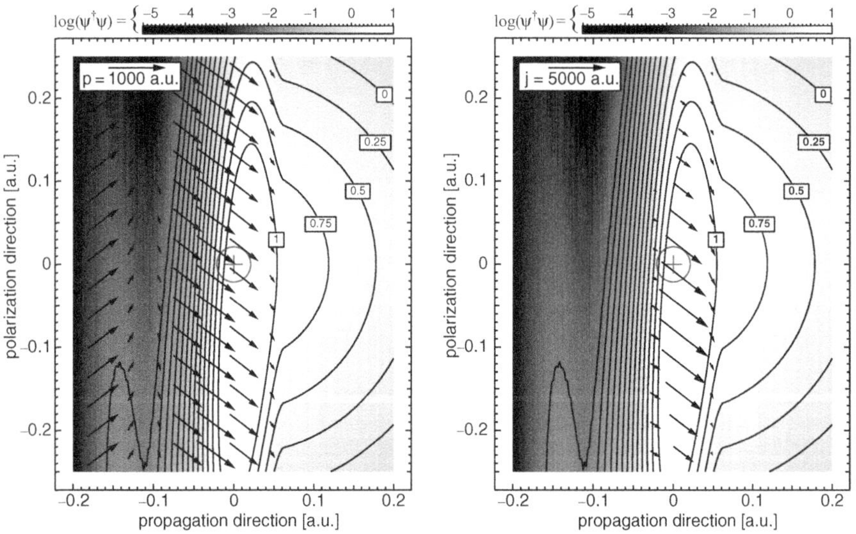

Fig. 6.4. Enlargement of (Fig. 6.3b) ($t = 0.00045$ a.u.) in the vicinity of the ionic core: The *contour lines* and the *grey shading* indicate the logarithm of the probability density $\rho = \psi^\dagger\psi$. The marker "+" indicates the ionic core's position right at the origin. In addition, the *arrows* depict the direction and magnitude ($p = |\boldsymbol{p}|$, $j = |\boldsymbol{\jmath}|$) of the local momentum $\boldsymbol{p}$ as defined by (6.20) (*left*) and the local current density $\boldsymbol{\jmath}$ as defined by (6.17) (*right*), respectively. Taken from [55] with permission

core is well above 387.6 a.u. (corresponding to 1.022 MeV) and the momentum flow passes through the vicinity of the core. The overall expectation value $\langle -i\hbar\boldsymbol{\nabla} - \frac{q}{c}\boldsymbol{A}\rangle_\psi$ of the kinetic momentum operator also reaches values that correspond to the MeV energies $E_{\text{kin.}}(\boldsymbol{p}) = \sqrt{(c\boldsymbol{p})^2 + (mc^2)^2} - mc^2$ that we require, i.e., up to 2.43 MeV. The greyshaded contour plot in the background shows that the probability density reaches 10 per unit area. The maximum of the probability current density is located a small distance away from the core, but with $|\boldsymbol{\jmath}| \simeq 4{,}848$ a.u., the probability of nuclear reactions should be high.

We conclude that, while the parameters needed for experimental realization are certainly still challenging, the counter-propagating setup of a pre-accelerated ion and a laser pulse is well-suited to induce electron-core collisions with kinetic energies in the MeV regime in principle. The collision flux is high and there especially is no difficult aiming or timing required.

6.3 Classical Dynamics

Laser field intensities in excess of $10^{18}\,\mathrm{W\,cm^{-2}}$ result typically in accelerating an electron to speeds comparable to that of light in vacuum [56–58]. A parameter of interest in this regard is

$$Q = \frac{eE_0}{mc\omega}. \tag{6.21}$$

Laser systems with $Q \geq 1$ are capable of accelerating electrons to relativistic speeds. This regime can be reached by high-power and/or high-frequency laser systems currently operational at a number of places around the world [38]. In this environment, ultra-fast processes take place and studies of the electron dynamics based on perturbation theory fail and must be sub-planted by more appropriate fully relativistic analysis. In this section, we review recent progress in the field of relevance to problems in which only the center-of-mass motion of the particle is of concern. In other words, where the quantum effects may be side-stepped, the (free) electron will be treated as a (relativistic) classical particle. The laser field will also be represented classically by vector and scalar potentials appropriate to the case at hand or, equivalently, by the electric and magnetic fields $\boldsymbol{E}$ and $\boldsymbol{B}$, respectively, derived from those potentials.

6.3.1 General Considerations

The electron's 4-momentum will be denoted by $p = (p_0, \boldsymbol{p})$, with $p_0 = \mathcal{E}/c$, where $\mathcal{E} = \gamma mc^2$ is the energy, $\gamma = (1-\beta^2)^{-1/2}$ is the Lorentz factor, $\boldsymbol{\beta} = \boldsymbol{v}/c$ is the electron velocity in units of c, and $\boldsymbol{p} = \gamma mc\boldsymbol{\beta}$ is its linear momentum. In principle, the dynamics of the electron may be discussed by solving the following energy-momentum transfer equations (SI units)

$$\frac{\mathrm{d}\boldsymbol{p}}{\mathrm{d}t} = -\mathrm{e}(\boldsymbol{E} + c\boldsymbol{\beta}\times\boldsymbol{B}); \quad \frac{\mathrm{d}\mathcal{E}}{\mathrm{d}t} = -ec\boldsymbol{\beta}\cdot\boldsymbol{E}, \tag{6.22}$$

which follow from the (covariant) Newton–Lorentz equation of motion as its space-like and time-like components, respectively [59]. To the best of our knowledge, analytic solutions to (6.22) exist under very specialized conditions and only for plane-wave fields. Provided the fields are appropriately modeled, a discussion of most aspects of the electron dynamics based on solutions, analytic or numerical, of these equations may then be obtained. In addition to that, from the obtained trajectories one may, in principle, obtain an accurate representation of the Liénard–Wiechert potentials of the accelerated electron. Radiation emitted by the electron in this environment may then be investigated by studying the emitted harmonics and their patterns.

6.3.2 Axicon (Radially Polarized) Laser Fields

The tremendous energy of present-day laser systems is made available for electrons to interact with when the beam is focused down to micron-size spatial dimensions. Recent experiments [60–64] have shown that radially polarized light, produced by submitting a Gaussian beam, for example, to an axicon (a mirror or lens with at least one conical side) of some variety, may be focused down to a spot of size smaller than the radiation wavelength. In cylindrical coordinates (r, θ, z) the fields turn out to have only three components [64–67]: two electric components (one axial, E_z and one radial, E_{r}) and a single magnetic component (polarized azimuthally, B_θ). Using the well known parameters of a fundamental Gaussian beam of wavelength λ, namely, its waist radius w_0, its Rayleigh length $z_{\mathrm{r}} = \pi w_0^2/\lambda$, and its diffraction angle $\varepsilon = w_0/z_{\mathrm{r}}$, the field components have recently been derived to order ε^5. In truncated series form, the fields are [65, 66]

$$\begin{aligned} E_{\mathrm{r}} = E \Bigg\{ & \varepsilon \left[\rho C_2\right] + \varepsilon^3 \left[-\frac{\rho C_3}{2} + \rho^3 C_4 - \frac{\rho^5 C_5}{4} \right] \\ & + \varepsilon^5 \left[-\frac{3\rho C_4}{8} - \frac{3\rho^3 C_5}{8} + \frac{17\rho^5 C_6}{16} - \frac{3\rho^7 C_7}{8} + \frac{\rho^9 C_8}{32} \right] \Bigg\}, \end{aligned} \tag{6.23}$$

$$E_z = E \left\{ \varepsilon^2 \left[S_2 - \rho^2 S_3\right] + \varepsilon^4 \left[\frac{S_3}{2} + \frac{\rho^2 S_4}{2} - \frac{5\rho^4 S_5}{4} + \frac{\rho^6 S_6}{4} \right] \right\}, \tag{6.24}$$

$$\begin{aligned} B_\theta = \frac{E}{c} \Bigg\{ & \varepsilon \left[\rho C_2\right] + \varepsilon^3 \left[\frac{\rho C_3}{2} + \frac{\rho^3 C_4}{2} - \frac{\rho^5 C_5}{4} \right] \\ & + \varepsilon^5 \left[\frac{3\rho C_4}{8} + \frac{3\rho^3 C_5}{8} + \frac{3\rho^5 C_6}{16} - \frac{\rho^7 C_7}{4} + \frac{\rho^9 C_8}{32} \right] \Bigg\}. \end{aligned} \tag{6.25}$$

In (6.23)–(6.25) $\rho = r/w_0$, and

$$E = E_0 \mathrm{e}^{-r^2/w^2}; \quad w = w_0 \sqrt{1+\zeta^2}; \quad \zeta = \frac{z}{z_{\mathrm{r}}}, \tag{6.26}$$

$$C_n = \left(\frac{w_0}{w}\right)^n \cos(\psi + n\psi_G); \quad n = 2, 3, \cdots, \tag{6.27}$$

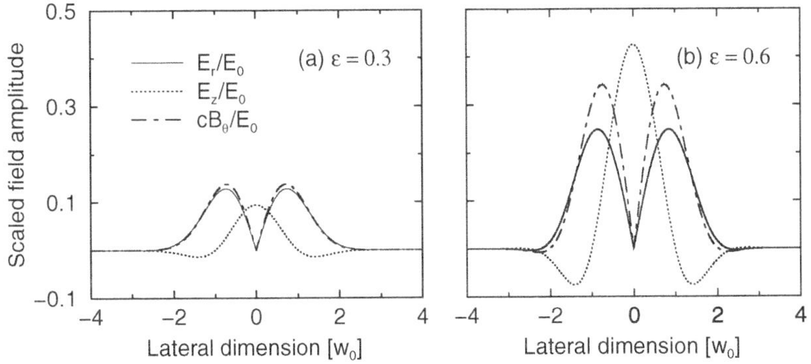

Fig. 6.5. Scaled amplitudes of the radially polarized fields given by (6.23)–(6.25) in the transverse plane through the beam focus at $z = 0$ as functions of the lateral dimension $\rho = r/w_0$. For a wavelength of $\lambda = 1\,\mu\text{m}$, the waist radii at focus are: (**a**) $w_0 = 10\lambda/3\pi \simeq 1.06\,\mu\text{m}$, and (**b**) $w_0 = 5\lambda/3\pi \simeq 0.58\,\mu\text{m}$. The legends in (**a**) apply to (**b**) as well

$$S_n = \left(\frac{w_0}{w}\right)^n \sin(\psi + n\psi_G); \quad \psi_G = \tan^{-1}\frac{z}{z_r}, \tag{6.28}$$

$$\psi = \psi_0 + \omega t - kz - \frac{kr^2}{2R}; \quad R = z + \frac{z_r^2}{z}, \tag{6.29}$$

where E_0 is a *constant* amplitude and ψ_0 is a *constant* initial phase. Note that the radial component of the electric field and the magnetic component vanish identically on the beam axis (the z axis, where $\rho = 0$) whereas E_z peaks there. This is shown in Fig. 6.5 quite clearly. The same figure also shows that, upon focusing, the maxima of all components increase in height, as expected, with E_z increasing the most. In fact, it has been shown in recent experiments that as much as three-quarters of the total beam power may be locked in the pencil-like axial focus of the output beam. This configuration is ideal for the particle confinement and acceleration scenario discussed recently. We review this acceleration configuration in the next subsection. For completeness, we quote here an expression for the output beam power derived from the above fields [65, 66]

$$P = \frac{\pi w_0^2}{2}\frac{E_0^2}{c\mu_0}\left(\frac{\varepsilon}{2}\right)^2\left[1 + 3\left(\frac{\varepsilon}{2}\right)^2 + 9\left(\frac{\varepsilon}{2}\right)^4\right], \tag{6.30}$$

where μ_0 is the permeability of free space. Finally, we mention that the field expressions given above have also been recently developed to order ε^{15}.

6.3.3 Electron Acceleration in Radially Polarized Fields

Features of the radially polarized fields alluded to briefly above make them ideally suited for electron laser acceleration to GeV energies. The fact that

E_r and B_θ both vanish on the z axis plays an important role in confining the electron motion to that axis [67,68]. With E_z having a maximum value along the same axis, this component works to accelerate the electron efficiently even from a position of rest at the origin.

In order to demonstrate acceleration we assume that an electron may be born at rest, like from above-threshold ionization, at the origin of coordinates. Equations (6.22) may be coupled into the following single first-order differential equation for the scaled velocity of the electron

$$\frac{\mathrm{d}\boldsymbol{\beta}}{\mathrm{d}t} = \frac{e}{\gamma mc}\left[\boldsymbol{\beta}(\boldsymbol{\beta}\cdot\boldsymbol{E}) - (\boldsymbol{E} + c\boldsymbol{\beta}\times\boldsymbol{B})\right]. \tag{6.31}$$

In principle, integration of this equation, subject to the initial conditions $x_0 = y_0 = z_0 = 0$ and $\boldsymbol{\beta}_0 = 0$, yields $\boldsymbol{\beta}$ which, in turn, may be used to calculate $\gamma(t)$. One typically integrates from $t_i = 0$ to $t_f \gg T$, where T is a laser-field cycle. The value of γ obtained in this way is then used to calculate the energy gain from

$$(\text{Energy Gain}) = [\gamma(t_f) - \gamma_0]mc^2, \tag{6.32}$$

where $\gamma_0 = \gamma(0)$.

Some results are displayed in Fig. 6.6 for acceleration by a 10 petawatt laser system. Integration has been carried out over more than 10^7 laser field cycles. In Fig. 6.6a the gain is shown as a function of the initial phase, as the latter is varied over the 360° range. The gain peaks at $\psi_0 \simeq \pi$, which seems to correspond to a situation of the electron riding with the radiation field and gaining energy maximally from it. Other points in the figure demonstrate maximum slippage and show zero gain. A maximum gain of about 3 GeV from a position of rest at the origin is obtained.

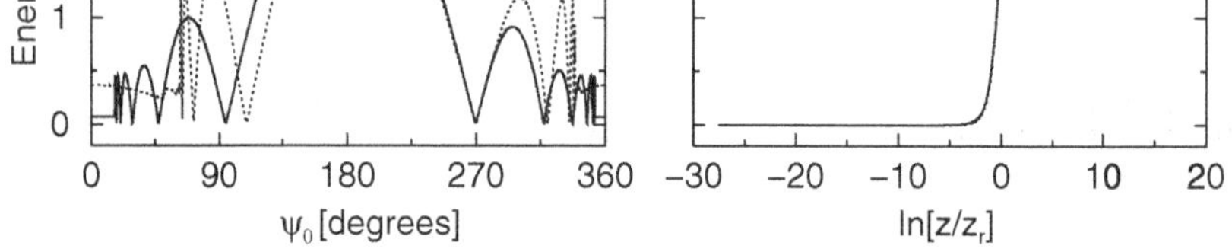

Fig. 6.6. Electron energy gain as a function of: (**a**) the initial phase angle, and (**b**) the logarithm of the axial distance of travel during acceleration (scaled by the Rayleigh length z_r). The laser output power is 10 PW, and the legends in (**b**) apply to (**a**) as well. See caption of Fig. 6.5 for the remaining parameters

The question of how far the electron can go over the time its motion is monitored (i.e., over the integration time interval) is addressed in Fig. 6.6b. The results shown there correspond to the situation of maximum gain (i.e., for $\psi_0 \simeq \pi$). The gain there is shown as a function of the axial distance of travel in units of the Rayleigh length (on a logarithmic scale). Note that the rise from a position of rest at the origin to an energy of 3 GeV is quite sharp and takes place over a small number of Rayleigh lengths. This also corresponds to a timescale equivalent to a few field cycles (a few femtoseconds for the parameter set used). After that, the electron seems to coast for a long time and over a long distance with the gained energy practically undiminished.

Further investigation has revealed that the axial electric field component is solely responsible for the dynamics of the electron subject to the initial conditions considered above. The radial electric and azimuthal magnetic components work to keep the electron motion confined along the beam axis.

References

1. Y. Gontier, M. Trahin, Phys. Rev. **172**, 83 (1968)
2. P. Agostini, F. Fabre, G. Mainfray, G. Petite, N.K. Rahman, Phys. Rev. Lett. **42**, 1127 (1979)
3. A. McPherson et al., J. Opt. Soc. Am. B **4**, 595 (1987)
4. M. Lewenstein, P. Balcou, M.Y. Ivanov, A. L'Huillier, P.B. Corkum, Phys. Rev. A **49**, 2117 (1994)
5. M. Protopapas, C.H. Keitel, P.L. Knight, Rep. Prog. Phys. **60**, 389 (1997)
6. C.J. Joachain, M. Dörr, N. Kylstra, Adv. At. Mol. Opt. Phys. **42**, 225 (2000)
7. H.F. Krause et al., Phys. Rev. Lett. **80**, 1190 (1998)
8. C. Brandau et al., Phys. Rev. Lett. **89**, 053201 (2002)
9. G.A. Mourou, T. Tajima, S.V. Bulanov, Rev. Modern Phys. **78**, 309 (2006)
10. P.A.M. Dirac, Proc. R. Soc. (London) Series A **117**, 610 (1928)
11. C.H. Keitel, Contemp. Phys. **42**, 353 (2001)
12. A. Maquet, R. Grobe, J. Mod. Opt. **49**, 2001 (2002)
13. W. Gordon, Z. Phys. **48**, 11 (1928)
14. C.G. Darwin, Proc. R. Soc. Lond., Ser. A **118**, 654 (1928)
15. D.M. Wolkow, Z. Phys. **94**, 250 (1935)
16. J.A. Fleck, J.R. Morris, M.D. Feit, Appl. Phys. **10**, 129 (1976)
17. R. Heather, Comput. Phys. Commun. **63**, 446 (1991)
18. O. Latinne, C.J. Joachain, M. Dörr, Europhys. Lett. **26**, 333 (1994)
19. J.R.V. de Aldana, L. Roso, Phys. Rev. A **61**, 043403 (2000)
20. R. Fischer, A. Staudt, C.H. Keitel, Comput. Phys. Commun. **157**, 139 (2004)
21. M. Protopapas, C.H. Keitel, P.L. Knight, J. Phys. B: At. Mol. Opt. Phys. **29**, L591 (1996)
22. S.X. Hu, C.H. Keitel, Phys. Rev. A **63**, 053402 (2001)
23. R. Taïeb, V. Véniard, A. Maquet, J Modern Opt. **50**, 365 (2003)
24. U.W. Rathe, C.H. Keitel, M. Protopapas, P.L. Knight, J. Phys. B: At. Mol. Opt. Phys. **30**, L531 (1997)
25. N.J. Kylstra, A.M. Ermolaev, C.J. Joachain, J. Phys. B: At. Mol. Opt. Phys. **30**, L449 (1997)

26. J.W. Braun, Q. Su, R. Grobe, Phys. Rev. A **59**, 604 (1999)
27. M. Casu, C.H. Keitel, Europhys. Lett. **58**, 496 (2002)
28. K. Momberger, A. Belkacem, A.H. Sørensen, Phys. Rev. A **53**, 1605 (1995)
29. Q. Su, B.A. Smetanko, R. Grobe, Opt. Express **2**, 277 (1997)
30. L.N. Gaier, C.H. Keitel, Phys. Rev. A **65**, 023406 (2002)
31. A. Pukhov, Rep. Prog. Phys. **66**, 47 (2003)
32. C. Müller, N. Grün, W. Scheid, Phys. Lett. A **242**, 245 (1998)
33. M. Frigo, S.G. Johnson, FFTW: An adaptive software architecture for the FFT, in *Proc. 1998 IEEE Intl. Conf. Acoustics Speech and Signal Processing*, vol. 3 (IEEE, 1998), pp. 1381–1384
34. G.R. Mocken, C.H. Keitel, J. Comput. Phys. **199**, 558 (2004)
35. L.L. Foldy, S.A. Wouthuysen, Phys. Rev. **78**, 29 (1950)
36. L.C. Biedenharn, L.P. Horwitz, Foundations Phys. **14**, 953 (1984)
37. V. Alonso, S.D. Vincenzo, L. Mondino, Eur. J. Phys. **18**, 315 (1997)
38. For a review, see Y.I. Salamin, S.X. Hu, K.Z. Hatsagortsyan, C.H. Keitel, Phys. Rep. **427**, 41 (2006) and references therein
39. L.S. Brown, T.W.B. Kibble, Phys. Rev. **133**, A705 (1964)
40. A.I. Nikishov, V.I. Ritus, Zh. Éksp. Teor. Fiz. **46**, 776 (1964)
41. A.I. Nikishov, V.I. Ritus, Sov. Phys. JETP **19**(2), 529–541 (1964)
42. J.S. Román, L. Roso, H.R. Reiss, J. Phys. B: At. Mol. Opt. Phys. **33**, 1869 (2000)
43. P. Strange, *Relativistic Quantum Mechanics*, 1st edn. (Cambridge University Press, Cambridge, 1998)
44. S. Zakowicz, J. Math. Phys. **46**, 032304 (2005)
45. J.S. Román, L. Plaja, L. Roso, Phys. Rev. A **64**, 063402 (2001)
46. V. Bargmann, L. Michel, V.L. Telegdi, Phys. Rev. Lett. **2**, 435 (1959)
47. M.W. Walser, C.H. Keitel, J. Phys. B: At. Mol. Opt. Phys. **33**, L221 (2000)
48. J.S. Román, L. Roso, L. Plaja, J. Phys. B: At. Mol. Opt. Phys. **37**, 435 (2004)
49. P.B. Corkum, Phys. Rev. Lett. **71**, 1994 (1993)
50. C.C. Chirila, C.J. Joachain, N.J. Kylstra, R.M. Potvliege, Phys. Rev. Lett. **93**, 243603 (2004)
51. G.R. Mocken, C.H. Keitel, J. Phys. B: At. Mol. Opt. Phys. **37**, L275 (2004)
52. C.H. Keitel, S.X. Hu, Appl. Phys. Lett. **80**, 541 (2002)
53. Y.I. Salamin, F.H.M. Faisal, Phys. Rev. A **54**, 4383 (1996)
54. B.C. Walker et al., Opt. Express **5**, 196 (1999)
55. G.R. Mocken, *Relativistische Quantendynamik in extrem starken Laserfeldern*, PhD Thesis, Universität Freiburg (2005)
56. Y.I. Salamin, C.H. Keitel, Phys. Rev. Lett. **88**, 095005 (2002)
57. S.X. Hu, A.F. Starace, Phys. Rev. Lett. **88**, 245003 (2002)
58. A. Maltsev, T. Ditmire, Phys. Rev. Lett. **90**, 0053002 (2003)
59. J.D. Jackson, *Classical Electrodynamics*, 3rd edn. (Wiley, New York, 1998)
60. S. Quabis, R. Dorn, M. Eberler, O. Glöckl, G. Leuchs, Appl. Phys. B **72**, 109 (2001)
61. R. Dorn, S. Quabis, G. Leuchs, Phys. Rev. Lett. **91**, 233901 (2003)
62. Y. Kozawa, S. Sato, Opt. Lett. **30**, 3036 (2005)
63. Y. Kozawa, S. Sato, Opt. Lett. **31**, 820 (2006)
64. P. Serafim, P. Sprangle, B. Hafizi, IEEE Trans. Plasma Sci. **28**, 1155 (2000)
65. Y.I. Salamin, Opt. Lett. **31**, 2619 (2006)
66. Y.I. Salamin, New J. Phys. **8**, 133 (2006), *ibid*, **10**, 069801 (2008)
67. Y.I. Salamin, Phys. Rev. A **73**, 043402 (2006)
68. I. Bialynicki-Birula, Phys. Rev. Lett. **93**, 020402 (2004)

7

Attosecond Scale Multi-XUV-Photon Processes

Dimitris Charalambidis, Paris Tzallas, Emmanouil P. Benis, and George D. Tsakiris

Abstract. Time domain investigations of dynamic processes involving electronic motion in all states of matter require temporal resolution of the scale of the atomic unit of time (24.1889 as). Intensive efforts in ultra-short radiation pulse engineering led recently to the breakthrough into the sub-femtosecond regime, assuring in the dawn of *attoscience*. Attosecond pulses are synthesized from XUV and/or X-ray waves, with higher order harmonic generation (HOHG) being so far the core process of the synthesizer. Intense attosecond pulse generation and the temporal characterization of such fragile against dispersion objects became a challenge for a number of international research campaigns. The collaboration between MPQ-Garching and FORTH-IESL led to the systematic development of one of the most promising approaches of attosecond metrology, based on pure non-linear XUV processes. The approaches are highly relevant to the temporal characterization of femtosecond and attosecond XUV and X-ray radiation, as well as to pump-probe applications. Hence, in addition to HOHG, they are highly pertinent to the diagnostics and time domain applications of the rapidly developing XFEL sources.

In the present manuscript, we review recent accomplishments of the campaign mentioned above. Two-XUV-photon ionization processes, including total and energy resolved ionization as also direct double ionization, are introduced in the context of non-linear detectors of non-linear autocorrelators for ultra-short XUV pulses. Experimental implementations of such processes, utilizing intense XUV generation, are presented. Promising extensions to two-X-ray-photon innershell ionization are discussed. We further review on the progress made in relevant XUV instrumentation. The design, development and exploitation of two complementary novel dispersionless, broadband XUV/X-ray, non-linear autocorrelators is summarized. Utilizing the achieved non-liner XUV processes and the new instrumentation, the first second order autocorrelation measurement of an attosecond pulse train and herewith the first direct observation of attosecond light localization is demonstrated. This measurement and its detailed through ab initio calculations assessment are further reviewed. The work presented opens up the venue for full four-dimensional investigations in light matter interactions, at extreme scales.

7.1 Introduction

High photon energy, high brilliance radiation sources, such as synchrotron installations, have been proven instrumental in structural studies in several disciplines including physics, chemistry, material sciences, biology and medicine. Recent sound developments of intense, coherent XUV/X-ray radiation sources, with ultra-short pulse duration are for the first time close to offering the possibility of complete four-dimensional (4D) studies in light matter interactions, at the unprecedented sub-Å^3 spatial and sub-fs temporal resolution. These sources may be classified in (a) free electron laser (XFELS) based installations [1,2] and (b) laser based sources utilizing non-linear laser frequency up-conversion processes, such as higher order harmonic generation (HOHG) in gas [3,4] or on solid surface [5–9] targets. XFEL and HOHG sources are at the moment complementary sources, with the XFELs providing today substantially higher brilliances, while HOHG sources hold by far the record in ultra-short pulse durations of 130 as (1 atto $= 10^{-18}$) [10].

As for the spatial resolution, two-dimensional (2D) sub-Å resolution can be obviously reached through near diffraction limited foci of tightly focused X-ray beams, even in linear photo-interactions. Confinement in the third dimension is a novel aspect that is offered by the non-linear interaction of the X-ray radiation with matter, provided that the required intensity is available. With the recently available intensities in the XUV spectral region, reaching values in the range of 10^{12}–10^{16} W cm^{-2} and targeting values as high as 10^{19} W cm^{-2} for the near future, non-linear XUV processes such as two-XUV-photon absorption, made their debut [11–17]. Although some of these works became controversial [18–21], non-linear XUV processes are today a fact in both XFEL and HOHG laboratories, fulfilling at last a long standing quest.

Confinement in the fourth dimension (time) is one of the most impressive developments in the last decade. While the striking success of laser pulse engineers in reducing the laser pulse duration from few picoseconds to a few fs has provided a unique tool for the study of a multitude of ultra-fast dynamical processes, opening up prominent new research areas such as femto-chemistry, several other processes in nature occur within characteristic times of one fs or even shorter. Some examples include the evolution of non-stationary electronic quantum states, charge migration in macro and biological molecules, electron–electron correlation effects in atoms molecules and solids; in the conjugation in aromatic rings or organic macromolecules in quantum confined systems such as quantum wells and quantum dots and in superconductivity; the early stages of energy redistribution in antenna complexes within the photosynthetic reaction centre. Real-time studies of the dynamics of such systems require at first place sub-fs temporal resolution [22,23]. This is the goal that the recently initiated attoscience is striving for. In fact, new and exciting developments during the last few years led to the laboratory breakthrough into the attosecond temporal regime [10,24–29].

Due to the initially low XUV intensities available, the metrology of attosecond pulses invoked smart cross-correlation approaches between the XUV radiation to be characterized and the driving IR field. Examples of such methods are the RABBIT [25,30], the "attosecond streak camera" [31,32] and the CRABFROG [33] approaches. Cross correlation is a powerful tool in short optical pulse metrology, but at the same time notorious in requiring a probe pulse that is considerably shorter than the pulse to be characterized. Otherwise, complicated indirect deconvolution procedures have to be applied based on an appropriate model that correctly describes the cross-correlation process. Since the calculations become intractable when using complete models, the above approaches utilize simplified models. Important effects such as spatiotemporal phase/amplitude distributions [34,35], correct spectral bandwidths, contributions of different electronic trajectories in the generation process and contributions of the IR field to the ionization process are not treated by these models.

In the well-established femtosecond metrology, the most widely used methods rely on a non-linear effect induced solely by the radiation to be characterized. Second order autocorrelation (AC) has been for many years the work horse in pico- and femtosecond laser laboratories, while techniques based on it, such as the FROG [36] technique, allow for nearly full reconstruction of ultrashort pulses. However, the extension of such approaches to sub-fs XUV pulses was far from trivial. The wavelength region below 100 nm is characterized by complete lack of transmitting optical components and even reflective optics is either of low reflectance or narrow bandwidth. This poses a formidable problem in the design of an XUV dispersionless autocorrelator. Interferometric stability and resolution is the next technological challenge in the development of autocorrelators operational at ultra-short wavelengths and ultra-short pulse durations. At the same time the IR/visible to XUV conversion efficiencies, were until recently too low to induce an observable second order process. Nonetheless, overcoming these obstacles is a secure path to rigorous attosecond pulse metrology, which is becoming the target of recent and ongoing research efforts of several research teams. It is worth noting that the experimental and conceptual arsenal of XUV second order AC based pulse metrology underlies concurrently time domain applications of attosecond pulses tackled by XUV-pump-XUV-probe techniques.

In the past 5 years we have systematically reached a number of milestones towards the above goals: (a) Different two-XUV-photon processes induced by HOHG superpositions forming attosecond pulse trains have been conclusively observed. These demonstrations evidently provide the means for the implementation of second order AC measurements of attosecond waveforms and attosecond XUV-pump-XUV-probe experiments in different spectral regions. (b) Dispersionless XUV auto-/cross-correlator instrumentation has been designed, constructed, evaluated and successfully exploited. (c) Utilizing the developments in (a) and (b) the first direct observation of an attosecond pulse train has been achieved by means of a second order XUV-AC. This project has been assessed through ab initio calculations and detailed numerical simulations, including spatiotemporal effects, that resulted in sub-fs pulse durations close to the experimental ones. (d) Very recently, a new approach for the generation of intense isolated attosecond pulses from many cycle pulses has been successfully applied, opening up the way to non-linear processes induced by isolated attosecond pulses. A review of the above accomplishments is presented in this manuscript.

7.2 The HOHG XUV Source

The XUV radiation source utilized in all the experiments illustrated below is a gas HOHG source. A picture of a typical workstation used, including the XUV generation, beam manipulation and application sectors, is shown in Fig. 7.1. Being interested in intense XUV generation, xenon gas is chosen as non-linear medium, as it provides highest conversion efficiency.

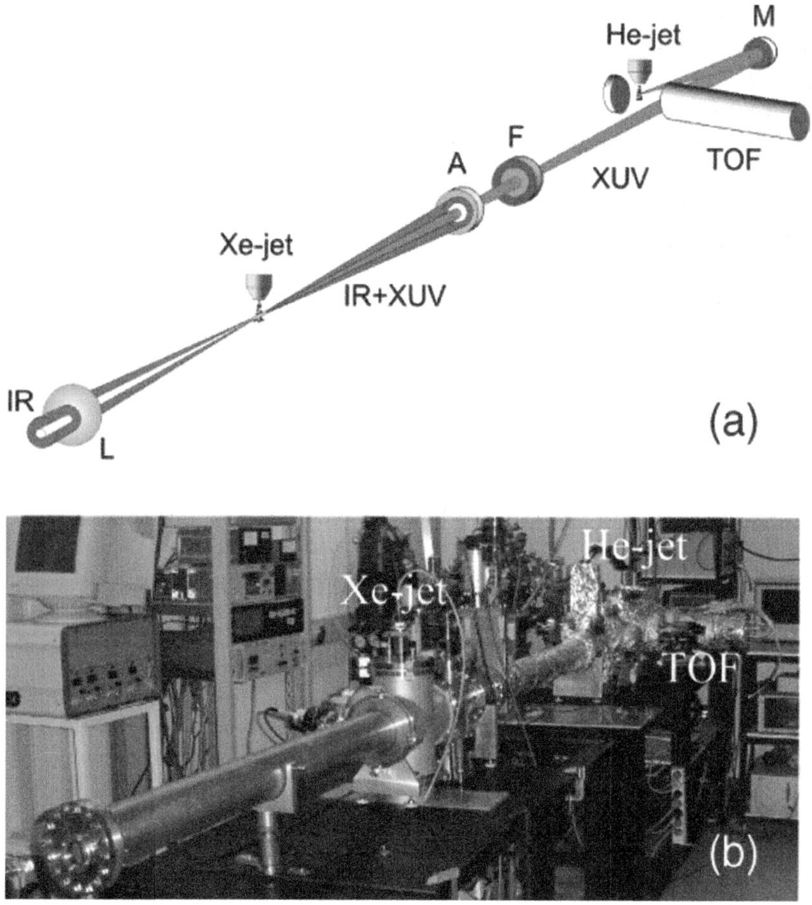

Fig. 7.1. Set-up of the XUV source including the generation, beam manipulation and application sectors. (**a**) L: lens ($f = 3$ m); A: iris; F: filter (In or Sn); M: spherical mirror; TOF: time of flight ion or electron spectrometer. XUV generation occurs at the Xe-jet by an annular IR beam focused before the Xe gas jet. The transmitted XUV radiation is interacting with the gas of a second jet (here He) for pulse diagnostics or applications. (**b**) Picture of the workstation

The laser systems used are 10 Hz Ti:sapphire systems emitting pulses with a central wavelength around 800 nm and durations of either 50 or 120 fs. Loose focusing geometries have been used utilizing 1.5 and 3 m focal lengths. This is a decisive parameter in increasing the interaction volume and thus the XUV photon number generated. Increasing the generating intensity far beyond 10^{14} W cm^{-2} is meaningless because the atomic target is depleted through ionization. In case of excessive laser power availability, looser focusing results in the desirable saturation intensity at an increased interaction volume. The thus increased conversion efficiency can reach values of the order of 10^{-5} for plateau harmonics [37]. For the laser intensities used, harmonics up to the 15th are generated, with an estimated total XUV radiation energy of few hundreds of nJ/pulse. The laser beam focus was placed at $b/2$ before the Xe gas jet, b being the confocal parameter, a position at which theoretical calculations predict optimum harmonic phase locking [38].

An annular IR beam is used for the harmonic generation. After generation the inner part hosts the harmonic emission, while an iris blocks the outer IR part. An appropriate filter selects the group of harmonics of interest and blocks the remaining scattered fundamental. 200 nm thick In and Sn filters have been used in this work selecting the set of harmonics 7–15 and 9–15, respectively. Extension in spectral regions of higher photon energies may be achieved replacing the xenon target by a lighter rare gas jet, i.e., with higher ionization energy [39], in combination with other selection filters, at the cost of the conversion efficiency. As no spatially dispersive elements are used in the apparatus, the harmonics transmitted through the filter are co-propagating. The XUV-beam divergence depends on the divergence of the driving IR field. In the used geometries it is measured to be of the order of 1 mrad and practically constant for the plateau harmonics [40]. A typical harmonic spectrum before and after an In filter is shown Fig. 7.2a. The comb of odd harmonic frequencies is indicative of a sequence of attosecond pulses with a separation of half the laser period [39]. Very recently, applying a novel optical switch approach, that is presented elsewhere [41], we have achieved to reduce the efficient emission to half of the driving laser cycle. The emitted spectrum becomes continuum, like the one shown in Fig. 7.2b.

The obtained single shot continuum spectra indicate emission of intense isolated attosecond bursts, starting from a many-cycle laser pulse. This is a vital development in intense isolated attosecond pulse generation that is presented in [41]

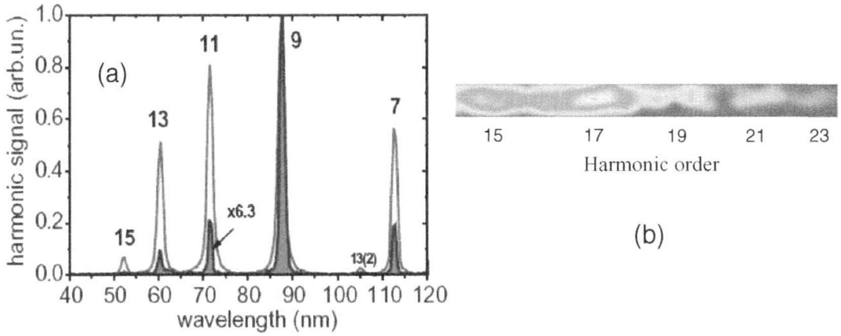

Fig. 7.2. XUV radiation spectra. (**a**) The harmonic "comb". Discrete higher order harmonic generation spectra produced in the Xe jet. The *red-line yellow-shaded spectrum* corresponds to the harmonics as measured without the In filter and the *blue-line green-shaded* to the ones transmitted through the filter. Both spectra are normalized to unity. The corrected relative intensities are 0.32(7th):1.0(9th):0.30(11th):0.11(13th):0.01(15th). (**b**) Quasi continuum XUV spectrum achieved through temporal confinement of the XUV emission process to the central half cycle of the driving many optical cycle field. The generation conditions are similar to those of (**a**)

Table 7.1. The attosecond XUV source

Gas target	Xe
Gas atomic density	10^{15}–10^{17} atoms cm^{-3}
Driving laser power	$\leq$0.5 TW
Driving laser intensity	$\leq 10^{14}$ W cm^{-2}
XUV spectral region of emission	$\leq$24 eV
XUV energy at the generation	~300 nJ
XUV beam divergence	~mrad
Filtered XUV focused intensity	$<10^{14}$ W cm^{-2}
XUV spectrum	Discrete $(2n+1)\hbar\omega_{\text{Laser}}$
XUV pulse duration[a]	(780 ± 80) as

[a]See Sect. 7.5

The transmitted XUV radiation can be steered by a removable metal mirror towards an XUV monochromator or be focused by spherical gold or tungsten (for increased reflectivity in the spectral region of interest) coated mirrors into a second jet for diagnostics or non-linear XUV applications. The products of the interaction of the XUV radiation with targets of interest are detected by (a) ion mass spectrometry, (b) energy (and angularly) resolved photoelectron spectroscopy by means of a TOF or an imaging spectrometer and (c) frequency or (d) time domain spectroscopy XUV facilities, attached to the source. Details about the diagnostic instrumentation will be given in Sect. 7.4.

The parameters of the presented XUV source are summarized in Table 7.1.

7.3 Non-Linear XUV Processes (As Non-Linear XUV Detectors)

The indisputable observation of non-linear XUV processes is pivotal to attosecond pulse metrology. In this section we summarize successful experimental efforts in demonstrating two-XUV-photon processes induced by superpositions of harmonics, appropriate for the implementation of second order AC measurements of as pulse trains.

7.3.1 General Considerations-Cross Sections-Rates

Non-linear interaction of XUV radiation with matter leads to ionization due to the large photon energy. The character of non-linear ionization of atoms is governed by the so called adiabaticity or Keldysh parameter

$$\gamma = \sqrt{\mathrm{IP}/2U_{\mathrm{P}}}, \tag{7.1}$$

where IP is the ionization potential and $U_{\mathrm{P}} = e^2E_0^2/4m\omega^2$ the ponderomotive energy, where e, m are the electron charge and mass and E_0, ω the field amplitude and frequency, respectively [42]. If $\gamma >> 1$ multiphoton ionization dominates, while if $\gamma << 1$ the interaction is quasi static and ionization is pronominally through tunneling. In the XUV spectral region the interaction is in the multiphoton regime unless the field intensity exceeds 10^{16} W cm^{-2}, an intensity not available yet in XUV laboratories. The multiphoton ionization rate, well below its saturation is given by:

$$\dot{N} = N\sigma^{(n)}F^{n}(t), \tag{7.2}$$

where N is the number of atoms in the interaction volume V, F the photon flux and $\sigma^{(n)}$ the generalized n-photon cross section. For a second order process $\sigma^{(2)}$ has small values ranging from 10^{-49} to 10^{-52} cm^4 s. Thus, for a realistic atomic density n_a (e.g., 10^{15}–10^{16} atoms cm^{-3}) and interaction volume, a non-saturated observable two-photon ionization yield

$$N_{\mathrm{ION}} = n_{\mathrm{a}}V\sigma^{(2)}\int_{-\infty}^{\infty} F^2(t)\mathrm{d}t \tag{7.3}$$

requires high XUV intensities, well exceeding 10^8 W cm^{-2}. This is why the observation of two-XUV-photon processes has been an unreachable target of intense experimental efforts for about two decades.

7.3.2 Two-XUV-Photon Ionization by an Attosecond Pulse Train

The first demonstration of a two-XUV-photon ionization (TXUVPI) of an atom by an attosecond pulse train has been achieved in 2003 [14]. He atoms have been ionized by a superposition of the 7th to the 15th harmonic. The ionization scheme, as shown in Fig. 7.3, consists of a non-resonant one harmonic photon absorption below the ionization threshold, followed subsequently by the absorption of a second photon of any of the harmonics available. The key parameter in these experiments has been the above discussed loose focusing conditions in the harmonic generation region. The thus increased interaction volume, leads to the required for a non-linear process XUV intensities.

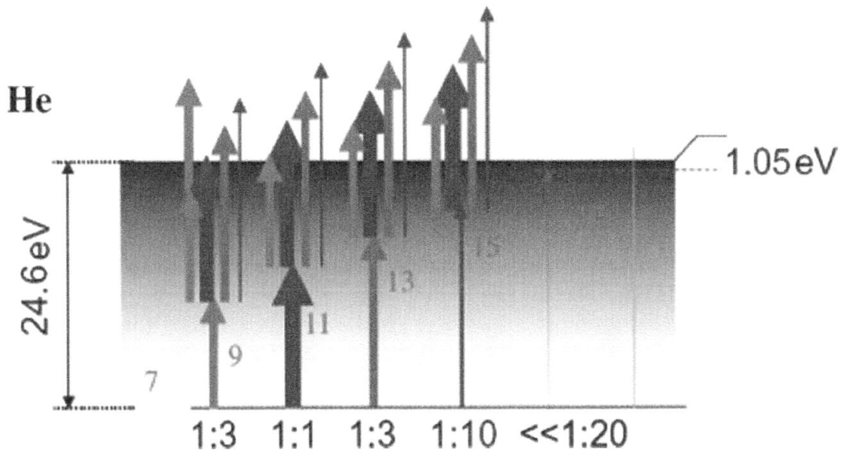

Fig. 7.3. Ionization scheme of He by the harmonic comp. The ionization occurs through non-resonant two-XUV-photon absorption. The ratios given are the relative harmonic intensities

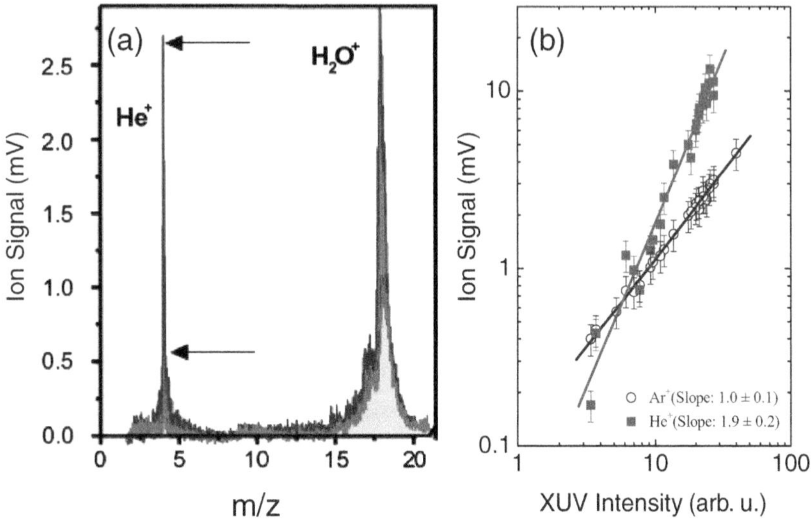

Fig. 7.4. He two-XUV-photon ionization by a superposition of harmonics. (**a**) Time of flight ion mass spectra measured at two different intensities. The *blue-cyan* shaded spectrum corresponds to a harmonic intensity twice as high as the one which is *red-yellow shaded*. The different intensity dependence of the He^+ and H_2O^+ yields is because the ionization of He is a two photon process, while H_2O has a much lower ionization energy and thus ionizes through one photon absorption [40]. (**b**) Ion yield dependence on the ionizing intensity. In log–log scale the slope for He is essentially 2 because He ionization is a second order non-linear process. Ar ionization is a linear process as it requires absorption of only one photon and thus the measured slope is one. The measured slopes provide the evidence of the observation of a two-XUV-photon ionization of He. Note that XUV intensity has been assumed to be proportional to the measured single-photon XUV ionization H_2O^+ signal [43]

He atoms have been detected by an ion time of flight mass spectrometer as shown in Fig. 7.1. Since He cannot be single photon ionized by this specific combination of harmonics, the presence of He ions in the mass spectra is indicative of the TXUVPI process. Further evidence is provided through the slope of the log of the ion yield dependence on the log of the ionizing intensity, below saturation. As it becomes apparent from Eq. (2) the slope for the TXUVPI is two. A number of experiments under different experimental conditions in the two collaborating laboratories have confirmed TXUVPI by the attosecond pulse train. Typical mass spectra including He ions and a measurement of the XUV intensity dependence of He ionization are shown in Fig. 7.4a, b, respectively. These results have established the platform for the breakthrough of the temporal characterization of attosecond pulse trains via the second order AC technique that followed [27].

7.3.3 Energy Resolved Two-XUV- Photon Ionization

An additional evidence of the observation of TXUVPI is provided through energy resolved photoelectron spectroscopy. The evidence comes from the position of the photoelectron peaks. At the same time, energy resolved TXUVPI is the basic requirement for the extension of a second order AC measurement to a FROG-like [36] approach in the XUV spectral region. This will allow for a detailed and rigorous temporal reconstruction of as waveforms, as well as of XFEL radiation, while the second order AC reveals simply averaged pulse durations.

Very recently we have succeeded in recording the photoelectron spectrum of the TXUVPI of He by an as pulse train, i.e., of the superposition of the 9th, 11th, 13th, and 15th harmonics [43]. In this experiment the TOF spectrometer was operated in its photoelectron energy mode, instead as a mass spectrometer. Figure 7.5 depicts measured (Fig. 7.5a) and calculated (Fig. 7.5b) spectra, where the harmonic combination channels $9 + 13 \oplus 11 + 11$ and $9 + 15 \oplus 11 + 13$ corresponding

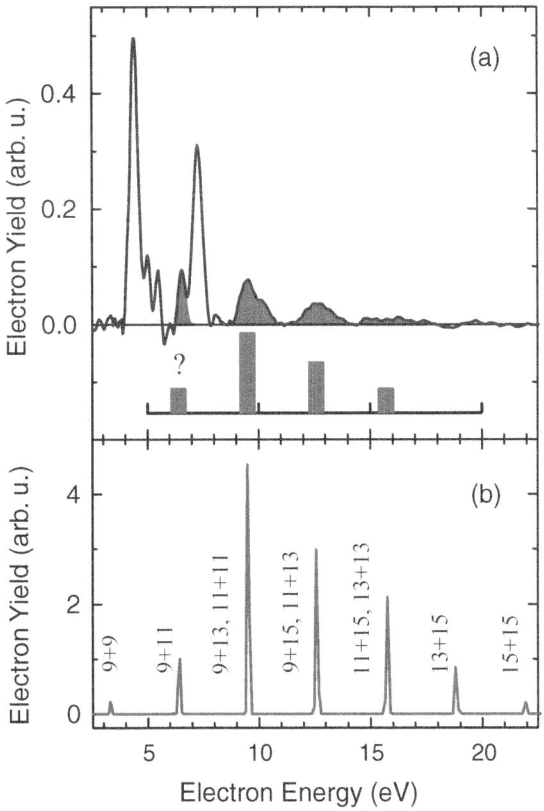

Fig. 7.5. TXUVPI energy resolved photoelectron spectra. (**a**) The measured spectrum of two-XUV-photon ionization of He is depicted by the shaded areas. The two peaks at energies 4.4 and 7.5 eV correspond to the single-photon ionization of Ar by the 13th and 15th harmonics, respectively. The red bars depict the integral of the ion mass peaks. (**b**) Theoretical energy photoelectron spectrum obtained by solving the He TDSE in the XUV field formed by the superposition from the 9th to the 15th harmonic. Harmonic phases were obtained by the three-step rescattering model for the laser intensity of 7×10^{13} W cm^{-2} used in the experiment

to photoelectron energies 9.5 and 12.6 eV, respectively, are clearly observed (peaks in red). There is also indication about the observation of two more channels, the $9 + 11$ and $11 + 15 \oplus 13 + 13$. The two additional peaks at energies 4.4 and 7.5 eV correspond to the single-photon ionization of Ar contaminants by the 13th and 15th harmonics, respectively, and were used for the TOF self-calibration.

Ab initio calculations by Nikolopoulos [35], based on a two electron basis expansion technique, have been utilized in numerically solving the six-dimensional (three for each electron) He time dependent Schrödinger equation (TDSE) including the electric dipole interaction with the field of the harmonics superposition. The measured (Fig. 7.5a) and calculated (Fig. 7.5b in green) photoelectron peak height distributions were found in very good agreement. As discussed above, this measurement is a decisive step towards FROG-like measurements. Here it should be noted that towards this goal, the energy resolution in the photoelectron spectra has to improve substantially (~10 meV) if attosecond pulse trains with fs envelopes are to be characterized. However, for the extremely broadband radiation of isolated attosecond pulses [10, 28], the available resolution is sufficient. A partially energy resolved second order AC study by Nabekawa et al. [29], nevertheless not a full FROG type measurement, has been reported almost concurrently with our work.

7.3.4 Direct Double Ionization

As our main motivation in investigating non-linear XUV processes is the development of precise and at the same time uncomplicated approaches for the temporal characterization of ultra short XUV pulses, and their XUV-pump-XUVprobe applications, it should be pointed out that TXUVPI provides such a tool as long as the spectrum of the radiation to be characterized is restricted to frequencies up to the ionization frequency of the ionizing atom. Otherwise single photon absorption causes ionization that is overwhelmingly stronger than the TXUVPI. As the highest atomic ionization energy is that of He with 24.6 eV, TXUVPI of He supports characterization of the broadest spectral distribution and thus of the shortest pulse durations than any other atom (or molecule). For beams involving photon energies larger than 24 eV an alternative non-linear process has to be used. A promising candidate is the so called direct double ionization.

For photon energies smaller than the double ionization energy, double ionization may occur either (a) sequentially, where the atom first ionizes leaving the ion in its ground or in an excited state, which subsequently doubly ionizes absorbing further photon(s) or (b) directly, where absorption of a number of photons ejects two electrons, which share the excess energy, ones reaching the double continuum. In the direct process the ejection of the two electrons is almost simultaneous but not exactly differing by few tens of as. Thus direct double ionization is a dynamic process the temporal investigation of which requires as resolution. In the multiphoton regime the sequential process is usually the dominant process, precluding the observation of the direct process. In the non-perturbative tunneling regime, the pioneering experiment by Walker et al. [44] put a clear signature of the direct double ionization process, opening up a series of electron–electron correlation studies in strong field double ionization [45–49]. In two-XUV-photon direct double ionization (TXUVPDDI) there are photon energy windows for which the sequential process is a three photon process, while the direct process is a two photon one.

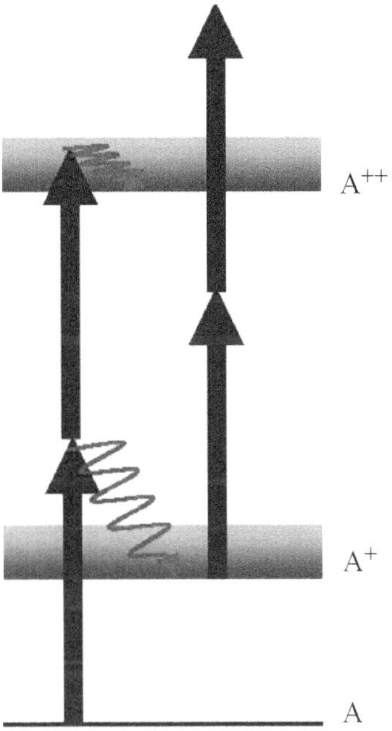

Fig. 7.6. Atomic two-photon double ionization. For photon energies satisfying the condition $\mathrm{IP} \leq \hbar\omega \leq \mathrm{IP}_2 - \mathrm{IP}_1$, IP_1, IP_2 being the single and double ionization energy, respectively, the direct process is a two-photon process, while the sequential becomes a three-photon process. Direct double ionization becomes then the dominant process for a certain range of ionizing intensities

For not too high intensities, the direct process becomes then the dominant one [50]. The scheme is illustrated in Fig. 7.6. The condition for the photon energy reads obviously

$$\mathrm{IP} \leq \hbar\omega \leq \mathrm{IP}_2 - \mathrm{IP}_1, \tag{7.4}$$

IP_1, IP_2 being the single and double ionization energies, respectively. Typical TXUVPDDI cross sections are 10^{-51}–$10^{-52}\,\mathrm{cm}^4\,\mathrm{s}$. The temporal evolution of the system as it results from the contributions of different competing channels can be evaluated at different intensities through rate equations [50,51]. Exploiting such an ionization scheme, evidence for TXUVPDDI of Ar and Kr, ionized by intense superpositions of the 11th, 13th and 15th harmonics, has been established [52]. The coupling schemes together with the XUV intensity dependence of the Ar^{2+} and Kr^{2+} are shown in Fig. 7.7a, b. The insets in Fig. 7.7b depict typical ion spectra. The Ar^{2+} and Kr^{2+} peaks are very small but clearly observable. The intensity dependence slopes are found close to two, indicative for a second order non-linearity. The observed double ionization can be assigned to the direct process [52] except maybe for the highest intensities used, where the sequential process may also contribute to the double ionization yield due to the depletion of the atomic ground state. However, in these cases the sequential process depicts also a second order non-linearity (two-photon ionization of the ion) and thus can be exploited in the implementation of second

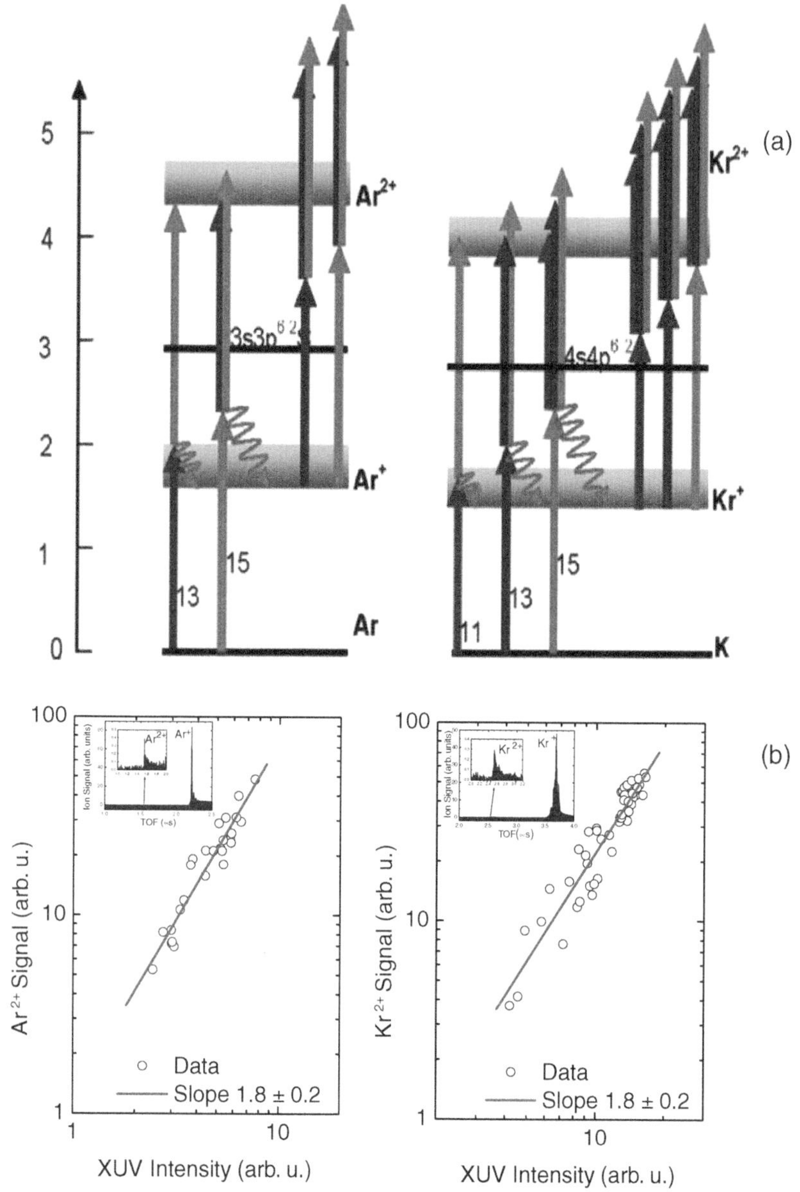

Fig. 7.7. Two-XUV-photon direct double ionization of Ar and Kr by a superposition of harmonics. (**a**) Ionization scheme. For the superposition of the 13th and 15th harmonic the sequential is a three photon, while the direct a two photon process in Ar. The same holds for the 11th, 13th and 15th harmonic in Kr. (**b**) Ion yield dependence on the ionizing intensity. The measured slope is indicative of a two photon process. The insets depict the measured ion spectra. Note that XUV intensity has been assumed to be proportional to the measured single-photon XUV ionization Ar^+ and Kr^+ signals, respectively

order AC measurements. These demonstrations indicate that second order AC runs may be extended up to 54 eV in He TXUVPDDI. Further extension to higher photon energies can be for certain energy intervals achieved through measurements of two-XUV-photon direct multiple ionization (triple, quadruple, etc.). Finally it should be noted here that alternatively two-XUV-photon ATI [29] allows continuous extension of the method to shorter wavelengths. Photoelectron spectroscopy is then mandatory and the process has a rapidly decreasing with photon energy cross section ($\sim(\hbar\omega)^{-5}$ for H atoms).

7.3.5 Extensions Towards the X-Ray Regime

An extension of the above described approaches in the X-ray spectral regime would be in principle possible through inner-shell multiphoton processes. This is a new unexplored area of non-linear physics, for which parameters are not known. However, realistic estimations can be made for the two-photon-ionization of the K-shell, which is known for H-like ions of Z nuclear charge

$$\sigma^{(2)}(Z,\omega) = \frac{1}{Z^6}\sigma^{(2)}\left(Z=1, \frac{\omega}{Z^2}\right). \tag{7.5}$$

Figure 7.8 shows K-shell generalized two-photon ionization cross sections for H-like ions (blue curve) for photon energies for which ionization is close to the threshold. The corresponding photon energies are shown with the red line. The light blue strip depicts the range relevant to existing and forthcoming XUV/X-ray sources such as HOHG, XFEL sources and the planned ELI European Research Infrastructure

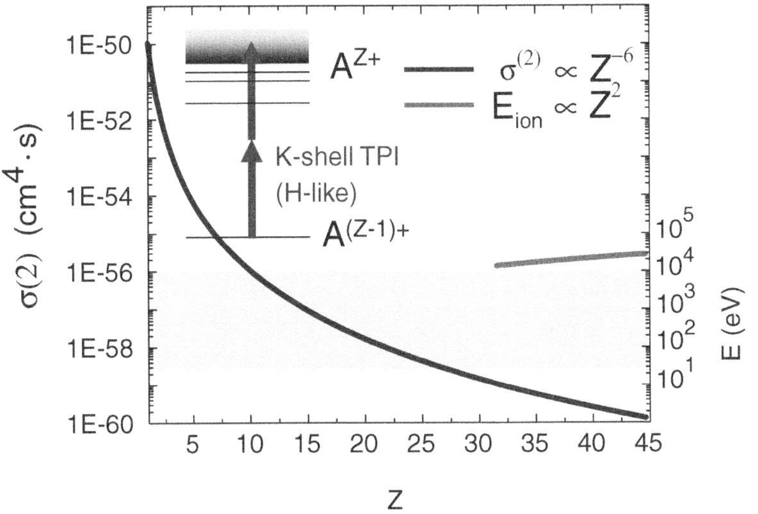

Fig. 7.8. Z dependence of the non-resonant two-photon ionization (TPI) of H-like ions. Photon energies are such that ionization is just above threshold. The *blue curve* depicts the generalized two-photon cross section. The *red line* is the photon energy of the two-photon ionizing radiation. The blue strip depicts the relevant region for existing and near future intense XUV/X-ray radiation sources

(see http://www.extreme-light-infrastructure.eu/). Required intensities for an observable signal would then range from 10^{12} to 10^{16} W cm^{-2}, for photon energies 0.3–10 keV and atomic targets with $Z = 5$–40. It is expected that the cross section will not be significantly affected by the presence of additional electrons either in the K-shell or in outer shells. They introduce though competing single photon ionization channels, which depending on the conditions might be the dominant ionization processes. This complication may be surmounted in different ways: (a) Measuring energy resolved photoelectron instead of ion yields. (b) For ultra short pulses with pulse duration of ∼50 as the two-X-ray K-shell ionization is estimated to become the dominant process at intensities between 10^{16} and 10^{17} W cm^{-2}. Such X-ray beam parameters have been predicted for surface harmonics produced at relativistic driving laser intensities [7]. (c) For pulses in the fs regime, the single-X-ray-photon process will dominate up to the saturation of ionization. Nevertheless, if photon energies are such that the ionization of the H-like ion is a two-X-ray-photon process, while all other ionization events are through a single photon process, a two photon process will be eventually observed detecting the bare nuclei. At such photon energies the atom is stripped mainly from inside. K-shell hole creation followed by the subsequent Auger decay, will be repeated until a H-like ion is formed, which will finally two-photon ionize. In this sequential multiple ionization the entire stripping sequence may be completed within the rising edge of the X-ray pulse. For 100–150 fs long pulses with photon energies 0.3–1.5 keV, atoms with $Z = 6$–10 will undergo such type of complete stripping at peak intensities in the range of 10^{16}–10^{17} W cm^{-2}.

7.4 Dispersionless XUV Auto- and Cross-Correlators

A key parameter in XUV attosecond metrology is the development of dispersionless interferometers to be used as non-linear correlators. The complete lack of refractive materials in the XUV spectral region sets a formidable barrier. The development of broadband XUV beam splitters became thus an alluring challenge. In the quest of non-dispersing set-ups for time domain measurements in the XUV spectral range, we have designed, developed and utilized two types of autocorrelators. The non-linear "split-mirror" volume autocorrelator and the "transmission grating" Michelson interferometer.

7.4.1 The Non-Linear Split-Mirror Volume Autocorrelator

This is a wavefront splitting arrangement [27, 53–55], consisting of a spherical mirror (SSM) cut into two halves (see Fig. 7.9), which is used to focus the harmonic radiation in the He gas jet. One of the halves can be translated by a piezo-crystal translation with a resolution of few nm over a total travel of about 10 μm. The two parts of the XUV radiation are brought into a common focus in the gas jet and can be delayed with respect to each other. The ionization products are analyzed by a TOF mass spectrometer. Assuming that the generated XUV radiation is cylindrically symmetric the measured TXUVPI signal as a function of the delay represents a second-order autocorrelation. For a given wavelength, a displacement of $D = \lambda/2$ between the two half mirrors divides the focal spot for zero displacement (Airy spot) with peak intensity $I_{\max}$ into two of equal size (see inset in Fig. 7.9). Since the total

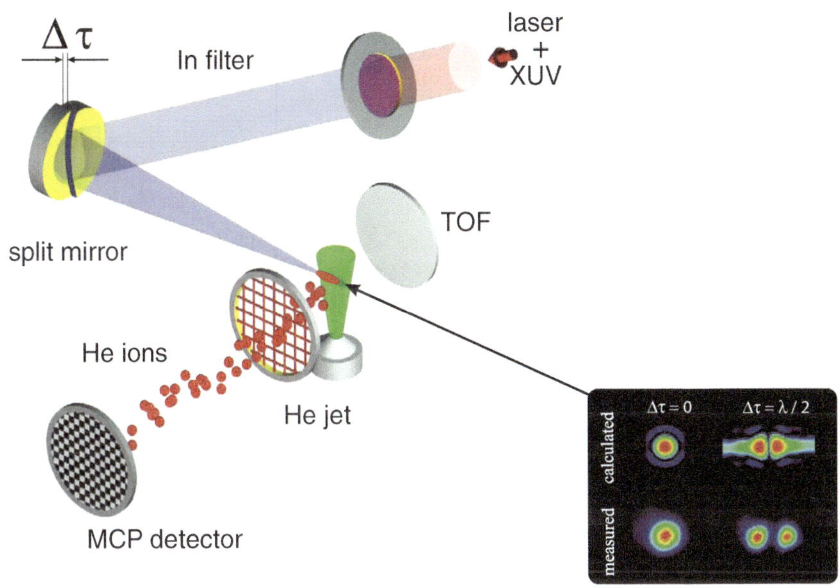

Fig. 7.9. The split mirror volume autocorrelator. At the focus there is an intensity redistribution as a function of the displacement of the two halves. This is shown for a monochromatic wave in the inset. For $\lambda/2$ displacement the Airy distribution is split into two distributions of equal intensity both in the experiment and the calculation. Assuming cylindrical symmetry the ionization yield measured as a function of the delay between the two wavefronts, corresponds to a second order autocorrelation measurement

energy in the interaction volume remains constant, the splitting of the focal spot would not affect the measured signal if the detector were linear, i.e., no modulation of the signal with the delay would be observed. For a non-linear (quadratic) detector instead, the rearrangement of the local intensity inside the interaction volume according to diffraction would produce a modulation. For a harmonic superposition consisting of four harmonics with equal amplitudes (7th, 9th, 11th, and 13th) the field distribution $E(D,x,y,z) = \sum_{n=7}^{13} E_n(D,x,y,z)$ at the focal spot of our spherical mirror has been calculated for various displacements (delays). The intensity distribution $I(D,x,y,z) = E(D,x,y,z)\ E^*(D,x,y,z)$ for three particular values of the displacement $D = 0$, $\lambda/4$ and $\lambda/2$ in the perpendicular and parallel to the propagation direction are shown in Fig. 7.10a.

The ionization signal $S(D) \propto \int\int\int_{\Delta V} I^2(D,x,z)\mathrm{d}x\ \mathrm{d}y\ \mathrm{d}z$ that results from the entire interaction volume ΔV, is depicted in Fig. 7.10c vs. delay. The blue curve is the interferometric equivalent of the volume autocorrelation for the superposition of the four harmonics. Comparing this result to that of a conventional planar interferometric auto-correlation, given by $S(\tau) = \int_{-\infty}^{+\infty} \left|[E_{\mathrm{T}}(t) + E_{\mathrm{T}}(t-\tau)]^2\right|^2 \mathrm{d}t$ where E_{T} is the total electric field of the harmonic superposition, the volume autocorrelation exhibits the same structure, but due to coarse sampling not all the peaks within

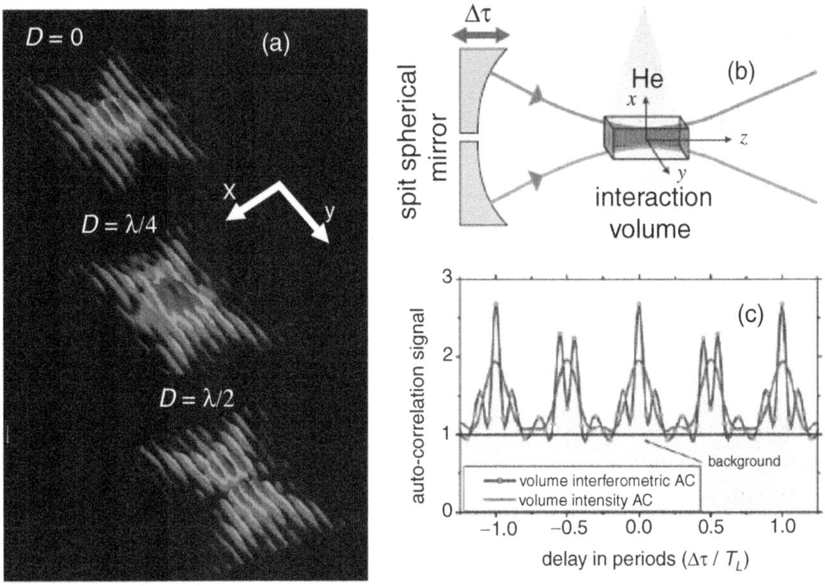

Fig. 7.10. The XUV volume autocorrelator. (**a**) Two-dimensional (x–z) plane color contours depicting the calculated intensity distribution of the harmonic superposition at the interaction region of the volume autocorrelator for the indicated mirror displacements. (**b**) Split mirror and interaction geometry. (**c**) Calculated interferometric (*blue line*) and intensity (*red line*) volume AC traces. T_{L} is the period and λ the wavelength of the laser

a period are visible in Fig. 7.10c. The main difference is the peak to background ratio, which instead of 8:1 is only ~2.75 : 1 for the volume autocorrelation. The corresponding intensity autocorrelation is obtained by filtering the fast oscillations (red line in the same graph Fig. 7.10c), for which the peak to background ratio from 3:1 (conventional) is reduced to ~2 : 1 (volume). The technique described is a unique dispersion-free characterization approach for XUV radiation waveforms of HOHG or XFEL based sources. As a second order AC it reveals the duration of the pulses within the approximations inherent in the second order AC technique. Substitution of the ion detection with high resolution energy resolved photoelectron spectroscopy, FROG type retrieval of spectral phase/amplitude distributions and hence pulse reconstruction, will be possible.

7.4.2 The Transmission Grating Interferometer

The second dispersionless interferometer developed is a Michelson type interferometer arrangement in which the beam splitter/combiner is a free standing transmission grating (Fig. 7.11). Rays of the same frequency split through zero and first order diffraction are retro-reflected and recombined at the grating through first and zero order diffraction, respectively. Correlation measurements can be performed by translating one of the two spherical mirrors (M1 or M2). For the ideal case of a single ray

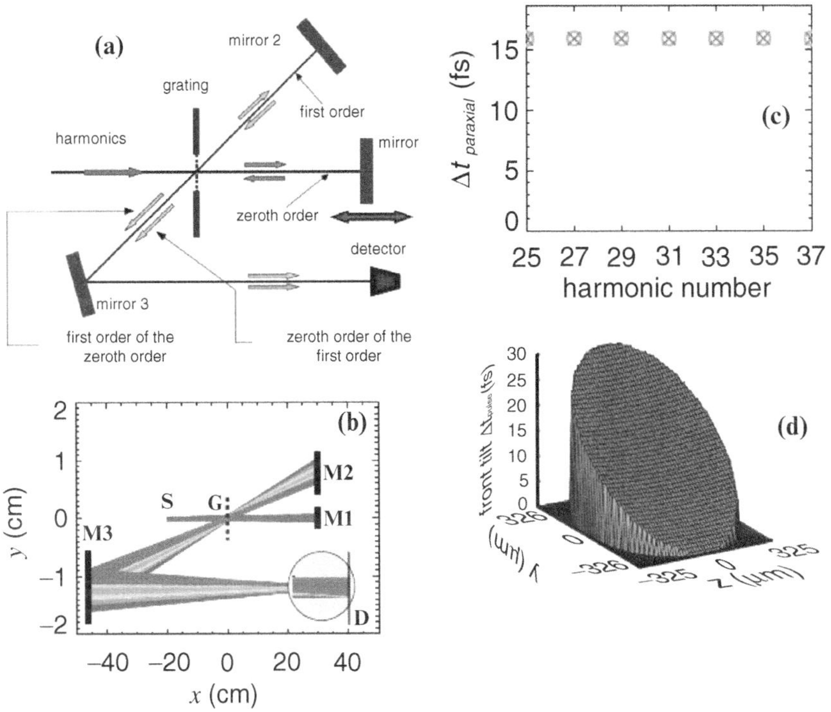

Fig. 7.11. The transmission grating interferometer. (**a**) The Michelson interferometer for operation in XUV spectral range using a free-standing transmission grating as a beamsplitter. (**b**) Optical paths of the 25th–37th harmonics for both arms (zero-first and first-zero order). S: source; G: grating and M1; M2, M3: mirrors. In the inset, a magnification is shown of the image of the grating for each harmonic at the detector plane D. (**c**) The difference in arrival time between the chief ray and the extreme paraxial ray for the harmonics 25–37. The symbols denote the two arms (circle: zero-first; cross: first-zero). (**d**) The pulse front for the central harmonic (31st) at the detector plane as calculated with the OPTICA package

(Fig. 7.11a) it becomes immediately apparent that equal optical paths guarantee dispersionless operation. For the realistic case of extended and diverging beams ray tracing analysis [56] has shown that by imaging the grating on the detector (Fig. 7.11b), the set up is dispersionless down to the 1 as regime (Fig. 7.11c, d).

The set up has a flat spectral response for a very large energy range, with lower and upper limits set by the grating constant and the absorption edge of the grating material, respectively. At the same time it allows for wavelength selection using slits or knife edges in the parts where the radiation is spatially dispersed. A serious drawback is its lower than 1% throughput, due to the double diffraction at the grating. Thus this arrangement is only appropriate for cross-correlation measurements of higher harmonics with the IR fundamental frequency or for non-linear AC measurements of intense XUV radiation like that foreseen for the advanced XUV/X-ray sources based XFELs or relativistic surface plasma harmonics [7]. The advanced properties of the apparatus have been so far demonstrated in: (a) performing a second order AC of a fs third harmonic field [57] and (b) the spectral phase distribution retrieval and pulse reconstruction of third harmonic fields [58], as well of an arbitrary lower order harmonic superposition [59] through an all optical version of a harmonic-IR cross correlation technique, that we have some years ago developed [60].

7.5 The Second Order Autocorrelation of an Attosecond Waveform

7.5.1 The Experiment

Utilizing the apparatus of Fig. 7.9 described in Sect. 7.4.1 the first genuine second order AC measurement of a train of XUV pulses, offering direct observation of attosecond temporal light confinement, has been realized. The results are summarized in Fig. 7.12. In particular, the graph in the upper panel gives the trace of the ion signal for a total temporal overlap near the peak of the laser pulse of ~18 fs in steps of ~37 as. Grey dots are averages of 10 data points laser shots, while the yellow points with errors bars the running average and standard deviations over 15 data points for the ion yield. A distinct modulation comprising two peaks within an optical cycle of the fundamental frequency is clearly perceived. The double peak per laser period is the result of having the driven electron revisiting twice per optical cycle the ion where it recombines emitting an XUV photon. The data averaging procedure we have used yields the equivalent of the intensity AC trace.

Following common practice of the second order AC technique, the mean individual pulse duration in the attosecond train is extracted from the trace by fitting from a sequence of Gaussian pulses. The red curve is the best fit and the individual Gaussian pulses are shown in light-blue color. This procedure has resulted to a 780 ± 80 as mean pulse duration, which is approximately twice the Fourier transform limited (FTL) value for the given bandwidth.

7.5.2 Theoretical Evaluation of the Approach

Attosecond metrology involves so far particularly delicate approaches, the validity and limitations of which should thus be subject to rigorous assessment. In particular the strong deviation of the measured pulse duration from its FTL value, described in the previous section, enhances the necessity of a detailed assessment of the approach employed.

The non-linear detector of a second order AC must fulfill the following two requirements. Its spectral response has to be practically constant and its temporal response instantaneous. For the broadband radiation (several eV if not tens of eV) of attosecond pulses and their ultra-short duration these conditions are not a priori fulfilled. These two aspects have been in depth analyzed through ab initio calculations performed by Nikolopoulos [35], solving the three-dimensional He time dependent Schrödinger equation (TDSE) [14, 61]. In this study the sufficient spectral response flatness has been established. The most important result is though the comparison of an interferometric second order AC trace as it results from the solution of the TDSE for a variable delay between the two autocorrelation XUV fields, with that resulting from a conventional ideal autocorrelator. The harmonic fields used have

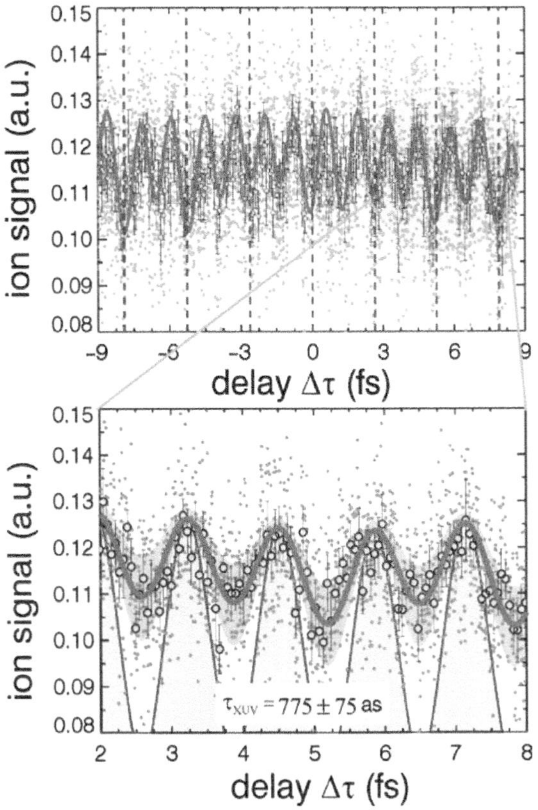

Fig. 7.12. The second order XUV measurement. Measured non-linear volume AC traces. The trace of the upper panel is a measured 18 fs long second order volume intensity AC trace of the superposition of the five harmonics 7th–15th, while the trace of the bottom panel a blow-up of the *green-shaded region* covering a delay range of about three laser cycles. The grey points are the measured data. The *yellow points* are their average and the *error bars* shown correspond to one standard deviation. The *red curve* is the best fit of an analytical function given by the sum of a sequence of Gaussian distributions. The *blue lines* in the expanded area are the individual (not superimposed) Gaussian distributions

amplitudes equal to those of the experiment and relative phases equal to zero, i.e., FTL pulses with a width of 315 as. The two AC traces sown in Fig. 7.13 are strikingly identical, indicating that the measurement is not at all distorted by the properties of the detector. An accurate extension of the method to other spectral regions through different two-XUV-photon processes requires a similar assessment.

Having assessed the performance of the non-linear volume autocorrelator, the origin of the strong departure of the measured pulse duration from its FTL value has been explored. It is found that this departure can be attributed to the phase difference $\Delta\varphi q$ between subsequent harmonics q, inherent to the dynamics of the harmonic emission process [30]. A further important parameter affecting the attosecond pulse

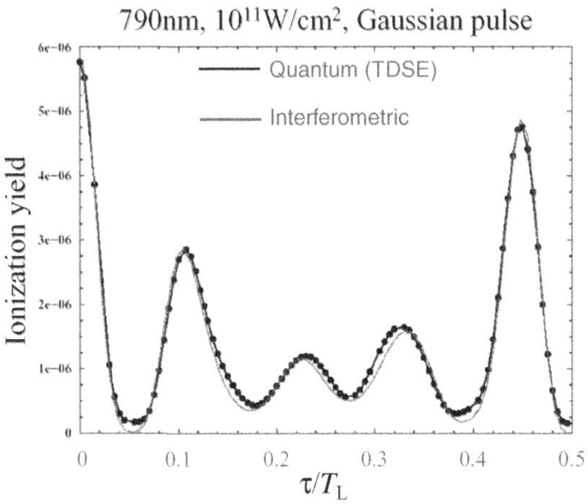

Fig. 7.13. Calculated second order AC interferometric traces. T_L is the laser period and τ the delay. *Black dots* are calculated by solving the TDSE for He interacting with two delayed replicas of the superposition of harmonics 7th–13th with no initial phase difference. The *red line* is the conventional AC trace $S(\tau) \propto \int_{-\infty}^{+\infty} (E_1(t) + E_2(t-\tau))^4 \, \mathrm{d}t$ of a perfect detector. $E_{1,2}$ are the two replicas of the harmonic superposition

duration is the dependence of the harmonic phase on the driving intensity. Since the driving laser field has a spatiotemporal intensity distribution, there is a spatiotemporal modulation of the harmonic phases and thus of the temporal characteristics of the as train. The radial modulation of harmonic phases causes at the same time a bending of the attosecond pulses fronts. An estimation of the attosecond pulse front shapes in our experiment is shown in Fig. 7.14.

This estimation relies on the three step model [39,61,62], using a Gaussian driving intensity distribution with the experimental peak intensity of 0.6×10^{14} W cm^{-2}. For driven electron dynamics, classical equations of motion have been used [61]. The model does not provide any information about the 7th harmonic, the photon energy of which is below the ionization energy of the generating Xe gas and thus an extrapolated phase value has been used for it. The upper panel of Fig. 7.14 depicts three temporal pulse sequences resulting from three different driving laser intensities across the radial laser beam profile. The corresponding radial positions are shown in the lower panel with straight color lines. The radial variation of the temporal profile of the as train is apparent. The spatially integrating autocorrelator detector "sees" the sum of the radial autocorrelation ion signals. The resulting burst duration is ~680 as. Inclusion of temporal integration increases this value to ~710 as. Due to the uncertainty in the phase of the 7th harmonic the above estimated durations have a large uncertainty with a lowest limit of 550 as in the averaged trace.

These estimations indicate that the measured pulse duration of (780 ± 80 as) is in good agreement with the expected XUV pulse duration, resulting from the atomic response taking into consideration the spatiotemporal intensity variation of

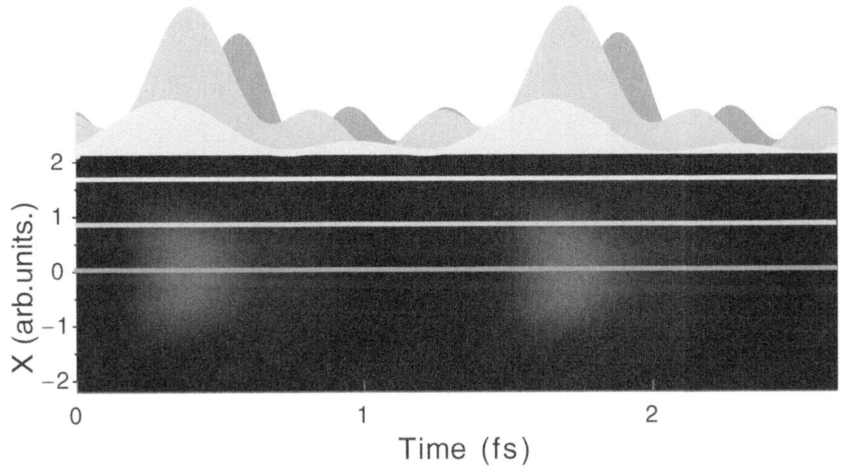

Fig. 7.14. Classically calculated radial intensity distributions of the attosecond train depicting the bending of the pulse fronts. The upper panel shows three attosecond pulse sequences corresponding to the three different driving laser intensities across the radial (x) laser beam profile (*green*: centre; *light blue*: middle; *yellow*: outer). The corresponding radial positions are shown with the *color straight lines*. The radial modulation of the temporal characteristics of the train is clear. Spatiotemporal driving intensity distributions affect the effective temporal characteristics of the attosecond pulses

the generating laser field. Propagation may possibly modify the result of the above analysis. Propagation effects for optimized phase matching conditions are taken into account by excluding the "long" trajectory due to the focusing conditions in the Xe jet [38]. Our result indicates that under these experimental conditions and for the set of harmonics used, spatiotemporal effects influence the pulse duration. In contrast, in approaches that measure spatiotemporally averaged spectral phases, like those using XUV-IR cross-correlations [25,30,58] the same effect results in a reduced phase difference between subsequent harmonics and therefore, to a shorter measured burst duration [34].

It is worth noting that the temporal resolution of the approach depends on both technical but also fundamental parameters. A technical restriction arises from the surface flatness and the reflectivity flatness requirements of the bisected mirror used. Pulses with few tens of as duration consist of wavelengths ranging from ~100 nm down to few nm and require precision mirrors with ultra-high surface flatness. Reflecting materials supporting such broadband radiation are lacking. From the fundamental point of view, the requirement of an instantaneous response may set the value of the temporal resolution to several tens of as. The response time of the TXUVPI detector is a very interesting open question, which relates to the life time of the virtual intermediate state of the non-resonant TXUVPI process and at the same time an experimental challenge as an immediate application of as pulses.

The results of Sects. 7.5.1 and 7.5.2 point out to phase correlated spectral components but with imperfect phase locking of the harmonics. They also make perceptible how precarious it is to infer pulse durations indirectly from partial knowledge of some of the parameters involved. The present study is highly pertinent to the metrology of other even more intense XUV radiation sources such as surface plasma harmonics and XFEL sources.

7.6 Conclusions and Outlook

We have reviewed recent developments and accomplishments of a research campaign on the generation, characterization and applications of intense attosecond pulses in the XUV spectral regime, based on the collaboration of scientists of the Max–Planck Institute für Quantenoptik (Germany) and FORTH-IESL (Greece). The achieved XUV intensity levels, through intense HOHG, are close to 10^{13} W cm^{-2}, with which a number of two-XUV-photon processes induced by harmonic superpositions have been demonstrated. Those include total and energy resolved two-photon single ionization, processes pivotal for the implementation of second order based autocorrelation measurements and XUV-pump-XUV-probe applications. Extension to higher intensities is feasible through a further increased generation volume. We have further reported extensions of these measurements to double ionization with evidence for the observation of direct double ionization by harmonic superpositions. These demonstrations are of significant importance to the generalization of the second order AC technique to a broader spectral region, since the single two-photon ionization is limited to photon energies up to 24 eV. Extension to shorter wavelengths has been quantitatively discussed, giving estimates for the required intensities. A proposal for further extensions to the X-ray regime, exploiting non-linear inner-shell processes, has been made. In addition to their importance for the temporal characterization of ultra-short X-ray pulses and X-ray time domain measurements, non-linear X-ray processes will provide a unique tool for 3D sample analysis with spatial resolution of the order of 1 Å or less.

Towards second order XUV autocorrelation based pulse diagnostics and time domain measurements in the attosecond regime, we have modeled, developed and successfully utilized two alternative dispersionless XUV/X-ray auto-/cross correlators, the transmission grating Michelson interferometer and the split-mirror non-linear volume autocorrelator. The two instruments provide complementary features, the first being extremely broadband, offering wavelength region selectivity but having a low throughput, while the later provides two orders of magnitude higher throughput with somewhat restricted wavelength selectivity. Both arrangements and their principles of performance have been presented in this manuscript.

Finally, we have summarized the essence of the first second order AC measurement of an attosecond pulse train and its detailed assessment trough ab initio theory and numerical simulations. This measurement establishes a new perspective towards rigorous attosecond pulse diagnostics and highest resolution time domain experiments. This has lured a number of other laboratories towards this approach. There are recent attempts to adopt the method for the temporal characterization of laser-surface plasma harmonics, while multi-XUV-photon processes have been included in the research programs of XFEL sources [63]. The proposed extension to the X-ray spectral region is a realistic near-future development, offering atomic scale 4D spatiotemporal resolution.

Acknowledgement

This work is supported in part by the European Community's Human Potential Programme under contract MRTN-CT-2003-505138 (XTRA); MIRG-CT-2004-506583 (CHARA) MTKD-CT-2004-517145 (X-HOMES); and the Ultraviolet Laser Facility (ULF) operating at FORTH-IESL (contract no. HPRI-CT-2001-00139).

References

1. R.H. Pantell, Y. Soncini, H.E. Putoff, Stimulated photon-electron scattering. IEEE Quant. Electron. **4**, 905–907 (1968)
2. D.A.G. Deacon, L.R. Elias, J.M.J. Madey, G.J. Ramian, H.A. Schwettman, T.I. Smith, First operation of a free-electron laser. Phys. Rev. Lett. **38**, 892–894 (1977)
3. A. McPherson, G. Gibson, H. Jara, U. Johann, T.S. Luk, I. McIntyre, K. Boyer, C.K. Rhodes, Studies of multiphoton production of vacuum-ultraviolet radiation in the rare gases. J. Opt. Soc. Am. B **4**, 595 (1987)
4. M. Ferray, A. L'Huillier, F.X. Li, L.A. Lompr, G. Mainfray, C. Manus, Multiple-harmonic conversion of 1064 nm radiation in rare gases. J. Phys. B **21**, L31–L35 (1988)
5. S.V. Bulanov, N.M. Naumova, F. Pegoraro, Interaction of an ultrashort, relativistically strong laser pulse with an overdense plasma. Phys. Plasmas **1**, 745–757 (1994)
6. R. Lichters, J. Meyer-ter-Vehn, A. Pukhov, Short-pulse laser harmonics from oscillating plasma surfaces driven at relativistic intensity. Phys. Plasmas **3**, 3425–3437 (1996)
7. G.D. Tsakiris, K. Eidmann, J. Meyer-ter-Vehn, F. Krausz, New J. Phys. 8, **19** (2006)
8. F. Quéré, C. Thaury, P. Monot, S. Dobosz, Ph. Martin, J.-P. Geindre, P. Audebert, Coherent wake emission of high-order harmonics from overdense plasmas. Phys. Rev. Lett. **96**, 125004 (2006)
9. B. Dromey, M. Zepf, A. Gopal, K. Lancaster, M.S. Wei, K. Krushelnick, M. Tatarakis, N. Vakakis, S. Moustaizis, R. Kodama, M. Tampo, C. Stoeckl, R. Clarke, H. Habara, D. Neely, S. Karsch, P. Norreys, High harmonicgeneration in the relativistic limit. Nature Phys. **2**, 456–459 (2006)
10. G. Sansone, E. Benedetti, F. Calegari, C. Vozzi, L. Avaldi, R. Flammini, L. Poletto, P. Villoresi, C. Altucci, R. Velotta, S. Stagira, S. De Silvestri, M. Nisoli, Isolated single-cycle attosecond pulses. Science **314**, 443–446 (2006)
11. D. Xenakis, O. Faucher, D. Charalambidis, C. Fotakis, Observation of two-XUV-photon ionization using harmonic generation from a short intense laser pulse. J. Phys. B **29**, L457–L463 (1996)
12. Y. Kobayashi, T. Sekikawa, Y. Nabekawa, S. Watanabe, 27-fs extreme ultraviolet pulse generation by high-order harmonics. Opt. Lett. **23**, 64–66 (1998)
13. H. Wabnitz, L. Bittner, A.R.B. de Castro, R. Dhrmann, P. Grtler, T. Laarmann, W. Laasch, J. Schulz, A. Swiderski, K. von Haeften, T. Möller, B. Faatz, A. Fateev, J. Feldhaus, C. Gerth, U. Hahn, E. Saldin, E. Schneidmiller, K. Sytchev, K. Tiedtke, R. Treusch, M. Yurkovk M, Multiple ionization of atom clusters by intense soft X-rays from a freeelectron laser. Nature **420**, 482–485 (2002)

14. N.A. Papadogiannis, L.A.A. Nikolopoulos, D. Charalambidis, P. Tzallas, G.D. Tsakiris, K. Witte, Two-photon ionization of He through a superposition of higher harmonics. Phys. Rev. Lett. **90**, 133902 (2003)
15. N. Miyamoto, M. Kamei, D. Yoshitomi, T. Kanai, T. Sekikawa, T. Nakajima, S. Watanabe, Observation of two-photon above-threshold ionization of rare gases by xuv harmonic photons. Phys. Rev. Lett. **93**, 083903 (2004)
16. Y. Nabekawa, H. Hasegawa, E.J. Takahasi, K. Midorikawa, Production of doubly charged helium ions by two-photon absorption of an intense sub-10-fs soft X-ray pulse at 42 eV photon energy. Phys. Rev. Lett. **94**, 043001 (2005)
17. T. Laarmann, A.R.B. de Castro, P. Grtler, W. Laasch, J. Schulz, H. Wabnitz, T. Möller, Photoionization of helium atoms irradiated with intense vacuum ultraviolet free-electron laser light. Part I. Experimental study of multiphoton and single-photon processes. Phys. Rev. A **72**, 023409 (2005)
18. D. Charalambidis, P. Tzallas, N.A. Papadogiannis, L.A.A. Nikolopoulos, E.P. Benis, G.D. Tsakiris, Comment on "Photoionization of helium atoms irradiated with intense vacuum ultraviolet free-electron laser light. Part I. Experimental study of multiphoton and single-photon processes". Phys. Rev. A **74**, 037402 (2006)
19. A. Maquet, B. Piraux, A. Scrinzi, R. Taeb, Comment on "Photoionization of helium atoms irradiated with intense vacuum ultraviolet free-electron laser light. Part II. Theoretical modeling of multiphoton and single-photon processes". Phys. Rev A **74**, 027401 (2006)
20. T. Laarmann, A.R.B. de Castro, P. Grtler, W. Laasch, J. Schulz, H. Wabnitz, T. Möller, Reply to "Comment on 'Photoionization of helium atoms irradiated with intense vacuum ultraviolet free-electron laser light. Part I. Experimental study of multiphoton and single-photon processes'". Phys. Rev A **74**, 037402 (2006)
21. A.R.B. de Castro, T. Laarmann, J. Schulz, H. Wabnitz, T. Möller, Reply to "Comment on 'Photoionization of helium atoms irradiated with intense vacuum ultraviolet free-electron laser light. Part II. Theoretical modeling of multiphoton and single-photon processes'". Phys. Rev. A **74**, 027402 (2006)
22. T.W. Hänsch, A proposed sub-femtosecond pulse synthesizer using separate phase-locked laser oscillators, Opt. Commun. **80**, 71–75 (1990)
23. G.Y. Farkas, C.S. Tóth, Proposal for attosecond light pulse generation using laser induced multipleharmonic conversion processes in rare gases. Phys. Lett. A **168**, 447–450 (1992)
24. N.A. Papadogiannis, B. Witzel, C. Kalpouzos, D. Charalambidis, Observation of attosecond light localization in higher order harmonic generation. Phys. Rev. Lett. **83**, 4289–4292 (1999)
25. P.M. Paul, E.S. Toma, P. Breger, G. Mullot, F. Augé, Ph. Balcou, H.G. Muller, P. Agostini, Observation of a train of attosecond pulses from high harmonic generation. Science **292**, 1689–1692 (2001)
26. R. Kienberger, M. Hentschel, M. Uiberacker, Ch. Spielmann, M. Kitzler, A. Scrinzi, M. Wieland, Th. Westerwalbesloh, R. Kleineberg, U. Heinzmann, M. Drescher, F. Krausz, Steering attosecond electron wave packets with light. Science **297**, 1144–1148 (2002)
27. P. Tzallas, D. Charalambidis, N.A. Papadogiannis, K. Witte, G.D. Tsakiris, Direct observation of attosecond light bunching. Nature **426**, 267–271 (2003)
28. R. Kienberger, E. Goulielmakis, M. Uiberacker, A. Baltuska, V. Yakovlev, F. Bammer, A. Scrinzi, Th. Westerwalbesloh, U. Kleineberg, U. Heinzmann, M. Drescher, F. Krausz, Atomic transient recorder. Nature **427**, 817–821 (2004)

29. Y. Nabekawa, T. Shimizu, T. Okino, K. Furusawa, H. Hasegawa, K. Yamanouchi, K. Midorikawa, Interferometric autocorrelation of an attosecond pulse train in the single-cycle regime. Phys. Rev. Lett. **96**, 083901 (2006)
30. Y. Mairesse, A. de Bohan, L.J. Frasinski, H. Merdji, L.C. Dinu, P. Monchicourt, P. Breger, M. Kovacev, R. Taïeb, B. Carré, H.G. Muller, P. Agostini, P. Saliéres, Attosecond synchronization of highharmonic soft X-rays. Science **302**, 1540–1543 (2003)
31. M. Hentschel, R. Kienberger, C. Spielmann, G.A. Reider, N. Milosevic, T. Brabec, P. Corkum, U. Heinzmann, M. Drescher, F. Krausz, Attosecond Metrology. Nature **414**, 509–513 (2001)
32. J. Itatani, F. Quéré, G.L. Yudin, M. Yu Ivanov, F. Krausz, Attosecond streak camera. Phys. Rev. Lett. **88**, 173903 (2002)
33. Y. Mairesse, F. Quéré, Frequency-resolved optical gating for complete reconstruction of attosecond bursts. Phys. Rev. A **71**, 011401(R) (2005)
34. S. Kazamias, Ph. Balcou, Intrinsic chirp of attosecond pulses: Single-atom model versus experiment. Phys. Rev. A **69**, 063416 (2004)
35. L.A.A. Nikolopoulos, E.P. Benis, P. Tzallas, D. Charalambidis, K. Witte, G.D. Tsakiris, Second order autocorrelation of an XUV attosecond pulse train. Phys. Rev. Lett. **94**, 113905 (2005)
36. R. Trebino, K.W. DeLong, D.N. Fittinghoff, J.N. Sweetser, M.A. Krumbügel, B. Richman, D.J. Kane, Measuring ultrashort laser pulses in the time–frequency domain using frequency-resolved optical gating. Rev. Sci. Instrum. **68**, 3277–3295 (1997)
37. J.-F. Hergott, M. Kovacev, H. Merdji, C. Hubert, Y. Mairesse, E. Jean, P. Breger, P. Agostini, B. Carré, P. Saliéres, Extreme-ultraviolet high-order harmonic pulses in the microjoule range. Phys. Rev. A **66**, 021801(R) (2002)
38. M.B. Gaarde, K.J. Schafer, Spacetime considerations in the phase locking of high harmonics. Phys. Rev. Lett. **89**, 213901 (2002)
39. P.B. Corkum, Plasma perspective on strong-field multiphoton ionization. Phys. Rev. Lett. **71**, 1994–1997 (1993)
40. P. Tzallas, D. Charalambidis, N.A. Papadogiannis, K. Witte, G.D. Tsakiris, Second-order autocorrelation measurements of attosecond XUV pulse trains. J. Mod. Opt. **52**, 321–340 (2005)
41. P. Tzallas, E. Skantzakis, C. Kalpouzos, E.P. Benis, G.D. Tsakiris, D. Charalambidis, Generation of intense continuum XUV radiation by many cycle laser fields. Nat. Phys. **3**, 846 (2007)
42. L.V. Keldysh, Sov. Phys. JETP **20**, 1307 (1965)
43. E.P. Benis, P. Tzallas, L.A.A. Nikolopoulos, M. Kovačev, C. Kalpouzos, D. Charalambidis, G.D. Tsakiris, Frequency resolved photoelectron spectra of two-photon ionization of He by an attosecond pulse train. New J. Phys. **8**, 92 (2006)
44. B. Walker, B. Sheehy, L.F. DiMauro, P. Agostini, K.J. Schafer, K.C. Kulander, Precision measurement of strong field double ionization of helium. Phys. Rev. Lett. **73**, 1227–1230 (1994)
45. B. Witzel, N.A.D. Papadogiannis, D. Charalambidis, Charge-state resolved above threshold ionization. Phys. Rev. Lett. **85**, 2268–2271 (2000)
46. R. Dörner, V. Mergel, O. Jagutzki, L. Spielberger, J. Ullrich, R. Moshammer, H. Schmidt-Böcking, Cold target recoil ion momentum spectroscopy: a & momentum microscope' to view atomic collision dynamics. Phys. Rep. **330**, 95–102 (2000)

47. Th Weber, H. Giessen, M.Weckenbrock, G. Urbasch, A. Staudte, L. Spielberger, O. Jagutzki, V. Mergel, M. Vollmer, R. Dörner, Correlated electron emission in multiphoton double ionization. Nature **405**, 658–661 (2000)
48. R. Lafon, J.L. Chaloupka, N. Sheehy, P.M. Paul, P. Agostini, K.C. Kulander, L.F. DiMauro, Electron energy spectra from intense laser double ionization of helium. Phys. Rev. Lett. **86**, 2762–2765 (2001)
49. R. Dörner, Th. Weber, M. Weckenbrock, A. Staudte, M. Hattass, R. Moshammer, J. Ullrich, H. Schmidt-Böcking, Multiple ionization in strong laser fields. Adv. Atom. Mol. & Opt. Phys. **48**, 1–36 (2002) (and references therein)
50. M.A. Kornberg, P. Lambropoulos, Photoelectron energy spectrum in 'direct' two-photon double ionization of helium. J. Phys. B **32**, L603–613 (1999)
51. P. Lambropoulos, L.A.A Nikolopoulos, M.G. Makris, Signatures of direct double ionization under xuv radiation. Phys. Rev. A **72**, 013410 (2005)
52. E.P. Benis, D. Charalambidis, T.N. Kitsopoulos, G.D. Tsakiris, P. Tzallas, Two-photon double ionization of rare gases by a superposition of harmonics. Phys. Rev. A **74**, 051402(R) (2006)
53. E. Constant, E. Mével, R.A. Zaÿ, V. Bagnoud, Salin, Toward sub-femtosecond pump-probe experiments: A dispersionless autocorrelator with attosecond resolution. J. Phys. IV (France) **11**, Pr2–537 (2001)
54. C.G. Durfee, S. Backus, H.C. Kapteyn, M.M. Murnane, Intense 8-fs pulse generation in the deep ultraviolet. Opt. Lett. **24**, 697–699 (1999)
55. H. Mashiko, A. Suda, K. Midorikawa, All-reflective interferometric autocorrelator for the measurement of ultra-short optical pulses, Appl. Phys. B **76**, 525–530 (2003)
56. E. Goulielmakis, G. Nersisyan, N.A. Papadogiannis, D. Charalambidis, G.D. Tsakiris, K. Witte, A dispersionless Michelson interferometer for the characterization of attosecond pulses. Appl. Phys. B **74**, 197–206 (2002)
57. N.A. Papadogiannis, G. Nersisyan, E. Goulielmakis, T.P. Rakitzis, E. Hertz, D. Charalambidis, G.D. Tsakiris, K. Witte, Temporal characterization of short-pulse third-harmonic generation in an atomic gas by a transmission-grating Michelson interferometer. Opt. Lett. **27**, 1561–1563 (2002)
58. E. Papalazarou, M. Kovacev, P. Tzallas, E.P. Benis, C. Kalpouzos, G.D. Tsakiris, D. Charalambidis, Spectral phase distribution retrieval through coherent control of harmonic generation. Phys. Rev. Lett. **96**, 163901 (2006)
59. P. Tzallas, E. Skatzakis, E.P. Benis, A. Bonarou, C. Kalpouzos, G.D. Tsakiris, D. Charalambidis, Full temporal reconstruction of a lower order harmonic superposition. New J. Phys. **9**, 232 (2007)
60. E. Hertz, N.A. Papadogiannis, G. Nersisyan, C. Kalpouzos, D. Charalambidis, T. Halfmann, G.D. Tsakiris, Probing attosecond pulse trains using "phase-control" techniques. Phys. Rev. A **64**, 051801(R) (2001)
61. M. Lewenstein, Ph. Balcou, M. YHu Ivanov, A. L'Hullier, P.B. Corcum, Theory of high-harmonic generation by low-frequency laser fields. Phys. Rev. A **49**, 2117–2132 (1994)
62. F. Lindner, W. Stremme, M.G. Schätzel, F. Grasbon, G.G. Paulus, H. Walther, R. Hartmann, L. Strüder, High-order harmonic generation at a repetition rate of 100 kHz. Phys. Rev. A **68**, 013814 (2003)
63. see e.g. 'Technical Design Report of the European X-Ray Free-Electron Laser' (2006)

8

Ultrashort Pulse Collapse in Quadratic Media

Jian Wu and Heping Zeng

Abstract. Experimental studies on spatiotemporal collapse of ultrashort pulse propagation in quadratic nonlinear media are reviewed. Two-dimensional multicolored transverse arrays are experimentally observed in a quadratic nonlinear medium based on cascaded non-collinearly quadratic couplings and spatial breakup of input pump beams. Furthermore, two-dimensional multicolored up-converted parametric amplifications are realized with preserved coherence. The observed two-dimensional arrays can be controlled by changing the boundary condition with a weak second harmonic beam along with the strong fundamental-wave pump pulses. As a result of spatiotemporal collapse of femtosecond pulses, colored conical emissions are observed under strong phase-matched fundamental pump, which are further amplified to support tunable broadband femtosecond pulse generation.

8.1 Introduction

In the ultrafast high-intensity laser interaction regime, strong optical fields may induce quite a lot of novel phenomena with many unrevealed intrinsic physics even only some basic nonlinear optical processes, such as second harmonic generation and optical parametric amplification, are considered. Recent studies on self-focusing behaviors of intense ultrashort pulses indicated that spatial and temporal degrees of freedom could not be treated separately [1]. Spatiotemporal modulational instability [1, 2] occurs in the presence of both diffraction and dispersion that act together with the material nonlinearities to defeat the spatial spreading and temporal broadening of optical wave-packet, which may induce spatiotemporal collapse of optical wave-packet, such as x-shaped light bullets [1], colored conical emission [2], and spatiotemporal solitary waves. Taking the well-known second harmonic generation process in quadratic nonlinear media as an example, strong mutual interaction and energy exchange between the involved optical fields at different frequencies can bring about some intriguing phenomena for the nonlinear wave propagations. Quadratic spatial solitons are typically observed as quadratic nonlinearities defeat transverse diffractions in quadratic nonlinear media. Furthermore, multiple quadratic spatial solitons are observed as reproducible one-dimensional spatial patterns in quadratic nonlinear media near phase-matched second harmonic generation by means of spatial modulational instabilities [3], or spontaneous amplification of small input beam asymmetries or material anisotropies [4–8]. All-optical control of one-dimensional quadratic spatial solitons has been realized by either seeding a weak second harmonic (SH) beam or changing the input beam asymmetries [7, 8]. However, pump intensities required for the direct observation of two-dimensional quadratic spatial solitons are usually much higher than the material damage thresholds [3]. Breaking the barriers to produce two-dimensional arrays needs new experimental approaches. From the practical application points of view, quadratic nonlinearities have been extensively applied to the generation of spatial and spatiotemporal solitary waves [1, 3–5], optical switching [9], pulse self-compression [10], mode-locking of femtosecond lasers [11], and so on. The generation of two-dimensional multicolored transverse arrays can help us to explore the fundamental physics of two-dimensional transient gratings and stimulate some applications such as multicolored amplifications, ultrashort pulse generation, and all-optical control such as carrier-envelope-phase control of tunable multicolored coherent optical source. Moreover, multiple three-photon processes could be involved in phase-matched second harmonic generation, and the accompanied spatiotemporal modulational instabilities could cause perturbations with certain frequencies to grow at given geometric angles governed by the phase-matching conditions, leading to colored conical emissions [2]. Such spatiotemporal collapse of ultrashort pulse is much more complicated than the typical case with only the spatial effects involved, and intensive experimental investigations are required to fully understand the dynamics and intrinsic physics of colored conical emissions in quadratic nonlinear media.

In this review article, we present the experimental observation of two-dimensional multicolored transverse arrays under the pump of two crossly overlapped femtosecond (fs) beams in a quadratic nonlinear medium [12]. The formation of two-dimensional arrays is based on the spatial breakup of an input elliptical beam, which interacts with the other pump beam non-collinearly through cascaded non-collinearly quadratic couplings. Two-dimensional transient gratings are formed, which could be probed with either a super-continuum pulse or a weak fundamental-wave (FW) beam, leading to two-dimensional multicolored up-converted parametric amplifications or enhanced two-dimensional multicolored transverse arrays, respectively. Seeding with a weak SH beam, the two-dimensional arrays could be suppressed on the basis of a controllable change of the boundary condition of the induced spatial breakup of the elliptic input beam. We also present the experimental observation of colored conical emissions in a thick quadratic nonlinear medium at a high-intensity pump, which extends the horizon of wave-packet collapse to the spatiotemporal domain [13]. Assisted by the spatiotemporal modulational instabilities, seeded amplification of colored conical emission exhibits a widely tunable range in wavelength and high energy conversion efficiency [14], which could be further used to generate carrier envelope phase stabilized pulse by nonlinear frequency mixing with the FW pump pulse.

8.2 Two-Dimensional Multicolored Transverse Arrays

The experiments reported throughout this article were carried out by using an amplified Ti:Sapphire system consisting of a mode-locked oscillator and a regenerative amplifier. The system generates 50 fs pulses of about 650 μJ per pulse at a 1 kHz repetition rate with central wavelength at 800 nm. The quadratic nonlinear medium used to generate two-dimensional multicolored transverse arrays is a 4 mm thick beta-barium borate (BBO, type I, 29.18° cut). As shown in Fig. 8.1a, the fs pulse from the laser system was split into an on-axis beam $\overrightarrow{k_1}$ and a slightly off-axis beam $\overrightarrow{k_2}$. The $\overrightarrow{k_1}$ beam was focused into the BBO crystal by a high-reflection spherical concave mirror of 2,000 mm curvature radius with a fold angle of 12°. This led to an elliptically spatial profile of $\overrightarrow{k_1}$ beam as $W_y/W_x = 1.58$, where W_x and W_y are the beam widths. While, the $\overrightarrow{k_2}$ beam was round and collimated down to the BBO crystal through a telescope system. At the input facet of the BBO crystal, the beam diameter along the long-axis direction was measured to be about 1.3 mm. The BBO crystal was placed about 200 mm before the focusing position for the sake of avoiding damage and adjusted to fulfill second harmonic generation for the on-axis $\overrightarrow{k_1}$ beam at 800 nm.

As presented in Fig. 8.1b, additional spatial patterns beside the original beam were observed after the BBO crystal as a result of wave-packet breakup of the spatially elliptical on-axis beam $\overrightarrow{k_1}$ with a self-induced spatially varying transverse phase properly balancing the effect of diffraction, where the mutual interactions between the FW and SH pulses played crucial roles [7, 8, 15–17]. The situation changed when a weak SH beam was seeded at the input of the crystal along with the on-axis $\overrightarrow{k_1}$ beam, which modified the induced spatially transverse phases and eventually influenced the wave-packet breakups. In order to demonstrate those effects, as shown in Fig. 8.1b, an additional 1 mm thick BBO crystal (labeled as BBO2) was placed before the original 4 mm BBO crystal (labeled as BBO1) to generate a weak SH beam collinearly with the on-axis FW beam. Here, BBO1 and BBO2 crystals are used for wave breakup observation and control, respectively. The pulse energy ratio SH/FW could be changed by adjusting the phase-matching condition of the 1 mm thick BBO crystal. As presented in Fig. 8.1b, the wave-packet breakup could be suppressed as the pulse energy ratio SH/FW reached 3.6%. Obviously, the generation of multiple one-dimensional quadratic spatial solitons was controllable by seeding along a weak SH beam at the input of the quadratic nonlinear crystal [8]. In order to verify our experimental observations, numerical simulations were carried out

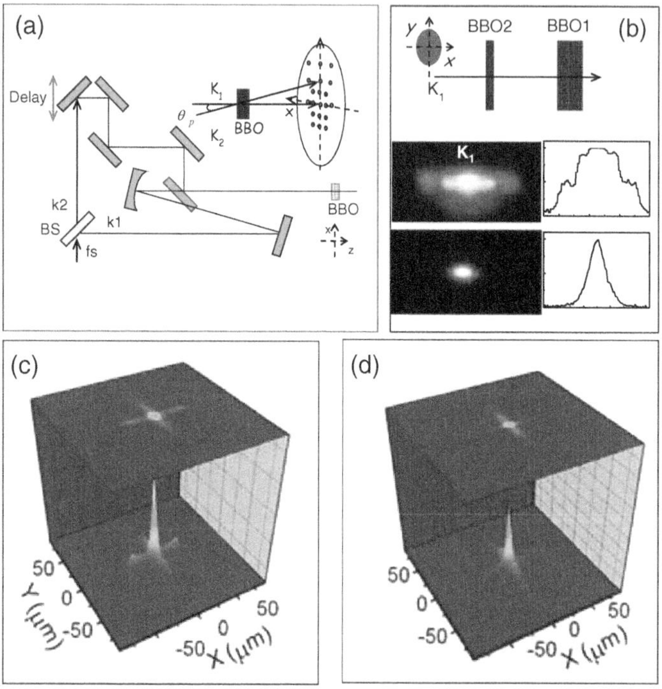

Fig. 8.1. (**a**) Schematic for generation of two-dimensional multicolored transverse arrays in a quadratic nonlinear medium. (**b**) Experimental scheme for generation and weak beam control of one-dimensional multiple quadratic spatial solitary waves. Experimentally observed pictures and corresponding transverse intensity spectra are presented as the insets. The wave breakup is suppressed as the pulse energy ratio SH/FW reached 3.6%. The numerical results of the quadratic nonlinearities induced wave breakup of FW pulse with the pulse energy ratios SH/FW as 0 and 3.0% are presented in (**c**) and (**d**), respectively. BS: beam splitter

by solving the (3 + 1)-dimensional nonlinear equations governing the second harmonic generation process in a BBO crystal (type I phase-matching) with a spatially elliptic FW beam as the initial input. As shown in Fig. 8.1c, d, after propagation in a 4 mm thick BBO crystal, quadratic nonlinearities induced a wave-packet breakup of the FW beam with a small input asymmetry, which could be further suppressed as a weak SH beam was initially seeded along the FW beam.

Two-dimensional multicolored transverse arrays were generated as both the on-axis $\overrightarrow{k_1}$ and off-axis $\overrightarrow{k_2}$ beams were synchronized in the 4 mm thick BBO crystal at a crossing angle θ_p and spatially overlapped each other in a scheme as shown in Fig. 8.1a. New patterns with a frequency shift (determined by the phase-matching condition) was generated based on the cascaded non-collinearly quadratic processes between the novel beams induced by spatial breakup of $\overrightarrow{k_1}$ and the input off-axis beam $\overrightarrow{k_2}$. The generated new multicolored patterns participated in cascaded non-collinearly quadratic processes in turn and eventually produced two-dimensional periodic patterns. As presented in Fig. 8.2a, reproducible two-dimensional multicolored transverse arrays were directly observed after the BBO crystal with pump intensities much lower than the damage threshold of the nonlinear medium. Overall, there are two sets of well-resolved red and blue arrays. For clarity of description later on, we label those two-dimensional patterns as $R(n,m)$ and $B(n,m)$, starting with $R(0,0)$ and $B(0,0)$ that correspond to the FW and SH of $\overrightarrow{k_1}$ beam, respectively. The other patterns are labeled in a reference frame of $(x,\ y)$ with the grid origin $R(0,0)$ for red patterns (and $B(0,0)$ for blue patterns), and some examples are presented in Fig. 8.2a. As a result of the required phase-matching in the cascaded non-collinearly quadratic processes, the spectra of the blue and red patterns varied at different rows and columns. The beam divergences of the red and blue patterns were measured, which showed typical values as $\Delta\theta_x \sim 1.2°$ ($\Delta\theta_y \sim 1.8°$) and $\Delta\theta_x \sim 0.7°$ ($\Delta\theta_y \sim 1.1°$), respectively. Two-dimensional multicolored transverse arrays with different conically angular spectra could be observed as the crossing angle θ_p was changed. Well-resolved red and blue patterns could be observed with enough large θ_p, and denser rows of transverse spots could be observed with smaller θ_p. However, due to the divergences of the transverse spots, the two-dimensional multicolored transverse arrays became blurred in the far-field observation screen for a small θ_p ($\theta_p \sim 1.26°$). As limited by the typical divergence of the observed two-dimensional patterns, the two-dimensional multicolored transverse arrays exhibited no resolved rows with $\theta_p < 2\Delta\theta_x$. Nevertheless, we note that the angular separation of adjacent columns in the observed multicolored transverse arrays did not change as θ_p varied.

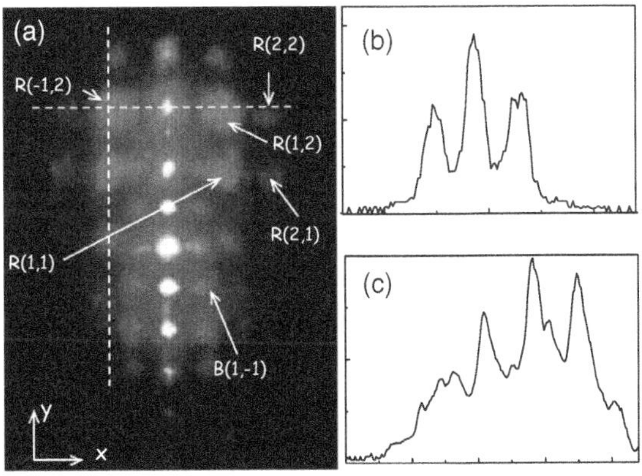

Fig. 8.2. The picture of the experimentally observed two-dimensional multicolored transverse arrays (**a**), and the corresponding intensity spectra of a row of multicolored transverse patterns crossing $R(1,2)$ (**b**) and a column of multicolored transverse patterns crossing $R(-1,2)$ (**c**) as labeled with *dashed line* in (**a**)

In the presence of the off-axis $\overrightarrow{k_2}$ beam, the spatial breakup of the elliptic input $\overrightarrow{k_1}$ beam was enhanced, as evidenced by the distinct columns in the two-dimensional multicolored transverse arrays. The enhanced spatial beam breakup was a direct result of energy transfer involved in the cascaded non-collinearly quadratic processes among different two-dimensional patterns. Fundamentally different from the two-dimensional discrete solitons in optically induced nonlinear lattices [18] or multiple filaments in air [6, 19], the approach demonstrated here can not only be served as a novel mechanism for the two-dimensional array generation with pump intensities below the material damage thresholds, but also provide us more degrees of freedom to control two-dimensional arrays of quadratic spatial solitons. This may stimulate further experiments to explore the fundamental dynamics of wave-packet breakup or collapse during wave propagation in quadratic nonlinear media, and may bring about various novel applications.

8.3 Transient Grating Probing

As the generation of two-dimensional multicolored transverse arrays was based on the cascaded non-collinearly quadratic couplings, the effective refractive index of the nonlinear crystal was modulated correspondingly. This established two-dimensional transient gratings. Experimentally, such two-dimensional transient gratings could be imaged and probed with synchronous coherent beams. In order to show this point clearly, as shown in Fig. 8.3a, a super-continuum pulse was produced by focusing another off-axis beam $\overrightarrow{k_3}$ (split from $\overrightarrow{k_2}$) into a 2 mm-thick sapphire plate. The residual FW pump pulses at 800 nm in the super-continuum were removed after passing through a short-pass filter with a cutoff wavelength around 750 nm. Here, the $\overrightarrow{k_3}$ beam is collimated down to the 4 mm thick BBO crystal and crossly overlapped with the $\overrightarrow{k_1}$ and $\overrightarrow{k_2}$ beams. As presented in Fig. 8.3b, the synchronously seeded super-continuum pulses were at first diffracted by the transient gratings and then amplified at the diffracted directions, leading to the generation of two-dimensional multicolored up-converted parametric amplifications.

Figure 8.3c presents the experimentally observed multicolored up-converted parametric amplifications as the super-continuum was seeded at $R(1, 2)$. The corresponding spectra of some selected spots were plotted in Fig. 8.3d. In order to identify the up-converted patterns from above multicolored transverse arrays, we labeled different spots in the two-dimensional multicolored up-converted parametric amplifications by $\mathrm{G}(n, m)$ with the seeded super-continuum beam as $G(0, 0)$. As the generated two-dimensional transient gratings were probed by the super-continuum beam at different positions with various geometric angles, the super-continuum beam would be diffracted and then amplified in different ways. These led to tunable multicolored up-converted parametric amplifications within different ranges. Figure 8.4 gives

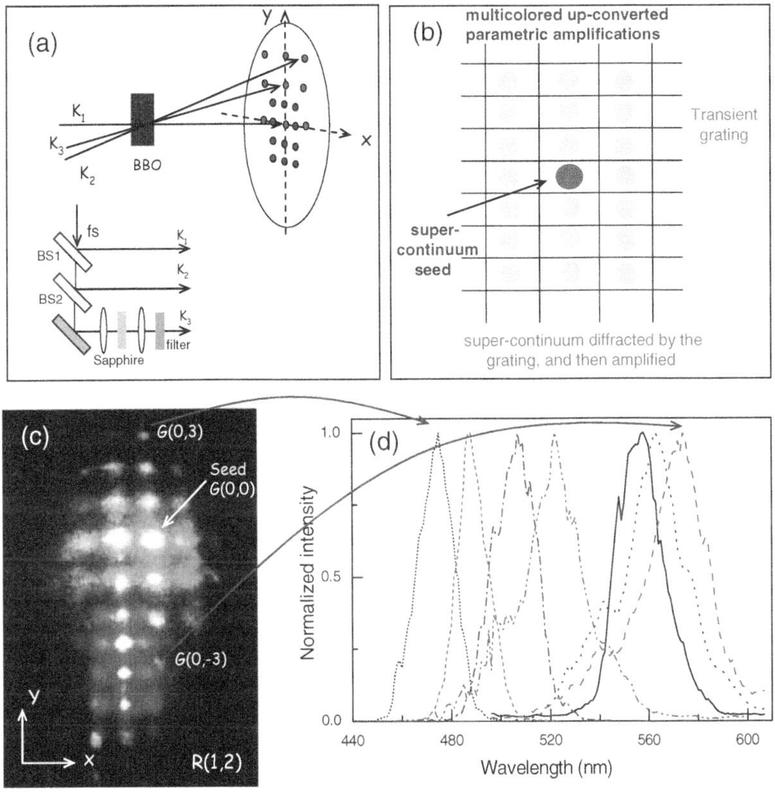

Fig. 8.3. (**a**) Experimental schematic for observation of multicolored up-converted parametric amplifications, where a super-continuum beam k_3 is used as a probe. (**b**) Diffraction and amplification of the seeded super-continuum pulse by the cascaded non-collinearly quadratic processes induced two-dimensional transient grating, which produces two-dimensional multicolored up-converted parametric amplifications. (**c**) Picture of the observed multicolored up-converted parametric amplification patterns as the super-continuum was seeded at $R(1,2)$. (**d**) Spectra of some selected spots $G(0,m)$ from multicolored up-converted parametric amplifications as shown in (**c**), where $m = 3,\ 2,\ 1,\ 0,\ -1,\ -2$, and -3, respectively, from left to right. BS: beam splitter

typical examples of two-dimensional multicolored up-converted parametric amplification patterns at different super-continuum seeding angles, where (a) and (b) correspond to the situations with super-continuum beams seeded at $R(2,1)$, and in the middle of $R(1,1)$, $R(1,2)$, $R(2,1)$, and $R(2,2)$, respectively. Figure 8.4c presents the spectra of some selected spots from the observed multicolored up-converted parametric amplifications as shown in Fig. 8.4b. The probe beam was not necessarily seeded at one of the spots of the two-dimensional multicolored transverse arrays for the observation of two-dimensional multicolored up-converted parametric amplifications since

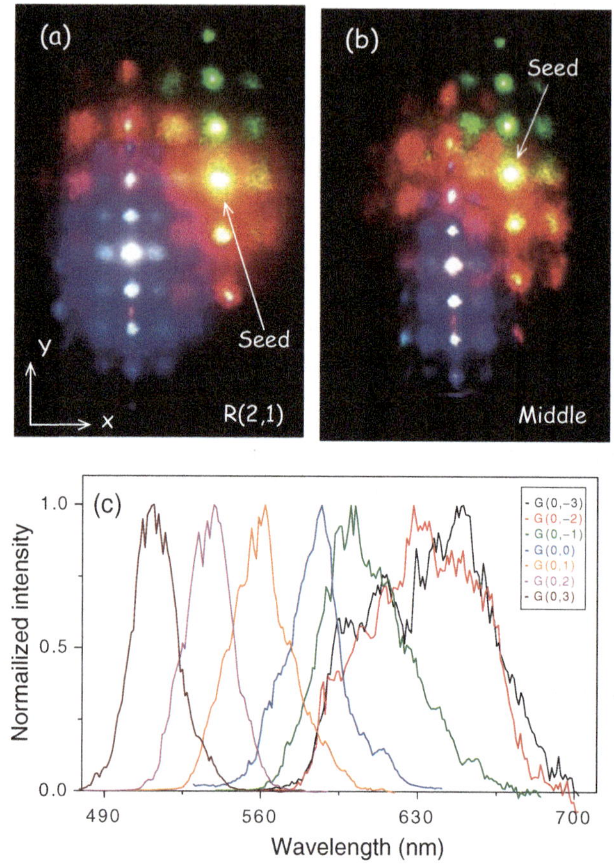

Fig. 8.4. Pictures of two-dimensional multicolored up-converted parametric amplifications observed with super-continuum seeded at (**a**) $R(2, 1)$ and (**b**) in the middle of $R(1, 1)$, $R(1, 2)$, $R(2, 1)$, and $R(2, 2)$, respectively. (**c**) The spectra of some selected spots of multicolored up-converted parametric amplifications as shown in (**b**)

the formed gratings existed in the whole interaction region, and the resulted multicolored up-converted parametric amplification patterns covered almost the whole optical spectrum in the visible region.

The dependence of two-dimensional multicolored up-converted parametric amplifications on the cascaded non-collinearly quadratic processes induced transient gratings could be tested by changing the relative delays among $\overrightarrow{k_1}$, $\overrightarrow{k_2}$ and $\overrightarrow{k_3}$. For example, Fig. 8.5 presents the dependence of multicolored up-converted parametric amplification pattern upon the delays of $\overrightarrow{k_1}$ and $\overrightarrow{k_2}$ while keeping the delay of $\overrightarrow{k_3}$ unchanged. Obviously, the two-dimensional multicolored up-converted parametric amplification patterns shrank dramatically as the delays of $\overrightarrow{k_1}$ and $\overrightarrow{k_2}$ walked away, which caused the degeneration of

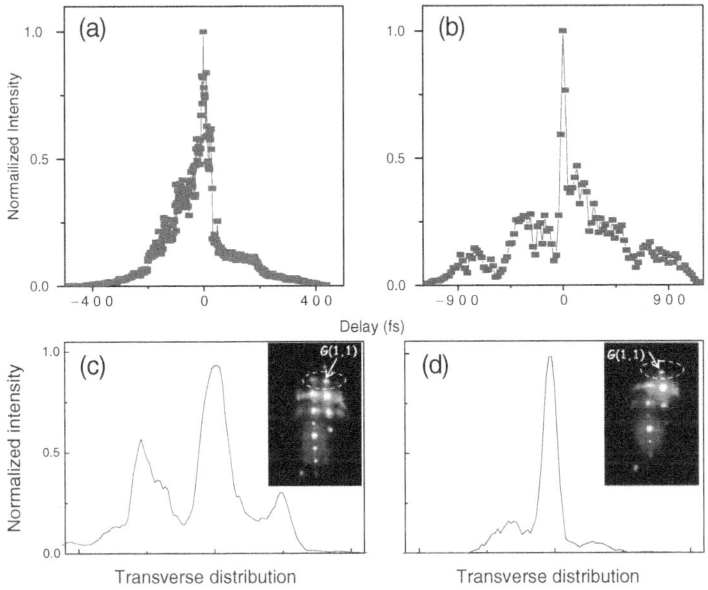

Fig. 8.5. Intensity dependences of multicolored up-converted parametric amplification patterns upon the delays of (**a**) k_1 and (**b**) k_2, respectively. Here, the k_3 is seeded at $R(1,2)$ and the up-converted pattern is selected to be $G(1,1)$ as labeled in the figures. (**c**) and (**d**) present the intensity spectra of a row of multicolored up-converted parametric amplification patterns crossing $G(0,1)$ as the delay of k_2 is changed to be 100 and 1,000 fs, respectively, while keeping k_1 and k_3 unchanged. The correspondingly observed pictures are also presented as the insets

the formed two-dimensional transient gratings. Furthermore, in the absence of transient gratings by blocking beam $\overrightarrow{k_2}$, the observed two-dimensional multicolored up-converted parametric amplifications degraded into a simple parametric amplification with patterns similar to that from beam $\overrightarrow{k_1}$. It clearly demonstrates that the multicolored up-converted parametric amplifications was produced by diffracting and amplifying seeded super-continuum pulse with the quadratic nonlinearities induced transient gratings.

Moreover, spectral coherence would be expected if the phase was preserved in the nonlinear amplifications along different directions, since they were from the same super-continuum pulse. We measured phase difference between two selected spots by using a so-called spectral interferometry technique. Figure 8.6a shows the setup for spectral interference technique. The beating signal on the charged coupled device could be written as

$$\begin{aligned} S(\omega) &= |E_1(\omega)e^{i\phi_1(\omega)} + E_2(\omega)e^{i[\phi_2(\omega)+\omega\tau]}|^2 \\ &= |E_1|^2 + |E_2|^2 + 2E_1E_2\cos(\omega\tau + \Delta\phi), \end{aligned} \tag{8.1}$$

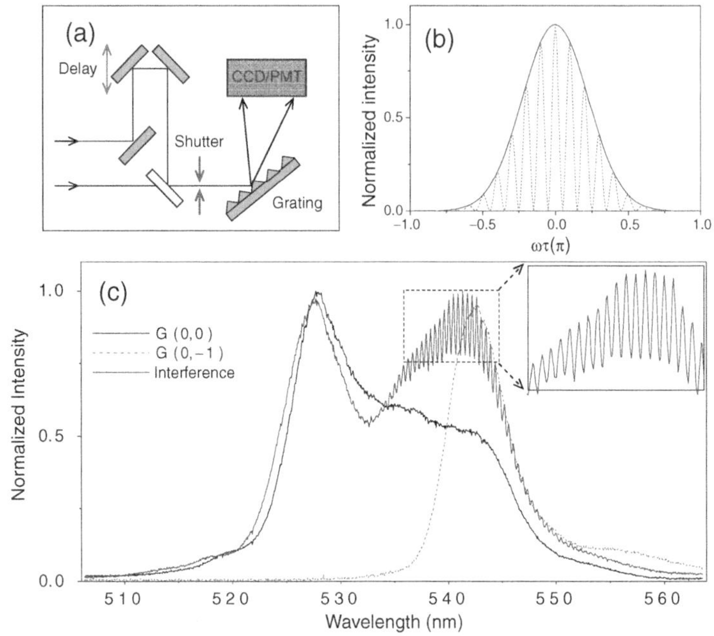

Fig. 8.6. (**a**) Experimental setup for spectral interference. (**b**) Simulated spectral interference patterns for pulses with constant (*dashed line*) and random (*solid line*) phase differences from shot to shot. (**c**) The measured spectral interference between two selected multicolored up-converted parametric amplification patterns with a preset delay and comparable intensities. CCD: charge coupled device; PMT: photomultiplier tube

where $\Delta\phi = \phi_2 - \phi_1$ is the phase difference between two interfering pulses of central frequency ω and preset delay of τ. As shown in Fig. 8.6b, a stable interference pattern can be observed if the beating pulses have a constant phase difference, otherwise the interference patterns will smear out after several laser shots if the phase difference between the beating pulses varies randomly from shot to shot. Experimentally, as shown in Fig. 8.6c, reproducible spectral interference pattern after an average of several thousand laser shots were observed between two selected spots with a preset delay and relatively comparable intensities. Obviously, phase coherence was preserved during the diffraction and amplification of the probe pulse at different directions, which led to a constant phase difference between the amplified up-converted patterns at different colors.

Under the pump of FW pulses at 800 nm, multiple frequency up-conversions with different colors distributed in two-dimensional transverse patterns were demonstrated as a direct result of crossly coupled cascaded non-collinearly quadratic and multiple parametric processes, which exhibited features quite different from the standard optical parametric amplifications. As the probing super-continuum pulses include different spatiotemporal chirps in different

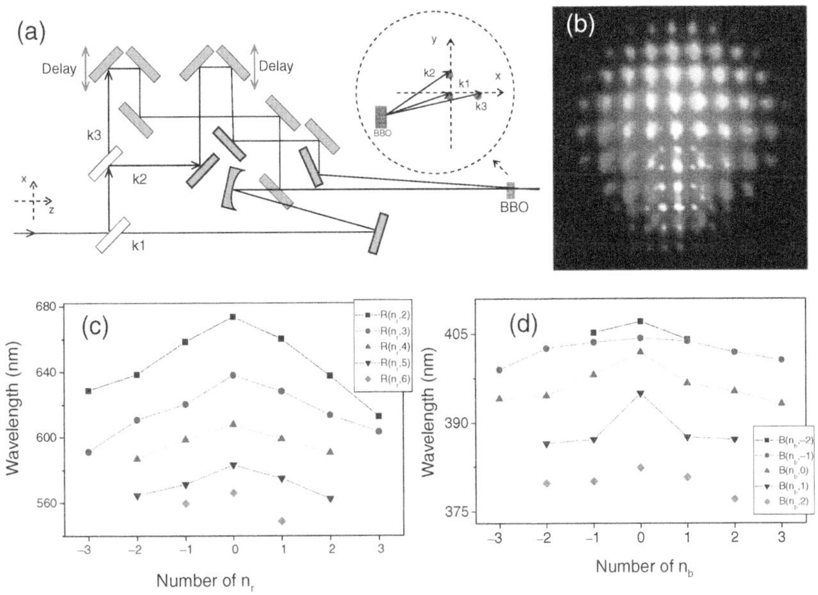

Fig. 8.7. (**a**) Experimental setup for observation of enhanced two-dimensional multicolored transverse arrays with a weak FW pulse k_3 split directly from FW pump pulse as the probe. (**b**) Picture of the observed enhanced two-dimensional multicolored transverse arrays. (**c**) and (**d**) plot the measured spectra of some selected spots as shown in (**b**)

spectral regions, the spatiotemporal dynamics of the involved cascaded non-collinearly quadratic processes could be investigated and the spatial distribution of the transient grating in the quadratic nonlinear medium could be dynamically imaged.

The transient gratings could be also observed by using a probing pulse directly split from the FW pulse. As shown in Fig. 8.7a, the $\overrightarrow{k_1}$, $\overrightarrow{k_2}$, and $\overrightarrow{k_3}$ beams were split directly from the FW pulses at 800 nm, and $\overrightarrow{k_3}$ pulse was very weak. Due to diffraction and amplification of the two-dimensional transient grating formed by $\overrightarrow{k_1}$ and $\overrightarrow{k_2}$, an significantly enhanced two-dimensional multicolored transverse arrays was observed as shown in Fig. 8.7b. More than 180 spots at different colors were actually produced by this novel technique, which could not be seen clearly in the picture due to the limitations of our digital camera. Differing from the observation of multicolored up-converted parametric amplifications with super-continuum pulse, here we synchronized $\overrightarrow{k_3}$ with $\overrightarrow{k_1}$ at the position where wave-packet breakup occurred, which also participated in the cascaded non-collinearly quadratic processes directly as $\overrightarrow{k_1}$ and $\overrightarrow{k_2}$ did. The wavelengths of different spots were measured. As shown in Fig. 8.7c, d, the measured central wavelengths of the adjacent spots agree well with the required phase-matching conditions of the corresponding cascaded non-collinearly quadratic processes. The emitting red spots cover a quite broad band spectral range from 550 to 800 nm. The spots farther away from the center occupy shorter wavelengths due to the cascaded non-collinearly quadratic processes. The observation of such enhanced multicolored transverse arrays further indicates that spatial wave-packet breakup and the followed cascaded non-collinearly quadratic processes play crucial roles in forming of our two-dimensional arrays. The phase coherence between different spots form the enhanced two-dimensional arrays was tested by using the spectral interferometry technique in the frequency domain. With a proper preset delay, a clear spectral interference pattern between the selected patterns was observed after an average of several thousand laser shots, which indicated the two-dimensional patterns were actually multicolored coherent optical sources.

The coherent two-dimensional patterns could be further combined into one beam by using a hollow fiber with a scheme as shown in Fig. 8.8a. In the experiment, all the spots from the enhanced two-dimensional multicolored transverse arrays were focused into a $L = 50\,\text{mm}$, $D = 1.0\,\text{mm}$ hollow fiber with a $f = 60$-mm achromatic lens. As shown in Fig. 8.8b, except the blue part around 400 nm, a quite broadband spectrum (590–820 nm) could be obtained at the end of the hollow fiber. The contributions from the outer spots with wavelengths shorter than 590 nm were almost eliminated due to their

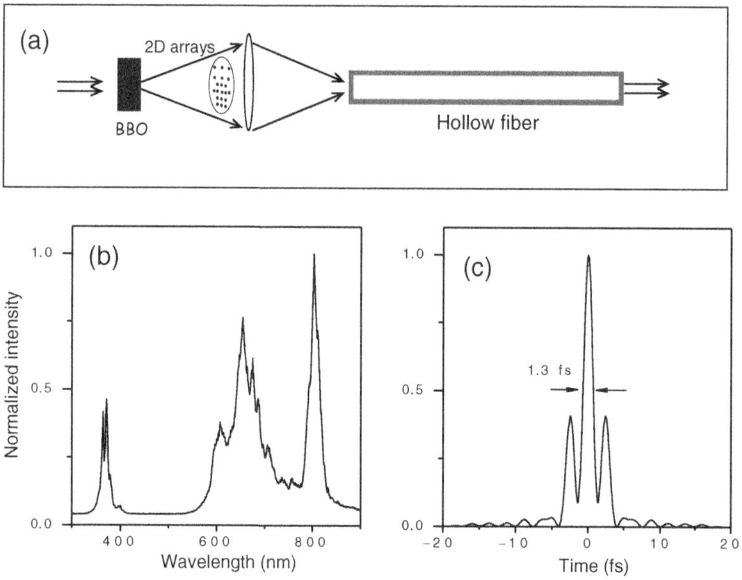

Fig. 8.8. (**a**) Experimental schematic for two-dimensional patterns combination using a hollow fiber. (**b**) The spectrum of the combined beam after a 50-mm hollow fiber, and (**c**) the corresponding transform-limited pulse by assuming the spectrum is chirped free with wavelength longer than 590 nm

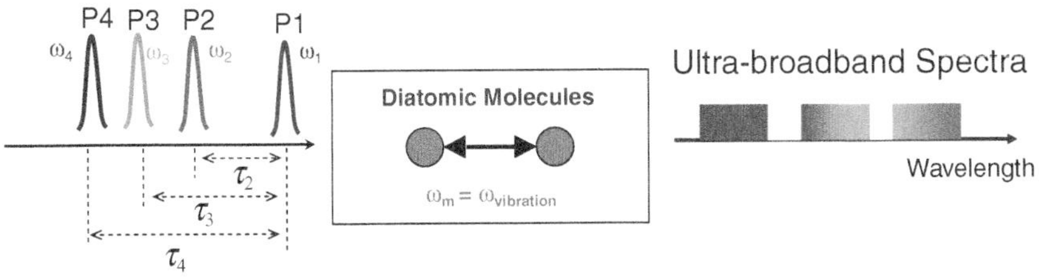

Fig. 8.9. Schematic for multicolored impulsive stimulated Raman scatterings to obtain ultra-broadband spectra with pump pulses at different frequencies selected from our two-dimensional arrays

relative weak intensities as compared with the inner spots and the waveguide effects of the hollow fiber. By using specially designed double chirped mirror pairs [20] or spatial light modulators [21] for dispersion compensation, the broadband spectrum can be compressed to produce ultrashort pulse. As shown in Fig. 8.8c, it can result in a transform-limited pulse in sub-5 fs range. Significantly, the combined beam has a high pulse energy (~44 μJ). The generation of high-peak-power optical pulse with duration down to sub-5 fs range is particularly important for a number of fundamental experiments, such as single attosecond pulse generation by high-order harmonic generation [22], ultrafast nonlinear spectroscopy, strong-field light-matter interactions, and various nonlinear processes.

Moreover, the coherent two-dimensional multicolored arrays could also be used as pumps for multicolored impulsive stimulated Raman scatterings [23]. As shown in Fig. 8.9, we could select one spot from the coherent two-dimensional arrays at frequency ω_1 with pulse duration shorter than the period of the Raman-active vibrational mode as the first pump beam, which leaved the molecules in a vibrational state. The modulated refractive index could be probed by the delayed coherent pulses at different frequencies (ω_2, ω_3, ω_4 ...) selected from the two-dimensional arrays. It was expected that each pulse itself would lead to a coherent broadened spectrum with central wavelength as the probe. Then, an ultra-broadband spectrum could be obtained to produce sub-fs pulses [24] when they were adjusted to walk together temporarily.

8.4 Weak Beam Control of Two-Dimensional Arrays

The strong nonlinear couplings between FW and SH fields are responsible for the collapses of optical waves during second harmonic generation processes in quadratic media [15–17]. It was demonstrated that one-dimensional multiple quadratic spatial solitons could be controlled by seeding a weak SH beam at the input of the nonlinear medium [8]. Since our two-dimensional arrays were formed on the basis of the generation of one-dimensional multiple

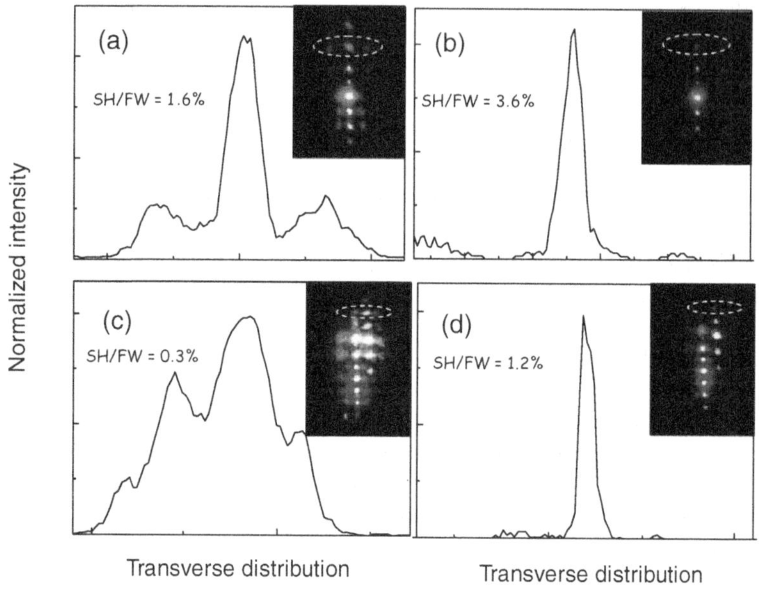

Fig. 8.10. Intensity spectra of a row of multicolored transverse patterns crossing $R(0,2)$ (**a, b**) and a row of multicolored up-converted parametric amplification patterns crossing $G(0,2)$ (**c, d**) under different pulse energy ratios SH/FW. The experimentally observed pictures are also presented as insets

quadratic spatial solitons, it is straightforward to expect that the observed two-dimensional arrays can also be controlled by a weak SH beam.

Figure 8.10a, b present the dependent results of two-dimensional multicolored transverse arrays on the SH control beam. The two-dimensional patterns were switched to one-dimensional ones as the energy ratio SH/FW reached 3.6%. In order to demonstrate the dependence of the control on the relative phase shift between the FW pump and weak SH control pulses, a 1-mm-thick CaF_2 plate was inserted between two BBO crystals to introduce a tunable relative phase shift. By tilting the CaF_2 plate, the relative phase shift could be varied from 0 to 2π readily. The experimental results showed that the control of two-dimensional multicolored transverse arrays was independent on the relative phase between the $\overrightarrow{k_1}$ FW and SH control pulses, which was consistent with the phase-insensitive control of one-dimensional multiple quadratic spatial solitons [8] caused by the spatial collapse of the spatially elliptic input beam. Obviously, the weak SH control beam brought about the same suppression of the two-dimensional patterns as it did on the one-dimensional multiple quadratic spatial solitons induced by the elliptic beam. This further suggests that the spatial collapse of the on-axis $\overrightarrow{k_1}$ beam is responsible for the formation of our two-dimensional multicolored transverse arrays by means of cascaded non-collinearly quadratic processes.

Figure 8.10c, d present the dependence of multicolored up-converted parametric amplifications with super-continuum seeding at $R(1,2)$ under different SH/FW ratios, demonstrating that the two-dimensional multicolored up-converted parametric amplifications could also be controlled by weak SH seeding pulses. Two-dimensional multicolored up-converted parametric amplification patterns were switched into one-dimensional ones when ratio SH/FW was about 1.2%. The threshold energy ratio SH/FW to suppress the two-dimensional multicolored up-converted parametric amplification patterns was consistent with those to suppress one-dimensional multiple quadratic spatial solitons [8] and two-dimensional multicolored transverse arrays.

Similar to one-dimensional multiple quadratic spatial solitons and two-dimensional multicolored transverse arrays, the control of two-dimensional multicolored up-converted parametric amplifications also exhibited a phase-independent feature. This implies that multiple coherent strong pulses could be controlled by using a weak SH beam without any predetermined phase relative to the pulses to be controlled, which may support a novel kind of phase-insensitive all-optical control with a weak beam to control multiple strong beams. As multicolored up-conversions in two-dimensional multicolored up-converted parametric amplifications could be coherent synthesized, a weak SH control of the two-dimensional multicolored up-converted parametric amplifications could facilitate an all-optical control of the relative intensities of different multicolored up-converted parametric amplification patterns, and consequently support an all-optical control of coherent pulses synthesis. This may find interesting applications in all-optical quantum coherent control.

8.5 Colored Conical Emissions

As shown in Fig. 8.11a, spatiotemporal modulational instabilities occur as a result of exponential growth of perturbations during propagations of wave-packets in nonlinear media as both dispersion and diffraction are present [1,2,13]. Assisted by spatiotemporal modulational instabilities, colored conical emissions by means of second harmonic generation involves multiple three-photon processes, which can be described by an intuitionistic picture as shown in Fig. 8.11b [2,13]. Under the pump of on-axis pulses at the FW frequency ω, spatiotemporal modulation instabilities induce decay of SH photons 2ω into photon pairs at frequencies $\omega_{\pm} = \omega \pm \delta$ traveling with opposite off-axis angles, resulting in colored conical emissions. Colored conical emissions could be clearly observed under a high-intensity pump, and it is intrinsically originated from exponential growth of noises or perturbations as a result of spatiotemporal modulational instability. The maximum nonlinear growth occurs at phase-matched directions for different wavelengths, with a characteristic of phase-matched parametric decay of the SH pulses. Accordingly, distinct colored conical emissions appear as the FW pump pulses are aligned at the second harmonic generation phase-matching condition. It is not necessary to

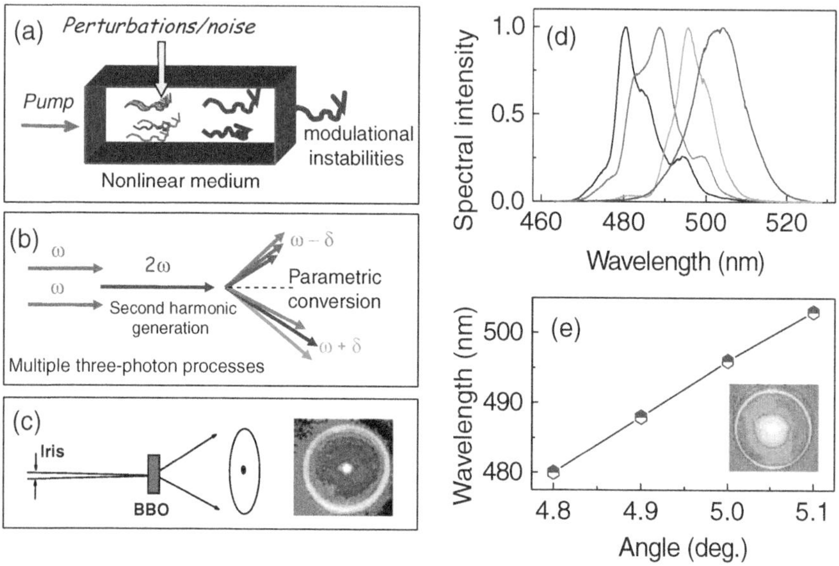

Fig. 8.11. (**a**) Exponential growth of perturbations during propagations of wave-packets in nonlinear media lead to the occurrences of modulational instabilities. (**b**) Multiple three-photon processes and (**c**) experimental schematic for observation of colored conical emissions during second harmonic generation in a thick quadratic nonlinear medium. (**d**) The spectra of the blue-green conical emission recorded with a small angular window. (**e**) The corresponding central wavelengths of the spectra as shown in (**d**) as a function of emission angles. The insets of (**c**) and (**e**) show the typical pictures of colored conical emissions as the FW pump pulses passing through a smaller iris (**c**), and at the optimum pumping chirp and angular spectra (**e**), respectively

have an input spatial asymmetry for the input beam as required for the generation of two-dimensional multicolored transverse arrays and multicolored up-converted parametric amplifications as discussed above [12].

In our experiments, as shown in Fig. 8.11c, the fs pulses from the amplified Ti:sapphire system with energy of 650 μJ/pulse were directly focused into a 6-mm-thick BBO crystal (type I, 29.18° cut) by using a lens with a focusing length of 1,000 mm. As shown in Fig. 8.11e, a bright blue-green cone with a conical angle of ∼ 5.0° emerged around the propagation axis when the pump reached a sufficiently high intensity and the BBO crystal was rotated to maximize the second harmonic generation. As plotted in Fig. 8.11d, e, the blue-green conical emission peaked near 500 nm and exhibited an angular width of about 0.4° (4.77°–5.17°). Spatiotemporal modulational instabilities and thus the induced colored conical emissions changed when the chirp (time) or beam divergence (space) of the FW pump pulse was varied. We used an iris placed before the concave mirror to vary the diameter and divergence of the laser beam focused into the BBO crystal. As presented in Fig. 8.11c, when

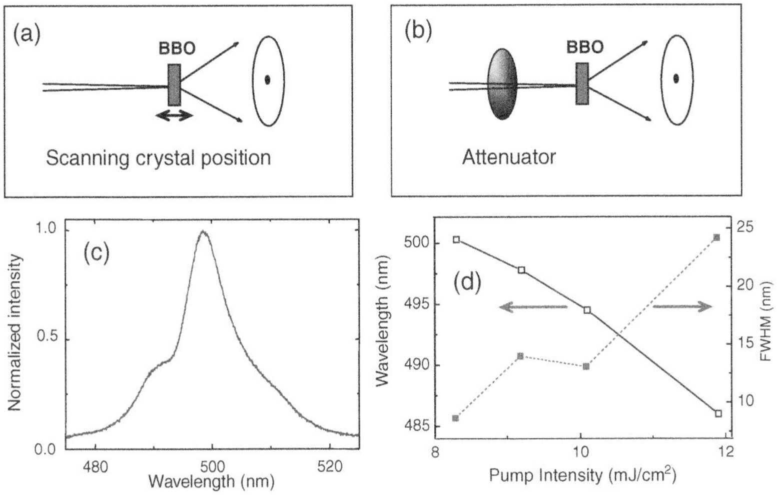

Fig. 8.12. Experimental schemes for change of pump intensities by (**a**) scanning the crystal along the focusing direction and (**b**) rotating a tunable attenuator before the BBO crystal. (**c**) The typical spectral profiles of blue-green conical emission. (**d**) The central wavelengths and FWHMs of the spectra of blue-green conical emission as a function of pump intensity

the iris was changed to a smaller diameter, the blue-green conical emission became less intense, and a red semiconical emission appeared inside the blue-green conical emission. On the other hand, red semiconical emissions could also be observed within the weakened blue-green conical emission as small negative chirps were intentionally introduced to the FW pump pulses. These observations clearly suggest that space and time degrees of freedom played the same role to entail colored conical emissions.

The pump intensity dependent spectral profile is one of the most important features of modulational instabilities [2, 3, 25]. As shown in Fig. 8.12a, b, the pump intensity on the facet of BBO crystal was changed by either scanning the position of the crystal along the focusing direction or rotating a tunable attenuator before the BBO crystal. The spectral features of the blue-green conical emission under different pump intensities are also plotted in Fig. 8.12. As the pump intensity increased, blue-shifted spectra were observed with increased full width at half maximum (FWHM). This implied strong nonlinear couplings among the involved multiple three-photon processes.

8.6 Seeded Amplification of Colored Conical Emission

Seeded amplification of colored conical emission could be realized by seeding with an accurately synchronized super-continuum pulse [14]. The injected seed could attain exponential gain along a certain conical angle of colored conical emissions and the nonlinear gain was assisted by spatiotemporal modulational instabilities, leading to a FW pump intensity dependence of the parametric gain. Intrinsically different from standard optical parametric amplification, seeded amplification of colored conical emission is pumped by the FW pulses, and up-converted amplification occurs as a direct result of multiple quadratic processes (second harmonic generation and parametric amplification). Rather than a simple combination, second harmonic generation and parametric amplifications are coupled each other. Nonlinear phase shifts induced by phase-matched second harmonic generation entail seeded amplification of colored conical emission output spectra dependent on the FW pump intensity. Strong couplings between FW and SH pulses may further balance the group velocity mismatching and dispersion, enabling large interaction length to output a quite high pulse energy of up-converted amplification. From the practical points of view, seeded amplification of colored conical emission supports a broadband fs up-conversion with a quite simple setup.

In the experiments, as shown in Fig. 8.13a, a weak FW pulse split from the output of the laser system was attenuated and then focused into a 2 mm thick sapphire plate to generate a super-continuum beam, which was further collimated to the BBO crystal as the seed pulses. The residual FW pulses were used to generate colored conical emissions with a scheme as discussed above. As shown in Fig. 8.13a, a sparkling green point appeared at point A just as a pearl over the blue-green conical emission ring. The seeded amplification of colored conical emission output beam showed a beam divergence about 0.7 mrad, and the corresponding M^2 parameter was measured to be 5.4 at a typical operation with the output pulse energy of 100 μJ under the fundamental pump intensity of 12 mJ cm^{-2}. The maximum seeded amplification of colored conical emission output pulse energy reached up to ~150 μJ as the FW pump pulse energy was increased to 650 μJ (~14 mJ cm^{-2}). This was at least one order of magnitude higher than those obtained with comparable FW pump pulses by using the standard non-collinear optical parametric amplification scheme.

Figure 8.13b, c presents the typical spectrum of seeded amplification of colored conical emission at a pump intensity of 10.6 mJ cm^{-2}, and the corresponding central wavelengths and FWHMs of the spectra versus the FW pump intensity, respectively. Seeded amplification of blue-green conical emission produced output pulses centered around 500 nm with FWHMs typically larger than 20 nm. Interestingly, seeded amplification of colored conical emission exhibited blue-shifted frequencies of amplification at higher FW pump intensities, implying that the corresponding photon-momentum conservation was modulated, which is actually one of the key features of seeded amplification of colored conical emission that differs from the standard optical parametric amplification. This can be qualitatively understood as follows. Since seeded amplification of colored conical emission consisted of multiple quadratic processes, FW and SH pulses in the second harmonic generation process strongly coupled with the signal and idler pulses in the parametric process. Interplay of

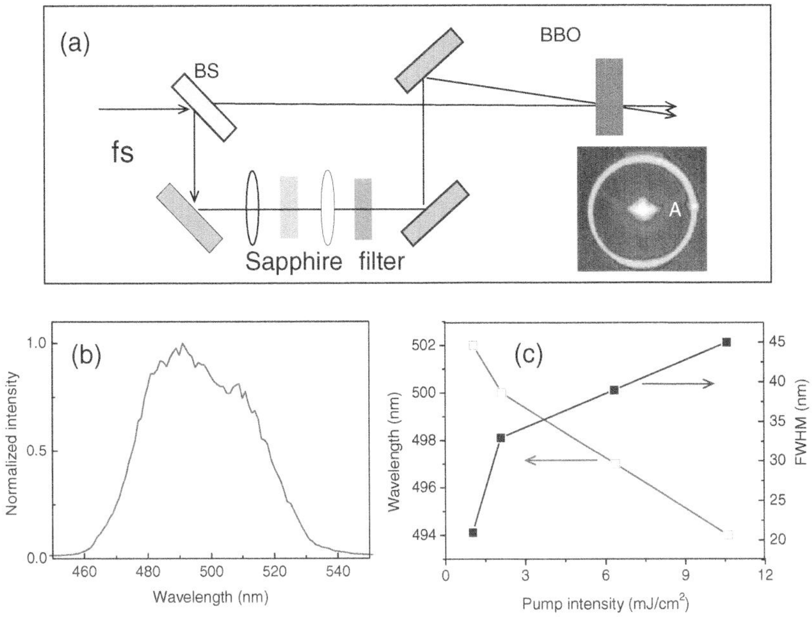

Fig. 8.13. (**a**) Experimental setup for seeded amplification of colored conical emission. (**b**) The typical spectrum of seeded amplification of colored conical emission at a pump intensity of 10.6 mJ cm^{-2}, and (**c**) the corresponding central wavelengths and FWHMs of the spectra of seeded amplification of colored conical emission on the ring as a function of pump intensity. The inset in (**a**) is the picture of the observed seeded amplification of colored conical emission as the seed angle is set to be 5°. BS: beam split

such multiple nonlinear processes caused nonlinear phase shifts to the pump pulses and modulated the effective nonlinear refractive index of the crystal. Consequently, the phase-matching condition and thus the photon-momentum conservation in seeded amplification of colored conical emission changed under different pump intensities. Therefore, for a given crossing angle and delay between the FW pump and super-continuum seed pulses, different spectral profiles could be observed as the FW pump intensities increase.

As colored conical emissions showed different spectra at different conical angles, seeded amplification of colored conical emission with different wavelengths could be obtained by varying the seed angle of the super-continuum beam. Figure 8.14 shows some typical pictures of seeded amplification of colored conical emission observed under different seeding angles. The detailed wavelength dependence of seeded amplification of colored conical emission on the seed angle is plotted in the right of Fig. 8.14, which shows a tunable range from 500 to 798 nm as the seed angle is varied from 5° to 0.68°. As a result of the self-phase-modulation processes in the sapphire plate, there existed a spectral chirp in the seeded super-continuum pulse. Therefore, the

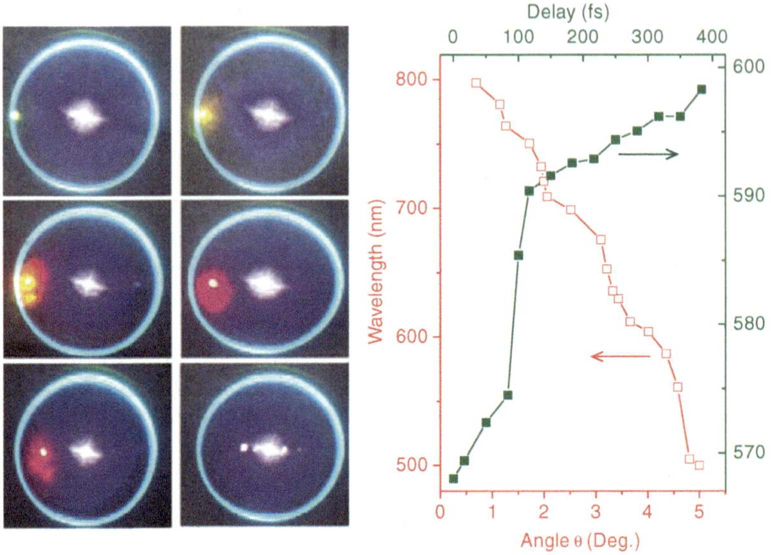

Fig. 8.14. Pictures of observed seeded amplification of colored conical emissions with super-continuum seeded at various angles (*left*). The dependences of the central wavelengths of seeded amplification of colored conical emissions as the seed angle and the delay between the FW and the super-continuum pulses are adjusted (*right*)

wavelength of seeded amplification of colored conical emission could also be tuned by adjusting the delay between FW pump and super-continuum seed pulses. The delay dependent tunable results are also presented in Fig. 8.14 as the seed angle was set as 4.5°. Obviously, tunable ultrashort pulses can be generated based on seeded amplification of colored conical emission, which supports significant amplification of up-converted pulses under infrared pump.

Carrier envelope phase stabilized fs pulses could be obtained by different frequency generation between the seeded amplification of colored conical emission and the FW pump pulses [26]. With a seed from self-phase-modulation broadened replica of a FW pump pulse and phase preservation in the amplification process, the amplified signal differs from FW pump pulse in carrier envelope phase by a constant value of $\pi/2$. Therefore, as shown in Fig. 8.15, the generated signal pulses by different frequency generation show a constant carrier envelope phase. In the experiments, we used a 1 kHz Ti:Sapphire regenerative amplifier to produce 1.4 mJ pulses of 40 fs duration at 800 nm. As shown in Fig. 8.15a, a part of output fs pulse with energy of 690 μJ was used to generate seeded amplification of colored conical emission based on a scheme as shown in Fig. 8.13, and the residual FW pulse with energy of 710 μJ was used for different frequency generation process. The different frequency generation signal in infrared optical region generated in a 2 mm thick BBO crystal (type-I, 31.5° cut) could be further frequency doubled to the visible

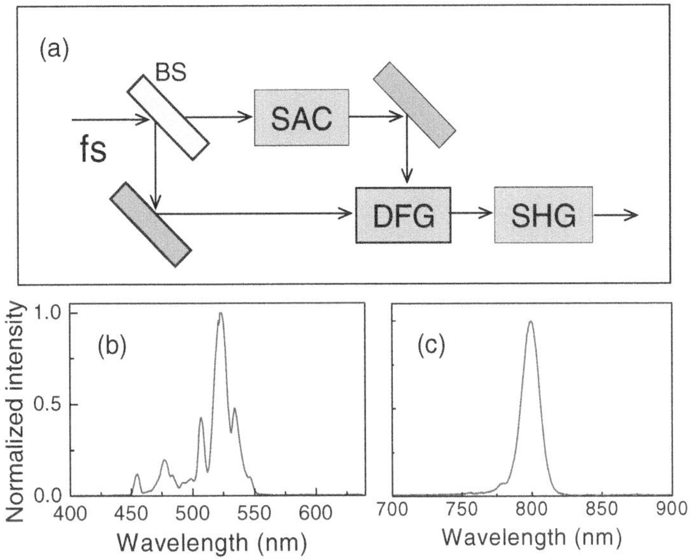

Fig. 8.15. (**a**) Experimental layout for generation of carrier envelope phase stabilized fs pulses by different frequency generation between the seeded amplification of colored conical emission and FW pump pulses. The spectral profiles of seeded amplification of colored conical emission and generated carrier envelope phase stabilized idler pulses are plotted in (**c**) and (**d**), respectively. BS: beam split; SAC: seeded amplification of colored conical emission; DFG: different frequency generation

region using another 2 mm thick BBO crystal (type-I, 19.8° cut). As shown in Fig. 8.15b, c, carrier envelope phase stabilized fs pulse centered at 800 nm could be generated after the second harmonic generation crystal as the central wavelength of the seeded amplification of colored conical emission pulse was tuned to 533 nm.

The carrier envelope phase stabilization of the generated different frequency generation pulse was verified by using a non-collinear optical parametric amplification based $f-2f$ spectral interference scheme [27,28]. As plotted in Fig. 8.16a, a stable interference pattern between the $f-2f$ beating pulses was observed, which indicated a stabilized carrier envelope phase of the different frequency generation signal pulse. The shot-to-shot drifts of the measured carrier envelope phase were calculated and plotted in Fig. 8.16b for the carrier envelope phase stabilized signal pulses. The carrier envelope phase drifts of the pulses emitted directly from the laser system were also plotted in Fig. 8.16c for comparison, which changed randomly from shot to shot without stabilization.

Since seeded amplification of colored conical emission exhibits a widely tunable range, the carrier envelope phase stabilized different frequency generation pulses can be tuned accordingly. As the central wavelength of seeded amplification of colored conical emission was tuned from 500 to 798 nm, as shown in Fig. 8.16d, the carrier envelope phase stabilized different frequency

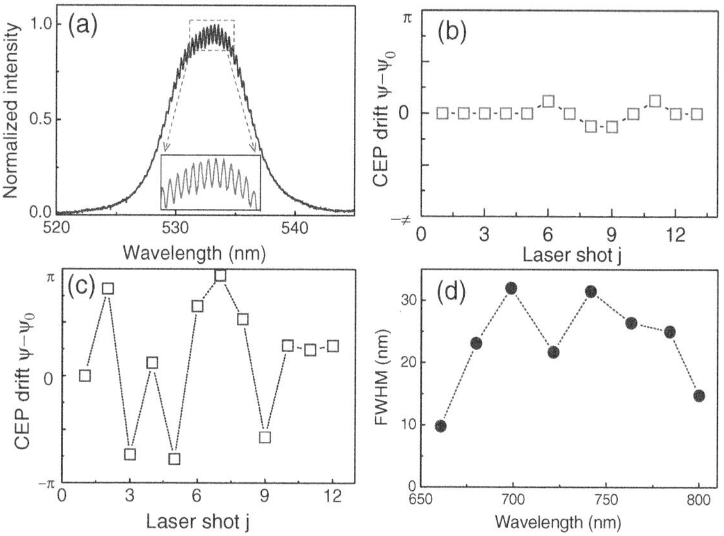

Fig. 8.16. (**a**) Spectral interference result of the generated carrier envelope phase stabilized pulses. (**b**) and (**c**) are the shot-to-shot carrier envelope phase drifts of the carrier envelope phase stabilized pulses and the pulses emitted directly from Ti:sapphire laser, respectively. (**d**) The measured FWHMs of the spectra of carrier envelope phase stabilized idler pulses as the central wavelength is tuned

generation pulses could be tuned from 660 to 800 nm with an FWHM of the spectrum varied between 10 and 32 nm, which correspond to a tunable range from 1,320 to 1,600 nm in the infrared region before the final second harmonic generation crystal.

8.7 Conclusions

As one of the most important and fascinating phenomena that have been observed so far in the propagation of intense laser pulses in nonlinear media, light wave collapse has attracted an ever-growing research interest, and has stimulated some interesting topics on fundamental physics and facilitated a variety of applications. Temporal, spatial, or spatiotemporal collapse of intense optical wave packets has been observed in various nonlinear systems. Nevertheless, the fundamental features of multidimensional wave-packet collapse have been left thus far widely unexplored due to the limit of material damage thresholds. We review here our experimental studies of two-dimensional multicolored solitary arrays with low pump intensities and colored conical emissions during second harmonic generation processes as a result of collapses of fs pulses in quadratic media, which showed various applications in different aspects. We summarize our main results as follows. (a) Two-dimensional multicolored transverse arrays were generated in a quadratic nonlinear medium under the pump of two crossly overlapped fs beams. The experimental observation of two-dimensional patterns demonstrated the spatial breakup of the input elliptic beam and cascaded non-collinearly quadratic processes as a novel mechanism to form two-dimensional patterns at low pump intensities below the material damage threshold. (b) Multicolored up-conversion parametric amplification was realized as a super-continuum probe pulse was diffracted by the transient grating established in the cascaded non-collinearly quadratic processes and then, the diffracted super-continuum pulses attained nonlinear growths with phase preservation along different directions. This presents further evidence of cascaded non-collinearly quadratic processes that were responsible for the formation of two-dimensional patterns. The synchronized super-continuum pulses provided a direct imaging of the spatial distribution of the transient grating in the quadratic medium and thus, diffraction from the transient grating could be used to probe the dynamics of the cascaded non-collinearly quadratic processes. Multiple parametric amplifications concurred with frequency up-conversion, interplay of cascaded non-collinearly quadratic and multiple parametric processes brought about unique features different from standard optical parametric amplification. (c) The two-dimensional arrays could be suppressed through weak beam control of the induced quadratic spatial solitary waves. This provides further evidence that two-dimensional patterns were originated from the spatial beam breakup of the elliptic pump and the induced quadratic spatial solitons equivalently functioned as new beams in the cascaded nonlinear couplings. (d) Colored conical emissions could be observed as a result of spatiotemporal collapse of fs pulses in a quadratic medium. Seeded amplification of colored conical emission was demonstrated to support ultra-broadband and tunable up-conversion, wherein interplay of multiple nonlinear processes makes seeded amplification of colored conical emission differ from standard optical parametric amplifications. Difference frequency generation between pump and seeded amplification of colored conical emission pulses was demonstrated to exhibit a constant carrier envelope phase, which can be used as a novel all-optical control of carrier envelope phase. The carrier envelope phase stabilization was verified and broadband tunable carrier envelope phase pulses were realized. Finally, we would like to point out that of ultrashort pulse collapses in quadratic media include abundant unrevealed features, further experimental and theoretical investigations are required to fully understand the ultrafast dynamics and fundamental physics related with ultrashort pulse propagation.

Acknowledgements

This work was supported in part by National Natural Science Fund (Grants 10774045, 10525416 and 10804032), Projects from Shanghai Science and Technology Commission (Grants 06JC14025, 06SR07102 and 08ZR1407100), National Basic Research Program of China (Grant 2006CB806005), Key projects from Chinese Ministry of Education (Grant 108058), Program for

Changjiang Scholars and Innovative Research Team in University, and Shanghai Educational Development Foundation (Grant 2008CG29). Contributions from Heping Zeng's group members and students (Jingxin Ding, Kun Wu, E Wu, Han Xu, Xuan Yang, Hua Cai, and Yan Peng) are highly appreciated.

References

1. C. Conti, S. Trillo, P.D. Trapani, G. Valiulis, A. Piskarskas, O. Jedrkiewicz, J. Trull, Nonlinear electromagnetic X waves. Phys. Rev. Lett. **90**, 170406 (2003)
2. S. Trillo, C. Conti, P.D. Trapani, O. Jedrkiewicz, J. Trull, G. Valiulis, G. Bellanca, Colored conical emission by means of second-harmonic generation. Opt. Lett. **27**, 1451 (2002)
3. R.A. Fuerst, D.M. Baboiu, B. Lawrence, W.E. Torruellas, G.I. Stegeman, Spatial modulational instability and multisolitonlike generation in a quadratically nonlinear optical medium. Phys. Rev. Lett. **78**, 2756 (1997)
4. S. Polyakov, R. Malendevich, L. Jankovic, G.I. Stegeman, C. Bosshard, P. Gunter, Effects of anisotropic diffraction on quadratic multisoliton excitation in noncritically phase-matched crystals. Opt. Lett. **27**, 1049 (2002)
5. H. Kim, L. Jankovic, G.I. Stegeman, S. Carrasco, L. Torner, D. Eger, M. Katz, Quadratic spatial solitons in periodically poled $KTiOPO_4$. Opt. Lett. **28**, 640 (2003)
6. A. Dubietis, G. Tamošauskas, G. Fibich, B. Ilan, Multiple filamentation induced by input-beam ellipticity. Opt. Lett. **29**, 1126 (2004)
7. S. Carrasco, S. Polyakov, H. Kim, L. Jankovic, G.I. Stegeman, Observation of multiple soliton generation mediated by amplification of asymmetries. Phys. Rev. E **67**, 046616 (2003)
8. S. Polyakov, H. Kim, L. Jankovic, G.I. Stegeman, M. Katz, Weak beam control of multiple quadratic soliton generation. Opt. Lett. **28**, 1451 (2003)
9. G. Assanto, G. Stegeman, M. Sheik-Bahae, E. Van Stryland, All-optical switching devices based on large nonlinear phase shifts from second harmonic generation. Appl. Phys. Lett. **62**, 1323 (1993)
10. X. Liu, L. Qian, F. Wise, High-energy pulse compression by use of negative phase shif ts produced by the cascade $\chi^{(2)}$: $\chi^{(2)}$ nonlinearity. Opt. Lett. **24**, 1777 (1999)
11. L.J. Qian, X. Liu, F.W. Wise, Femtosecond Kerr-lens mode locking with negative nonlinear phase shifts. Opt. Lett. **24**, 166 (1999)
12. H. Zeng, J. Wu, H. Xu, K. Wu, Generation and weak beam control of two-dimensional multicolored arrays in a quadratic nonlinear medium. Phys. Rev. Lett. **96**, 083902 (2006)
13. H. Zeng, J. Wu, H. Xu, K. Wu, E. Wu, colored conical emission by means of second harmonic generation in a quadratically nonlinear medium. Phys. Rev. Lett. **92**, 143903 (2004)
14. H. Zeng, K. Wu, H. Xu, J. Wu, Seeded amplification of colored conical emission via spatiotemporal modulational instability. Appl. Phys. Lett. **87**, 061102 (2005)
15. M. Segev, G.I. Stegeman, Self-trapping of optical beams: spatial solitons. Phys. Today **51**, 42 (1998)
16. A.V. Buryak, Y.S. Kivshar, Spatial optical solitons governed by quadratic nonlinearity. Opt. Lett. **19**, 1612 (1994)

17. K.D. Moll, A.L. Gaeta, Mutual spatio-temporal collapse of femtosecond pulses, quantum electronics and laser science conference, Long Beach (California), USA (2002)
18. J.W. Fleischer, M. Segev, N.K. Efremidis, D.N. Christodoulides, Observation of two-dimensional discrete solitons in optically induced nonlinear photonic lattices. Nature (London) **422**, 147 (2003)
19. G. Méchain, A. Couairon, M. Franco, B. Prade, A. Mysyrowicz, Organizing multiple femtosecond filaments in air. Phys. Rev. Lett. **93**, 035003 (2004)
20. E.J. Mayer, J. Mbius, A. Euteneuer, W.W. Rhle, R. Szipcs, Ultrabroadband chirped mirrors for femtosecond lasers. Opt. Lett. **22**, 528 (1997)
21. S. Guo, Z.Y. Rong, H.T. Wang, Y.R. Wang, L.Z. Cai, Phase-shifting with computer-generated holograms written on a spatial light modulator. Appl. Opt. **32**, 6514 (2003)
22. I.J. Sola, E. Mével, L. Elouga, E. Constant, V. Strelkov, L. Poletto, P. Villoresi, E. Benedetti, J.-P. Caumes, S. Stagira, C. Vozzi, G. Sansone, M. Nisoli, Controlling attosecond electron dynamics by phase-stabilized polarization gating. Nat. Phys. **2**, 319 (2006)
23. A. Nazarkin, G. Korn, Raman self-conversion of femtosecond laser pulses and generation of single-cycle radiation. Phys. Rev. A **58**, R61 (1998)
24. J. Wu and H. Zeng, Subfemtosecond pulse generation and multiplicative increase of pulse spacing in high-order stimulated Raman scattering. Opt. Lett. **28**, 1052 (2003)
25. K. Tai, A. Hasegawa, A. Tomita, Observation of modulational instability in optical fibers. Phys. Rev. Lett. **56**, 135 (1986)
26. K. Wu, X. Yang, H. Zeng, All-optical stabilization of carrier-envelope phase by use of difference frequency generation with seeded amplification of colored conical emission. Appl. Phys. B **88**, 189 (2007)
27. P. Baum, S. Lochbrunner, E. Riedle, Carrier-envelope phase fluctuations of amplified femtosecond pulses: characterization with a simple spatial interference setup. Appl. Phys. B **77**, 129 (2003)
28. K. Wu, Y. Peng, S. Xu, H. Zeng, All-optical control of the carrier-envelope phase with multi-stage optical parametric amplifiers verified with spectral interference. Appl. Phys. B **83**, 537 (2006)

9

Femtosecond Laser Induced Various Luminescent Phenomena in Solid Materials

Jianrong Qiu, Li Wang, and Bin Zhu

Abstract. In this chapter, we review femtosecond laser induced various luminescent phenomena in solid materials. The phenomena of multiphoton excited upconversion luminescence in glasses, crystals and semiconductors, enhanced multiphoton excited upconversion luminescence due to energy transfer between second harmonic generation micro-crystallites and rare-earth ions, different emissions due to the valence state change of active ions induced by femtosecond laser, long-lasting phosphorescence induced by femtosecond laser, and unusual polarization-dependent luminescence induced by femtosecond laser were introduced and their mechanism were discussed. Promising applications of the observed phenomena were also demonstrated.

9.1 Introduction

Compared with CW and long pulsed lasers, femtosecond laser has two apparent features: (1) elimination of the thermal effect due to extremely short energy deposition time, and (2) participation of various nonlinear processes enabled by highly localization of laser photons in both time and spatial domains. Therefore, it is expected that femtosecond laser may induce different phenomena when the femtosecond laser interacts with matter compared with CW and long pulsed lasers.

We know that many glasses and crystals have no intrinsic absorption at 800 nm (the wavelength of the used Ti:sapphire laser). Linear absorption of the laser radiation does not occur when the glasses and crystals are irradiated by the laser beam. This is because the energy gap between the valence and conduction bands is larger than the energy of the single photon. Thus the energy of a single photon is not sufficient to excite an electron at the ground state. At sufficiently high laser intensities, however, an electron can simultaneously absorb the energy from the multiple photons to exceed the band gap. This nonlinear process is called multiphoton absorption. It is a highly intensity-dependent process, with the rate $P(I) = \sigma_k I^k$, where σ_k is the multiphoton absorption coefficient for k-photon absorption.

Once an electron is promoted to the conduction band, it serves as a seed to a process called avalanche ionization. Seed electrons can also arise from the other processes such as electron tunneling and thermal excitation from impurity states. An electron in the conduction band can absorbs sufficient energy $nh\nu \geq E$ (band gap), where n is the number of photons absorbed sequentially, it can then use the excess energy to ionize another electron via direct collision, known as impact ionization. The resulting two electrons in the conduction band can then continue the process of the linear absorption and impact ionization to achieve an exponential growth of the free electrons. Such avalanche ionization produces a highly absorptive and dense plasma, facilitating the transfer of energy from the laser to the glasses and crystals. Many phenomena occur during and after the irradiation of femtosecond laser due to the various nonlinear interaction between laser and glasses and crystals. In this chapter, we introduce various luminescent phenomena in solid materials induced by the femtosecond laser.

9.2 Multiphoton Excited Upconversion Luminescence

Upconversion luminescence is a process that materials absorb longer wavelength light to emit shorter wavelength light, which is in contrast to the process of conventional photoluminescence. The first upconversion luminescence phenomenon was observed in rare-earth doped material ($Eu^{2+}:CaF_2$) in 1960s. From then, it has attracted considerable attention. The mechanisms of upconversion luminescence includes: ETU (energy transfer upconversion), ESA (excited-state absorption, cooperative upconversion), photon avalanche, and multiphoton simultaneous absorption. The upconversion luminescence based on multiphoton simultaneous absorption mechanism has some advantages in practical applications compared to the upconversion luminescence based on the other mechanisms. Multiphoton simultaneous absorption requires the electrons in ground state absorbing two or more pumping photons to transit to excited state directly, while the other mechanisms requires the help of interbands, which means the absorption is a more complicated stepwise process.

The relationship between intensity of fluorescence induced by multiphoton absorption and the power of incident light usually shows exponential relation: $I \propto P^n$, n is the number of photons absorbed. We can obtain the value of n through the measurement of the relationship between intensity of fluorescence induced by multiphoton absorption and the power of incident light. Multiphoton simultaneous absorption is not a brand new phenomenon in photonics filed. Early in 1930s, Gopper–Mayer had predicted this phenomenon theoretically. To induce multiphoton simultaneously absorption requires pump source with very high electric field. And pulsed laser can satisfy with this requirement.

Most solid lasers work at the infrared wavelength range. The conversion of infrared radiation to visible light has received much attention for the increasing applications of UV and visible lasers in optical storage, three-dimensional display, infrared detection and biological imaging. We can convert the infrared ray to visible light through nonlinear process. The usual methods are using high harmonic generation and optical parameters amplification technology, which are confined by the rigorous conditions, such as the direction of polarization, temperature of the crystal and the direction of the crystal axis, etc. However, the confinement of upconversion laser is much less. Most upconversion luminescence induced by multiphoton absorption is pumped by nanosecond laser or picosecond laser. They can hardly generate upconversion luminescence effectively in solid inorganic materials due to low nonlinear effects induced when they interact with the materials. The solvent of the problem is to adopt pulsed laser with higher peak power and shorter pulse duration as pump source.

In recent years, the Ti:sapphire femtosecond laser whose peak power is high has become the appropriate pump source for upconversion luminescence. The advantages of femtosecond laser are as follows: (1) it can offer higher repetition rate, shorter pulse duration and more intensive intensity within the range of wavelength from 267 nm to 2.5 μm. It is possible to induce nonlinear effects more easily duo to its ultrahigh intensity. As early as the year of 1999 and 2000, Kazansky and Qiu observed the phenomena of multiphoton excited upconversion luminescence pumped by infrared femtosecond laser based on multiphoton simultaneous absorption mechanism in Ge-doped SiO_2 glass [1] and rare-earth doped glass [2]. Nowadays, there have been intensive investigations on multiphoton excited upconversion luminescence induced by infrared femtosecond laser.

9.2.1 Multiphoton Excited Upconversion Luminescence in Various Inorganic Glasses

Kazansky et al. observed multiphoton excited upconversion luminescence in Ge-doped silica glass (Ge:SiO_2) in 1999 [1]. Ge:SiO_2 glass is the main constituent material of optical fibers for optical communication and which has an interesting property, strong photosensitivity, associated with defects in glass such as germanium oxygen deficient centers (Ge-ODC or Ge–Si wrong bonds). These centers produce a strong absorption band at 5 eV (240 nm, singlet–singlet transition), a weak absorption band at 3.7 eV (330 nm, forbidden singlet–triplet transition) and blue triplet luminescence at 3.1 eV with a decay time of about 100 ms. Many studies of the blue luminescence in silica glass under the excitation with ultraviolet light have been carried out. These studies have been motivated by the important role of defects responsible for the blue luminescence in two kinds of phenomena. The first one includes refractive index changes and gratings induced via one- or two-photon absorption of ultraviolet or visible light. The second one involves gratings of second-order optical nonlinearity induced via coherent photocurrent in glass (modulation of the angular distribution of photoelectrons) or coherent photoconductivity in glass under applied dc electric field (modulation of the total cross section of ionization) as a result of quantum interference between coherent light fields at two different frequencies. More recently, strong refractive index charges were induced in glass by femtosecond laser pulses via multiphoton absorption of infrared light. On the other hand, the luminescence of silica glass under excitation with intense (10^{11-12} W cm^{-2}) ultrashort (100 fs) infrared radiation has received little attention. It should be pointed out that high optical damage threshold of silica pumped with ultrashort pulses offers unique possibilities for the studies of optical excitations at high light intensities.

The laser radiation in Gaussian mode produced by regeneratively amplified mode-locked (120 fs pulse duration, 200 kHz repetition rate) Ti:sapphire laser operating at a wavelength of 800 nm was used in the experiments. Glass samples were placed on a stage under the optical microscope. The infrared laser radiation reflected by a dichroic mirror inside of the microscope was focused via a 20× objective onto the sample. Simultaneously, the irradiated spot was imaged in the visible spectral range via the microscope by using a color CCD (charge-coupled device) camera.

During the experiments on Ge-doped (GeO_2, 8 mol%) silica glass strong blue luminescence (with a center wavelength at 410 nm) of defect states (Ge-ODC) was observed (Fig. 9.1). Using a cross-sectional area for Ge-ODC of 5×10^{-18} cm^2, the absorption value at 240 nm for our sample of ~15 cm^{-1} gives the Ge-ODC concentration of about 10^{19} cm^3. This luminescence (triplet luminescence) can be excited via the singlet–singlet transition by absorption of three pump photons or one UV photon of the third harmonic of the pump followed by quick nonradiative decay (with a decay time 1 ns) to the long-lived triplet level (Fig. 9.2).

Figure 9.3 shows the absorption spectrum of the $1EuF_2 \bullet 14YF_3 \bullet 10MgF_2 \bullet 20CaF_2 \bullet 10SrF_2 \bullet 10BaF_2 \bullet 35AlF_3$ (mol%) fluoroaluminate glass sample [2]. No absorption peak in the wavelength region from 380 nm to 2.5 μm was observed. A broad peak can be observed at 250 nm, ranging from 190 to 380 nm. The peak can be assigned to the absorption of the $5d$–$4f$ transition of Eu^{2+} [3]. No apparent absorption due to the $4f$–$4f$ transition of Eu^{3+} was observed at 394 nm [3, 4] Therefore, most of the Eu ions existed as a divalent state in the glass.

Figure 9.4 shows the photoluminescence spectra of the glass sample when excited by a femtosecond laser beam using a focusing lens ($f = 100$ mm). A small peak exists at 360 nm due to the $^6P_{7/2}$–$^8S_{7/2}$ transition of Eu^{2+} ions and a broad peak exists at 460 nm due to the $5d$–$4f$ transition of Eu^{2+} ions. Calculations showed that the intensity of the blue luminescence at 460 nm was nearly proportional to the three units of power of the excitation power of the femtosecond laser. Therefore, the blue luminescence at 460 nm was due to three-photon absorption and subsequent relaxation from the $5d$ level to the $^8S_{7/2}$ ground state of the Eu^{2+} ions.

You et al. observed the near-infrared to visible upconverted fluorescence in the Al_2O_3–SiO_2 glass containing Eu^{3+} ions during 800 nm femtosecond laser

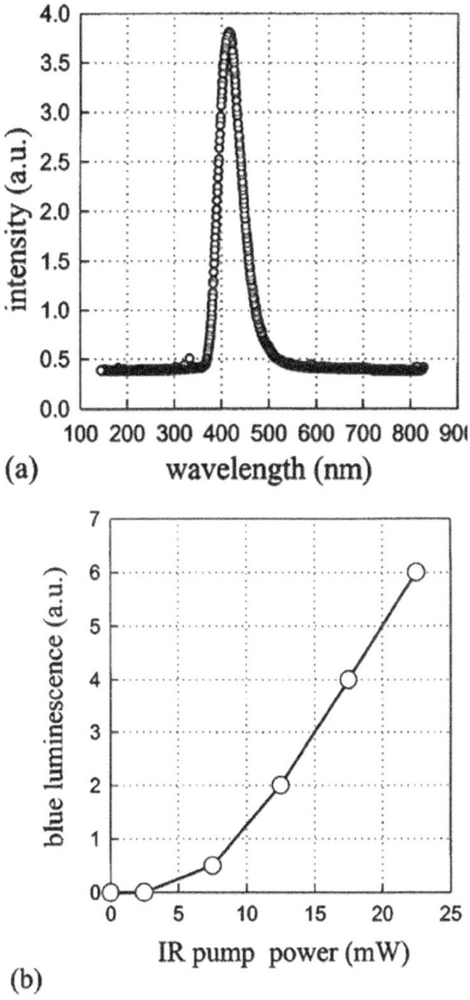

Fig. 9.1. Spectrum (**a**) and pump power dependence (**b**) of the blue luminescence

irradiation [5]. The glass with the composition Si: Al: Eu = 80: 20: 2 (in molar ratio) was prepared by the sol–gel process. The dependence of the intensity of the Eu^{3+} emission on the pump power reveals that the three-photon excitation is dominant in the upconversion process from near-infrared into the visible fluorescence. The analysis of the upconversion mechanism reveals that the three-photon simultaneous absorption leads to the population of the $4f$ levels via the excitation of the charge transfer state of the Eu^{3+} ions, then relaxes into the ^5D_o level of the Eu^{3+} ions from which the fluorescence occurs.

You et al. further used Ce^{3+} ions instead of Eu^{3+} ions, and prepared Al_2O_3–SiO_2 glass with the composition Si: Al: Ce = 80: 20: 0.4 (in molar ratio) by sol–gel method [6]. Near-infrared to visible blue upconversion luminescence

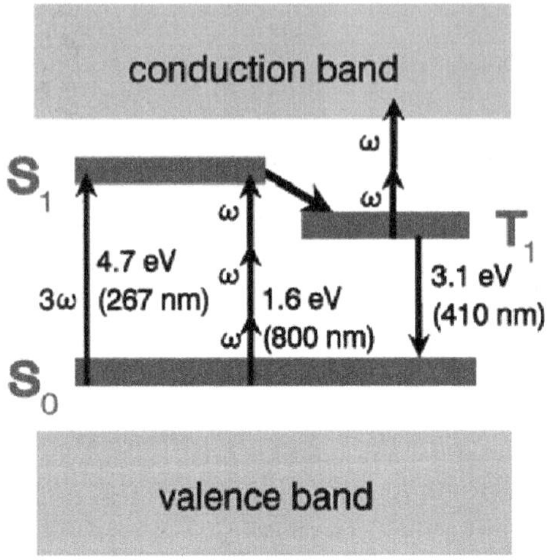

Fig. 9.2. Energy-level diagram of Ge–Si defect in silica glass with the possible channels of excitation

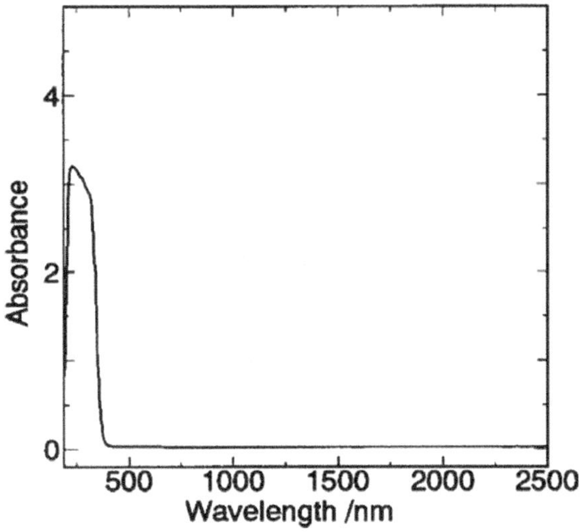

Fig. 9.3. Absorption spectrum of Eu^{2+}-doped fluoroaluminate glass

comes from the $5d$–$4f$ transition of the Ce^{3+} ions was observed in the glass under femtosecond laser irradiation. The relationship between the intensity of the Ce^{3+} emission and the pump power reveals that a three-photon absorption predominates in the conversion process from the near-infrared into the blue luminescence. The analysis of the upconversion mechanism suggests that the upconversion luminescence may come from a three-photon simultaneous absorption that leads to a population of the $5d$ level in which the characteristic luminescence occurs.

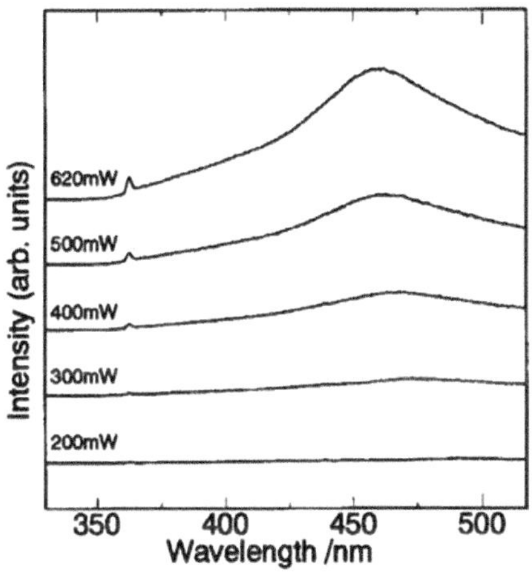

Fig. 9.4. Photoluminescence spectra of Eu^{2+}-doped fluoroaluminate glass excited by a femtosecond laser beam

Meng et al. observed strong blue upconversion luminescence centered at about 420 nm under irradiation with a 780 nm femtosecond pulsed laser [7]. They suggested that the electron was excited into the Ta^{5+} $5d^0$ energy level by absorbing two photons of the short-wavelength part of the supercontinuum of the femtosecond laser and then relaxes to the lowest vibrational level of Ta^{5+} $5d^0$. Finally, the recombination of electron–hole pairs contributed to the blue upconversion. This discovery facilitated understanding the mechanism of blue emission from tantalum-doped glasses and demonstrates the great potential of transition-metal ions without d electrons in the fields of optics.

Zhang et al. reported he near infrared to visible upconversion luminescence in a Tb^{3+}-doped ZnO–B_2O_3–SiO_2 glass excited with 800 nm femtosecond laser irradiation [8]. The upconversion luminescence is attributed to $^5D_4$ to ^7F_j (j = 3, 4, 5, 6) transitions of Tb^{3+}. The relationship between upconversion luminescence intensity and the pump power indicates that a three-photon simultaneous absorption process is dominant in this upconversion luminescence. The calculated value of the three-photon absorption cross section σ_3 of the glass is 1.832×10^{-81} cm^6 s^2. Also, three-dimensional display is demonstrated based on the multiphoton absorption upconversion luminescence for the first time.

As the Tb^{3+}-doped glass has shown highly localized green emission when excited with focused femtosecond laser, this observed phenomenon can be used for three dimensional solid state displays. Figure 9.5 presents the photograph of a simple mark through scanning the 800 nm femtosecond laser into the Tb^{3+}-doped glass at a scanning rate of 300 pixels s^{-1} via computer controlled scanning type galvanometer. The femtosecond laser beam was focused onto the glass by a lens with focal length of 2 cm. They have observed a three

Fig. 9.5. Photograph of a simple mark through scanning the femtosecond laser into the glass

dimensional green ellipsoidal shape image with 13 horizontal lines inside the glass. It should be possible to realize three-dimensional, solid state, three color display based on multiphoton absorption upconversion luminescence if one can design and control the combination of doping species and microstructure of the glass.

9.2.2 Enhanced Multiphoton Excited Upconversion Luminescence to Due to Energy Transfer Between Second Harmonic Generation Micro-Crystallites and Rare-Earth Ions

Zhu et al. observed enhanced near-infrared to visible red upconversion luminescence in transparent Eu^{3+}-doped SrO–TiO_2–SiO_2 glass–ceramics induced by near-infrared femtosecond laser (Fig. 9.6) [9]. X-ray diffraction analysis indicates that $Sr_2TiSi_2O_8$ crystallites of good second-order optical nonlinearity are precipitated in the glass after heat treatment at 850°C for 2 h. They suggested that the enhanced emissions can be attributed to the absorption of the second-harmonic generation as a result of precipitated $Sr_2TiSi_2O_8$ microcrystalline particles with second-order optical nonlinearities. The 790 nm femtosecond laser was converted to a 395 nm laser in the glass–ceramic at first, and then the latter was reabsorbed by Eu^{3+}. It should be pointed out that the half-width at middle maximum of the femtosecond laser is about 18 nm, and the wavelength of the second-harmonic generation covers the wavelength region from 386 to 404 nm. Second-harmonic generation is re-absorbed effectively by Eu^{3+} via the ${}^7F_0 \rightarrow {}^5L_6$ transition at 394 nm. The intensity of the upconversion luminescence was proportional to the square of the excitation power. Furthermore, the efficiency of luminescence due to multiphoton excitation is far less than that of the second-harmonic generation. The Eu^{3+}-doped glass–ceramics show strongly enhanced upconversion luminescence and greatly improved damage threshold compared with the as-prepared glass.

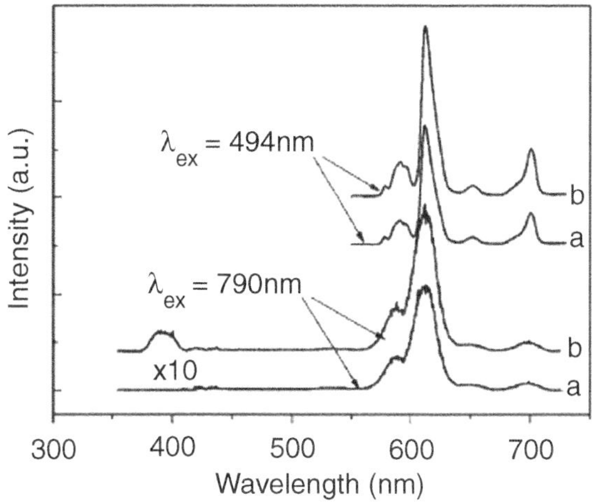

Fig. 9.6. Emission spectra of the Eu^{3+}-doped glass (**a**) and glass–ceramic (**b**) under femtosecond laser irradiation (the excitation power is 32 mW) and 394 nm light excitation

9.2.3 Multiphoton Excited Upconversion Luminescence in Various Crystals

Dong et al. observed blue frequency-upconversion fluorescence emission in Ce^{3+} doped Gd_2SiO_5 single crystals, pumped with 120-fs laser 800 nm IR laser pulses (Fig. 9.7), while no visible emission was observed under irradiation of 800 nm picosecond laser [10]. The observed fluorescence emission peaks at about 440 nm is due to $5d \rightarrow 4f$ transition of Ce^{3+} ions. The intensity dependence of the blue fluorescence emission on the IR excitation laser power obeys the cubic law, demonstrating three-photon absorption process. Analysis suggested that three-photon simultaneous absorption induced population inversion should be the predominant frequency upconversion mechanism. The process of the direct three photon induced fluorescence of $Ce^{3+}:Gd_2SiO_5$ by femtosecond laser irradiation may be as follows: first, the simultaneous absorption of three photons by $Ce^{3+}:Gd_2SiO_5$ leading to population of the upper $5d$ excited state. Second, the excited state relaxes nonradiatively to the zeroth vibronic level – the bottom of the $5d$ state before returning to the $^2F_{5/2}$ ground state via emission of a single photon. The population inversion is created between the lower emitting state, $5d$, and the ground state, $^2F_{5/2}$.

They also found that purple upconversion luminescence in $Ce^{3+}:Lu_2Si_2O_7$ single crystals excited at 800 nm by using a femtosecond Ti:sapphire laser [3]. The emission spectra of $Ce^{3+}:Lu_2Si_2O_7$ crystals were the same for one-photon excitation at 267 nm as for excitation at 800 nm (Fig. 9.8). The emission intensity of $Ce^{3+}:Lu_2Si_2O_7$ crystals was found to depend on the cube of the laser power at 800 nm, consistent with simultaneous absorption of three 800 nm photons.

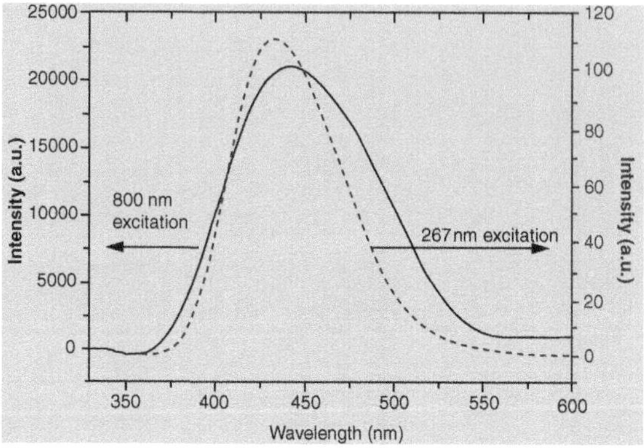

Fig. 9.7. Emission spectra of Ce^{3+}: Gd_2SiO_5 irradiated by femtosecond laser at 800 nm (*solid line*) and xenon lamp at 267 nm (*dashed line*)

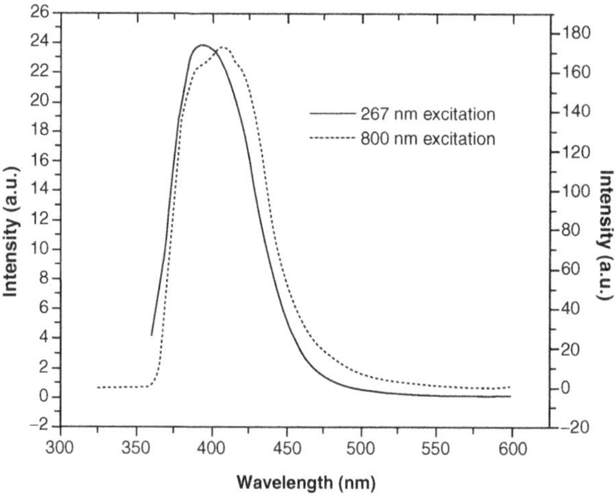

Fig. 9.8. Emission spectra of Ce^{3+}: $Lu_2Si_2O_7$ single crystals for excitation at 800 and 267 nm

The reports on upconversion luminescence in transparent solids doped with transition metal ions are not too much. Among these ions, Cr^{3+} ion is particular because its $^2E_3 \rightarrow {}^4A_2$ transition is typical in solid spectroscopy. The $^2E_3 \rightarrow {}^4A_2$ transition in crystal and glass is studied extensively to utilize its emission at infrared wavelength and explore the crystalline field around Cr^{3+} ions.

Yang et al. reported visible red upconversion luminescence in Cr^{3+}: Al_2O_3 crystal under focused femtosecond laser irradiation [4]. The luminescence spectra showed that the upconversion luminescence originates from the $^2E \rightarrow {}^4A_2$ transition of Cr^{3+}. The dependence of the fluorescence intensity of Cr^{3+} on the pump power revealed that a two-photon absorption process dominates in

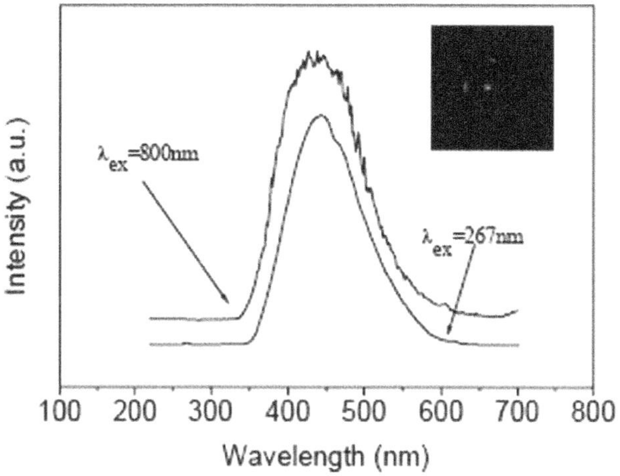

Fig. 9.9. Emission spectra of the YVO_4 crystal sample under focused femtosecond laser irradiation and 267 nm monochromatic light excitation. The inset was the photograph of the YVO_4 crystal sample irradiated by focused femtosecond laser, which was taken in darkroom

the conversion of infrared radiation to the visible emission. It was suggested that the simultaneous absorption of two infrared photons produces the population of upper excited states, which leads to the characteristic visible emission from ^2E state of Cr^{3+}. They also reported on blue upconversion luminescence of a pure YVO_4 single crystal excited by an infrared femtosecond laser [11]. The luminescent spectra (Fig. 9.9) showed that the upconversion luminescence comes from the transitions from the lowest excited states $^3T_1$, $^3T_2$ to the ground state $^1A_1$ of the $VO_4{}^{3-}$. The dependence of the fluorescence intensity on the pump power density of laser indicated that the conversion of infrared irradiation to visible emission is dominated by three-photon excitation process. They suggested that the simultaneous absorption of three infrared photons promotes the $VO_4{}^{3-}$ to excited states, which quickly cascade down to lowest excited states, and radiatively relax to ground states, resulting in the broad characteristic fluorescence of $VO_4{}^{3-}$.

They observed infrared to ultraviolet and visible upconversion luminescence in Ce^{3+} doped $YAlO_3$ crystal (Ce^{3+}: YAP) under focused infrared femtosecond laser irradiation [12]. The fluorescence spectra showed that the upconverted luminescence comes from the $5d - 4f$ transitions of Ce^{3+} ions. The dependence of luminescence intensity of Ce^{3+} ions on infrared pumping power reveals that the conversion of infrared radiation is dominated by three-photon excitation process. It is suggested that the simultaneous absorption of three infrared photons pumps the Ce^{3+} ion into upper $5d$ level, which quickly nonradiatively relax to lowest $5d$ level. Thereafter, the ions radiatively return to the ground states, leading to the characteristic emission of Ce^{3+}.

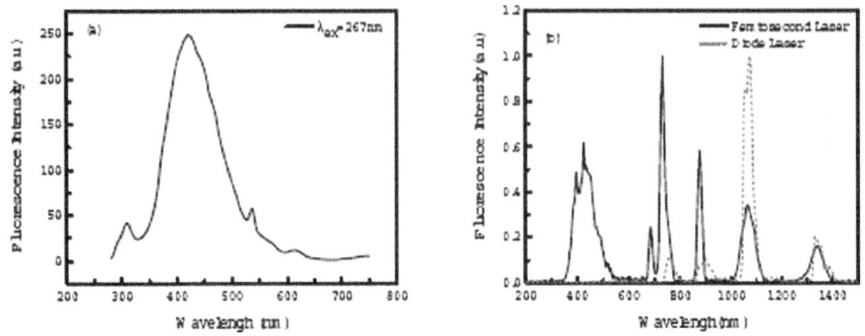

Fig. 9.10. Emission spectra of the $Nd^{3+}: YVO_4$ crystal sample under (**a**) 267 nm monochromatic light, (**b**) femtosecond laser and CW diode laser excitation

Wang et al. observed upconversion luminescence in Nd: YVO_4 under excitation of a continuous wave (CW) diode laser and a femtosecond laser at 800 nm (Fig. 9.10) [13]. It was found that Nd: YVO_4 shows different upconversion and downconversion luminescencent behaviors when excited by the diode laser and the femtosecond laser. The log–log relationship between pump power density of the femtosecond laser and fluorescence intensity of Nd:YVO_4 showed that the upconversion processes for the 424, 731 and 684 nm signals are three-photon-absorption processes at low pump power, while at high pump power, four- or five-photon excitation process might occur. They also found that the upconversion fluorescence intensity of Nd^{3+} excited by the femtosecond laser is higher than the downconversion one, which is consistent with the calculation result based on Bloch equations, and that the upconversion luminescence efficiency pumped by the femtosecond laser is higher than that by the diode laser.

9.2.4 Multiphoton Excited Upconversion Luminescence Inside Semiconductors

The nonlinearity of the semiconductor has received considerable attention, because it can be applied in devices for optical power limitation. The semiconductor used in femtosecond laser inducing upconversion luminescence is IV–VI semiconductor.

Dushkina et al. reported on green upconversion luminescence centered at 525 nm in CdS crystal after irradiation of 804 nm femtosecond laser [14]. At room temperature, CdS can hardly emit green light. Sometimes, one may observe red emission duo to the defects and sulfur vacancy. In this case, femtosecond laser induced two-photon absorption which could excite the electron–hole pairs of CdS. Then the green light was emitted duo to the recombination of electron sand holes. Pan et al. observed upconversion luminescence centered at 500 nm in CdS nanoribbons which was irradiated by femtosecond laser at room temperature [15]. These ribbons have single-crystal structures and uniform rectangular cross sections and grow along the (100) direction. The upconversion luminescence was attributed to photo-induced bipolaronic excitons.

Compared to CdS, the forbidden band of CdTe is more narrow, and the emission quantum efficiency is higher. Wu et al. measured the PL spectra of CdTe nanocrystals of size of 3.8, 4.4 and 5.4 nm irradiated by 400, 800 and 1,208 nm femtosecond laser, respectively [16]. The upconversion luminescence spectra had slopes of 2.05 ± 0.03 and 1.88 ± 0.02 at 800 and 1,208 nm excitation, respectively, while the normal luminescence spectra had a slope of 0.85 ± 0.004. The results indicate that normal luminescence intensity was linearly dependent on excitation power, whereas upconversion luminescence intensity varied quadratically with pump power. It was therefore concluded that upconversion luminescence was attributed to two-photon excitation or an Auger recombination process.

Recently, ZnO has attracted renewed interest as a promising optoelectronic material in the UV for laser diodes and light-emitting diodes operating at room temperature owing to its wide bandgap and extremely large binding energy of free excitons. Dai et al. found that there are two emission bands in representative MPA-induced PL spectra of ZnO excited by a femtosecond laser with tunable wavelength range of 700–1,000 nm at room temperature: a narrow UV band at 395 nm (3.14 eV) and a broad visible (VIS) band centered at ~492 nm (2.52 eV) (Fig. 9.11a) [17]. The peaks marked by vertical arrows were the second-harmonic generation signals of the laser lines. The UV emission band was due to electron–hole plasma emission, which was 230 meV below the bandgap, ~3.37 eV (368 nm) of ZnO at room temperature and was a result of bandgap renormalization that happened immediately after intense excitation. The broad VIS emission band was due to the deep luminescent centers in ZnO. As the energies of the excitation photons were well below the bandgap of ZnO, the observed luminescence bands were certainly due to

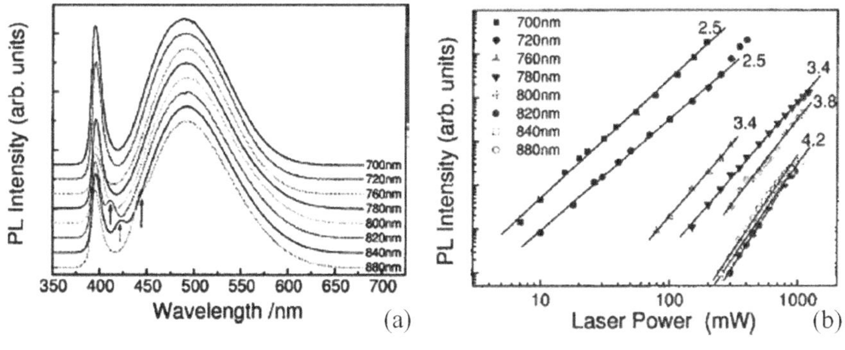

Fig. 9.11. (**a**) MPA-induced PL spectra of ZnO at room temperature. (**b**) Dependence on excitation power of the integrated intensity of the UV emission band. *Solid curves* are the results of fitting

multiphoton absorption. They measured the dependence of the UV emission band on excitation power for several laser wavelengths, and the results are shown in Fig. 9.11b on a logarithmic scale. The slopes of the solid curves which were the results of fitting proved that the upconversion luminescence was due to two-photon absorption, three-photon absorption and four-photon absorption. In particular, the observation of efficient 3PA in ZnO was of scientific significance because this phenomenon was rarely reported in wide bandgap semiconductor materials. These results showed that ZnO was a promising material for nonlinear optical applications.

Prasanth et al. carried out two-photon absorption measurement for ZnO nanowires using femtosecond laser pulses in the wavelength range of 700–800 nm [18]. They observed deep-level green emission around 530 nm due to surface defects and the near band-edge ultraviolet emission around 360 nm due to the exciton under the irradiation of 700 and 753 nm femtosecond laser.

Li et al. reported an observation of efficient two-photon photoluminescence (TPL) of InGaN/GaN multi-quantum-well (MQW) structures using broadband femtosecond near-infrared excitation laser at low temperature of 77 K [19]. The excitation wavelengths were 710, 760 and 780 nm, respectively. The inset showed the logarithmic excitation-power dependence of the integrated TPL intensity for excitation wavelength of 760 nm. The power index obtained from the least-squares fitting was 2.12 ± 0.07. It was proved that the blue upconversion luminescence was due to two-photon absorption. These results show that InGaN/GaN MQWs structures are promising heterostructures for applications of nonlinear optics.

Yang et al. observed second-harmonic generation (SHG) and yellow nonlinear photoluminescence (NPL) in GaN film grown on sapphire simultaneously or individually by using a tunable broadband femtosecond laser as excitation source at room temperature [20]. They carefully measured dependence of the SHG and NPL signals on polarization of the excitation light. The results revealed that the reabsorption of the SHG photons with energies higher than the fundamental gap of GaN significantly contributed to generation of the efficient NPL signal.

Wu et al. reported visible luminescence from SiO_x formed by microstructuring silicon surfaces with 800 nm femtosecond laser pulses in air. SiO_x, $x < 2$, consisted of silicon nanoclusters embedded in a SiO_2 matrix [21]. At room temperature the luminescence was visible to the naked eye with less than 10 mW excitation at 488 nm. The peak wavelength of the primary luminescence band varies between 540 and 630 nm. They attributed the green band to recombination of carriers at defects ("defect luminescence") and the red band to recombination of confined excitons in silicon nanoclusters ("quantum confinement"). The behavior of the luminescence upon annealing was one way to distinguish these two mechanisms. Annealing increased the size of Si clusters within the SiO_2 matrix by diffusion of Si, a process known as Ostwald ripening. Luminescence resulting from quantum confinement should therefore show a wavelength shift to lower energies upon annealing, as the cluster size increased and the excitonic energy decreased. By contrast, for defect luminescence, annealing might eliminate radiative or nonradiative defect sites (or both, potentially at different rates), and thus might alter the intensity, but should not alter the peak wavelength substantially. The luminescence wavelength was dependent on the average number of laser shots per unit area $<N>$. The luminescence intensity increased with increased oxygen content, unlike the luminescence from porous silicon, which degraded upon oxidation. The amount of oxygen incorporated into the silicon increase with the laser fluence used for microstructuring. Laser production of luminescent SiO_x offered the possibility of precise, localized SiO_x production on silicon chips without the use of masks or multistep processing; other methods of producing luminescent SiO_x lack this versatility.

9.3 Different Emissions to Due to the Valence State Change of Active Ions Induced by Femtosecond Laser

Qiu et al. have carried out systematic studies on the valence state change of rare-earth ions and noble ions in glasses induced by femtosecond laser [22–27]. They observed that the color of the focused area of the Sm^{3+} ions doped alumina borate glass became orange after 800 nm femtosecond laser irradiation [22]. Under the excitation of 514.5 nm Ar^+ laser, orange fluorescence attributed to $4f \rightarrow 4f$ transition of Sm^{2+} ions was observed. Sm^{2+} ions are due to the photoreduction of Sm^{3+} induced by femtosecond laser. The mechanism of photoreduction is: when the femtosecond laser is focused into the sample, free electrons are generated by a series of ionization processes of multiphoton absorption, thermal absorption, collision and so on. The holes are trapped by non-bridging oxygens while free electrons are trapped by Sm^{3+} to form Sm^{2+}.

Recently, they examined the possibility of achieving 3D optical data storage inside glass by using a focused femtosecond laser to permanently photoreduce Sm^{3+} to Sm^{2+} [26]. The composition of the glass samples doped with samarium ions used in this study was $35AlF_3$–$10BaF_2$–$10SrF_2$–$20CaF_2$ –$10MgF_2$–$14.5YF_3$–$0.5SmF_3$ (mol%). Samarium was present in the glass in the Sm^{3+} ionic state. As a recording source. They employed a regeneratively amplified 800 nm Ti:sapphire laser emitting 20 Hz or 250 kHz mode-locked pulses. Femtosecond laser was tightly focused inside the bulk glass to locally photoreduce Sm^{3+} to Sm^{2+}. A 3D recording was made by laser irradiation through a water-immersion objective (63× magnification, NA = 1.2) in the interior of the glass samples with an XYZ stage. Photoluminescence images were obtained with a three-dimensional nanometer-scanning spectroscopic microscope (Nanofinder TII) using 488 nm light from an Ar^+ laser as the readout (excitation) source. All experiments were carried out at room temperature.

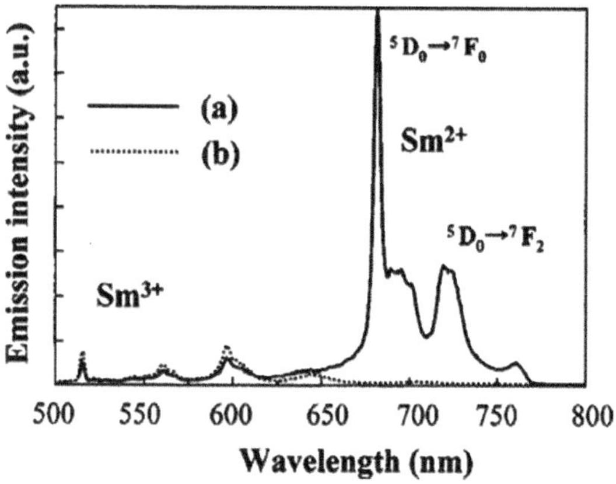

Fig. 9.12. Photoluminescence spectra obtained by excitation at 488 nm for a laser-irradiated area (**a**) and non-irradiated area (**b**) in the interior of the glass

Figure 9.12 shows photoluminescence spectra obtained by excitation at 488 nm for a laser-irradiated area (a) and non-irradiated area (b) in the interior of the glass. Comparing (a)–(b) shows that the emission in the 650–775 nm region differed appreciably. The broadbands observed around 560, 600, and 645 nm can be attributed to the $^4G_{5/2} \rightarrow ^6H_{5/2,\ 7/2,\ 9/2}$ transitions, respectively, of the Sm^{3+} ions. On the other hand, the emissions at 680, 700, and 725 nm are attributed to the $^5D_0 \rightarrow ^7F_{0,\ 1,\ 2}$ transitions, respectively, of the Sm^{2+} ions. This means that laser-irradiated areas (photo-reduced areas) recorded inside glass can be detected only by emissions at 680, 700, or 725 nm. Although the mechanism is not fully clear at present, it is suggested that the mechanism of the photoreduction is as follows. Active electrons and holes were created in the glass through a multiphoton process. Holes were trapped in the active sites in the glass matrix while some of the electrons were trapped by Sm^{3+}, leading to the formation of hole-trapped defect centers and Sm^{2+}.

By using the photoreduction of Sm^{3+} to Sm^{2+}, alphabetical characters were recorded in the form of sub-micron-size bits in a three-dimensional (layered) manner in the glass samples. Here, one recorded character consisted of 300–500 photoreduction bits recorded with 5,000 laser shots per bit with a repetition rate of 250 kHz (pulse energy: 40 nJ). The spacing between alphabetical characters was 2 μm. The bits had a diameter of 400 nm, which was significantly smaller than the focal-beam size and the wavelength of the recording laser. Figure 9.13 shows photoluminescence images of alphabetical characters recorded on different layers, which were observed by using a 40× objective lens and the 680 nm emission from Sm^{2+} with confocal detection implemented. They confirmed that the spacing of 2 μm between alphabetical character planes was sufficient to prevent cross talk in the photoluminescence images. Although the 3D memory bits (photo-reduced areas) were recorded with a femtosecond laser, they could be read with a cw laser at 0.5 mW.

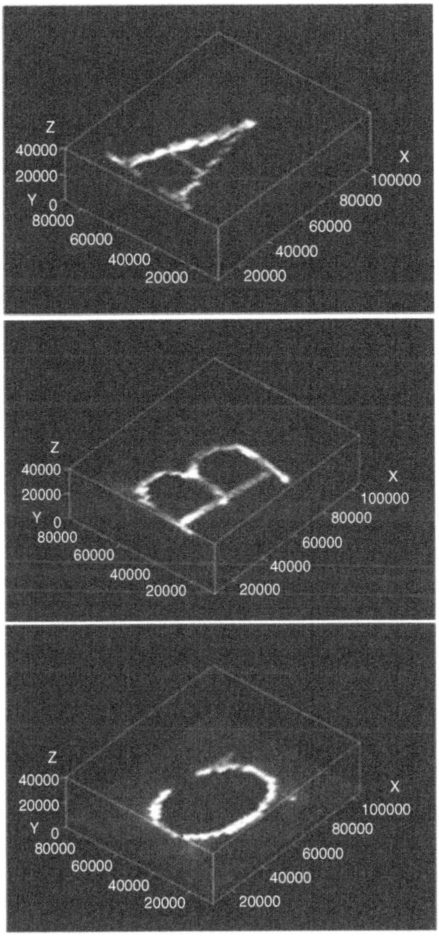

Fig. 9.13. Photoluminescence images of alphabetical characters recorded on different layers, which were observed by using a 40× objective lens and the 680 nm emission from Sm^{2+} with confocal detection implemented (excitation at 488 nm, 1 mW Ar^{+} laser)

Recording and readout in a three-dimensional memory are possible without any influence from bits in upper or lower layers recorded previously. Although obtaining a multilayer of several hundred layers has been difficult in a conventional 3D memory, it can be achieved by applying photoreduction bits to current optical memory technology. In addition, as shown in Fig. 9.14, photoreduction bits with a 200 nm diam could also be read out clearly by detecting the fluorescence as a signal. Since photoreduction bits can be spaced 150 nm apart on a layer in glass, a memory capacity of as high as 1 Tbit could be achieved for a glass piece with dimensions of 10 mm × 10 mm × 1 mm.

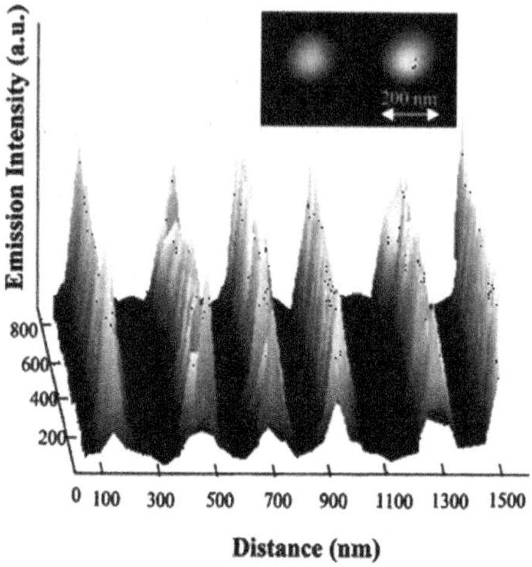

Fig. 9.14. Signal readout by detecting the fluorescence for photoreduction bits with a 200 nm diam

Another outstanding characteristic of the photoreduction of Sm^{3+} to Sm^{2+} is that stable Sm^{2+} at room temperature can be changed to Sm^{3+} by photo-oxidation with a cw laser, such as an argon-ion laser or a semiconductor laser. An example of a photoreduction bit erased by irradiation with an Ar^+ laser (10 mW at 514.5 nm) is shown in Fig. 9.15. Figure 9.16 shows the contrast achieved in readout of bits (I, II) in Fig. 9.15. These results indicate the possibility of achieving a three-dimensional optical memory with rewrite capability. More than ten times rewriteable capability was confirmed recently.

By using the valence-state change of samarium ions to represent a bit, data can be read out as fluorescence information and erased by irradiation with a cw laser. The ratio of the fluorescence intensity at 680 nm after and before irradiation with the femtosecond laser is limitless, so a high signal-to-noise ratio can be achieved. Moreover, it is possible to create a multilayer record of data over several hundred layers, because the recording, readout, and erasure processes for the data are not affected by layers recorded nearby. Therefore, this technique will be useful in fabricating next-generation 3D optical memory devices with an ultra-high storage density.

Qiu et al. found that Eu^{3+} ions were changed into Eu^{2+} in the Eu^{3+} ions doped fluoroaluminate [27] glass by 800 nm femtosecond laser irradiation. Blue fluorescence attributed to $5d$–$4f$ transition of Eu^{2+} ions could be observed under the excitation of UV light (Fig. 9.17).

Besides the rare-earth ions, Qiu et al. also reported noble metal ions can be reduced or oxidized in various glasses after infrared femtosecond laser irradiation, and resulting in different emission under the excitation of UV light [22–27].

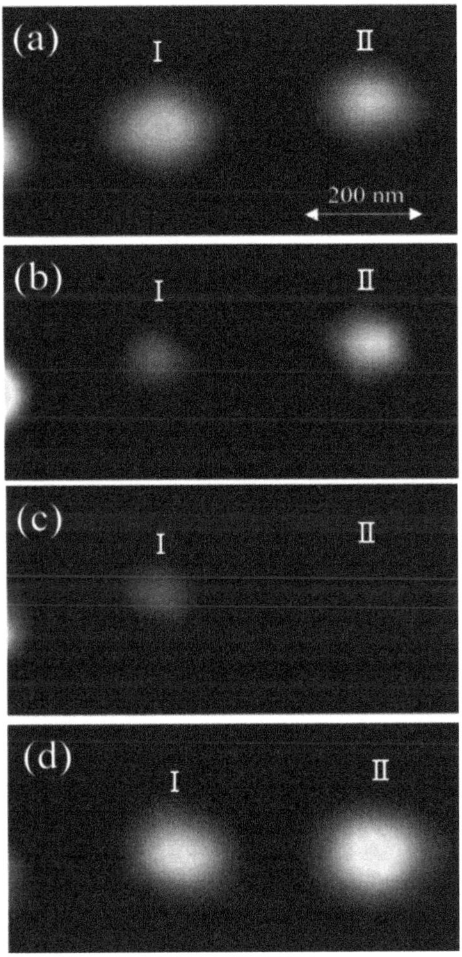

Fig. 9.15. Example of erasing and rewriting by irradiation with an Ar^+ laser (5 mW at 514.5 nm) and a femtosecond laser. (**a**) Photoluminescence image before the erasure. (**b**) Image after Ar^+ laser irradiation to photoreduction bit I. (**c**) Image after Ar^+ laser irradiation to bit II. (**d**) Image after femtosecond laser irradiation to areas I and II

Watanabe et al. reported a bright orange fluorescence from the irradiated area inside soda-alumino-phosphate glass doped with silver ions Ag^+ by near-ultraviolet femtosecond laser pulses (Fig. 9.3) [28]. The orange fluorescence is attributed to the formation of color centers such as Ag^0 and Ag^{2+} in the glass, which is found to be associated with intensity-dependent nonlinear optical process induced by extremely high irradiance up to $\sim$TW cm^{-2} of the femtosecond pulses. They demonstrate a bright orange fluorescence from the induced color centers inside the glass should be applicable to a functional optical device.

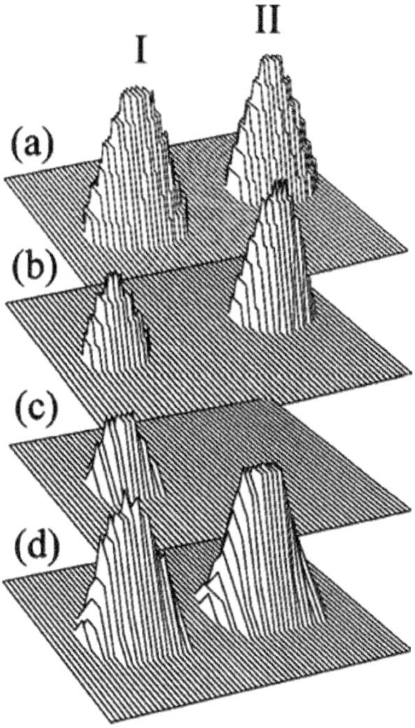

Fig. 9.16. Intensity distribution of the fluorescence (680 nm) for bits of I and II in Fig. 9.4a–d

Podlipensky et al. studied the luminescence of spherical Ag nanoparticles doped soda-lime glass upon irradiation by intense femtosecond laser pulses [29]. The results give evidence that the nanoparticles are ionized by the laser pulses, and rather high local concentrations of Ag^+ ions are created near to the clusters, apparently playing an important role for the processes of shape deformation. Samples with Ag nanoparticles show, after irradiation and only moderate temperature treatment at 100 °C, a strong broad band luminescence (excitation wavelength 400 nm) centered at 600 nm, which can be attributed to small silver complexes ${Ag_2}^+$, ${Ag_3}^+$, ${Ag_3}^{2+}$.

Gleitsmann et al. observed strong orange luminescence from silver clusters generated by 800 nm femtosecond-laser-induced reduction of silver oxide nanoparticles embedded in a polymeric gelatin matrix upon exposure to Hg lamp [30, 31]. Actually, fluorescence could be observed from the pure silver oxide nanoparticles film after irradiation of 800 nm femtosecond laser under excitation of Ar^+/Kr^+ laser (488 nm blue light or 568 nm green light). Green fluorescence was observed under excitation of blue light, and red fluorescence was observed under green light. However, there was no fluorescence observed under the same excitation source without irradiation of femtosecond laser. They suggested that the femtosecond laser activate fluorescence from silver oxide nanoparticles, and the process of activation including

Fig. 9.17. Valence state change of Eu ion induced by femtosecond laser

Fig. 9.18 Fluorescence observed by excitation with an Hg lamp

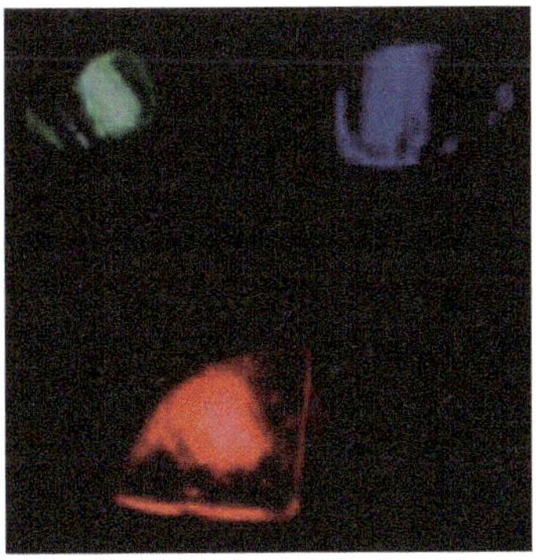

Fig. 9.19 Photograph of the emission states of phosphorescence in glass samples 5 min after the removal of the exciting laser. The Ce^{3+}-doped calcium aluminosilicate glass sample shows *blue light* emission. The Tb^{3+}-doped calcium aluminosilicate glass sample shows *green light* emission. The Pr^{3+}-doped calcium aluminosilicate glass sample shows *red light* emission

multiple sequential steps such as chemical reduction, diffusion, cluster formation, as well as cluster decomposition is most likely to prevail. But the fluorescence from the femtosecond-laser-activated matrix areas considerably exceeds the luminescence intensity of similarly activated bare silver oxide nanoparticle films.

The favorable mechanical properties of the polymeric gelatin matrix material, as well as the measured long-time stability of the luminescent structures in the matrix, suggest possible applications as versatile light-emitting materials for all-optical logic devices or data-storage media. In addition, the stabilization of the formed luminescent clusters by interaction with the organic matrix might determine gelatin encapsulated fluorescent silver clusters as potential biocompatible luminescent pharmacological probes.

Takita et al. observed a phenomenon of increase of fluorescence intensity at the structural changes of human fingernail induced by focused 800 nm femtosecond laser pulse [32]. The fluorescence intensity is higher than its surrounding auto-fluorescence of a fingernail. The increase of fluorescence intensity was also observed with the fingernail heated by a drying oven. It is suggested that the fluorescence increase of the structural change, which is useful to read out three-dimensionally recorded data inside human fingernail, is most likely caused by locally heating caused by femtosecond laser pulse irradiation.

9.4 Long-Lasting Phosphorescence Induced by Femtosecond Laser

Usually, luminescence is divided into two categories: fluorescence and phosphorescence, according to the duration of luminescence. The luminescence whose duration is less than 10^{-6} s is called fluorescence, while the luminescence whose duration is more than 10^{-6} s is called phosphorescence. And phosphorescence can be divided into short-lasting phosphorescence and long-lasting phosphorescence according to the decay time of phosphorescence.

Qiu et al. observed bright long-lasting phosphorescence in various glasses after femtosecond laser irradiation [33–39]. The phosphorescence was observed from the focused area inside the Tb^{3+} ions doped ZBLAN fluoride glass under near-infrared femtosecond laser irradiation after the removal of the activating laser. The intensity of the phosphorescence decreases with the time increasing. They wrote various colors of three-dimensional image into glasses via designing the compositions of glasses and the species of rare-earth ions. The mechanism of long-lasting phosphorescence was as follows. The femtosecond laser induced the defects like V_k centers in glass sample. The unstable defect centers could be excited again when they were thermal stimulated. At the time, the electrons and holes trapped by defect centers were released, and then recombined giving out energy. The energy excited the electrons located in ground state of Tb^{3+} ions near the defect centers to the excited state. The phosphorescence due to $5d \rightarrow 4f$ transition of Tb^{3+} was emitted when the electrons returned to ground state from excited state. The duration of the phosphorescence was very long because the energy level distribution was very broad.

They also reported on a long-lasting phosphorescence phenomenon in calcium aluminosilicate glasses doped with Ce^{3+}, Tb^{3+}, and Pr^{3+} ions [35]. After irradiation by an 800 nm femtosecond pulsed laser, the focused part of the laser in the glasses emits bright and long-lasting phosphorescence able to be clearly seen with the naked eye in the dark even one hour after the removal of the activating laser. Moreover, by selecting appropriate glass compositions and species of rare earth ions, optional three-dimensional image patterns emitting long-lasting phosphorescence in various colors, including blue, green, and red, can be formed within glass samples by moving the focal point of the laser (Fig. 9.3). Based on absorption spectra, the long-lasting phosphorescence is considered to be due to the thermo-stimulated recombination of holes and electrons at traps induced by the laser irradiation, which leave holes or electrons in a metastable excited state at room temperature.

Qiu et al. reported on three-dimensional bright and long-lasting phosphorescence in a Ce^{3+}-doped $Ca_2Al_2SiO_7$ crystal [36]. After they scanned with a focused 800 nm femtosecond pulsed laser, the path traversed by the focal point of the laser in the crystal emits bright blue and long-lasting phosphorescence that can be seen with the naked eye in the dark even 10 h after the removal of the activating laser. Absorption spectra of the crystal show that defect centers have formed after the laser irradiation, and the absorption that is due to the defect centers decays with time. It is suggested that a mechanism of the long-lasting phosphorescence consists of a thermo-stimulated recombination of holes and electrons at traps induced by the laser irradiation at room temperature. They [1] observed on a novel phenomenon in oxygen-deficient Ge-doped silica glasses at room temperature. Irradiation of focused 120 fs laser pulses at 800 nm induced long-lasting phosphorescence with peaks at 290 and 390 nm for oxygen-deficient Ge-doped silica glass. The phosphorescence persisted for not less than 1 h after the removal of the irradiating light. The intensity of the phosphorescence at 390 nm increased with an increase in the concentration of oxygen-deficiency associated with Ge ions. Based on the time dependence of the intensity of the phosphorescence, the long-lasting phosphorescence in these glasses is considered to be due to the thermally activated electron and hole recombination at shallow traps. Femtosecond laser induced long-lasting phosphorescence has also been observed in various glasses.

Qiu et al. also reported on the observation of photo-stimulated long-lasting phosphorescence in a Tb^{3+}-doped zinc–boron–silicate glass sample [39]. After irradiation by an ultraviolet (UV) light at 254 nm or an 800 nm femtosecond pulsed laser, they found that the glass sample emits bright and long-lasting phosphorescence. Long-lasting phosphorescence was observed once again after further excitation by 365 nm UV light when the 254 nm UV light or the femtosecond laser-induced long-lasting phosphorescence cannot be detected. The intensity of the photo-stimulated long-lasting phosphorescence decreased in inverse proportion to the time. Based on absorption, emission spectra and thermo-luminescence curves, they considered the photo-stimulated long-lasting phosphorescence to be due to the light-induced re-arrangement and thermo-stimulated recombination of electrons and holes in traps induced by the 254 nm UV light or the femtosecond laser at room temperature. They infer that the observed phenomenon is useful in the fabrication of rewriteable three-dimensional optical memory with ultrahigh storage density and ultrafast storage speed.

9.5 Polarization-Dependent Luminescence Induced by Femtosecond Laser

In 1999, Kazansky et al. observed bright polarization-dependent luminescence phenomenon in Ge-doped silica glass irradiated by 800 nm femtosecond laser [1]. The blue emission was attributed to germanium oxygen deficient centers (Ge-ODC). When the pump (10 mW average power, 0.4 MW peak power, 2.5×10^{12} W cm^{-2} intensity in the focus of a beam) was focused slightly (~50 μm) above the surface of the sample the shape of the spot of the blue luminescence imaged via the microscope and CCD camera was circular (Fig. 9.20a). Unexpectedly, it had been discovered that when the pump

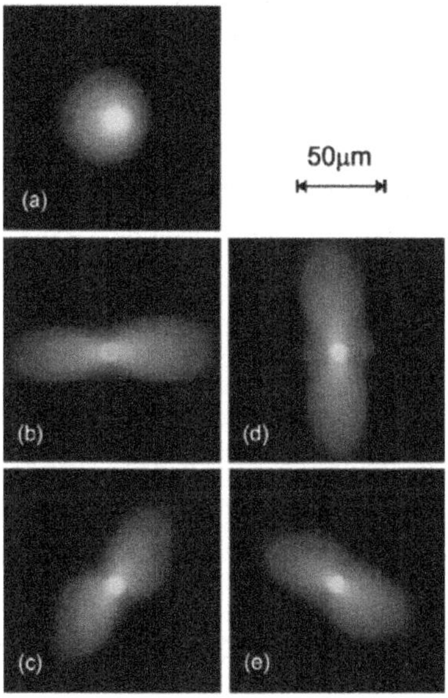

Fig. 9.20. Photographs of the blue luminescence spots in the focus of the linearly polarized pump. Pump is focused above the sample (**a**). Pump with four different orientations of polarization is focused inside the sample [(**b**)–(**e**)]. Notice that the *blue luminescence spot* is elongated along the pump polarization. The *pink spot* is produced by the pump leaking through the dichroic mirror

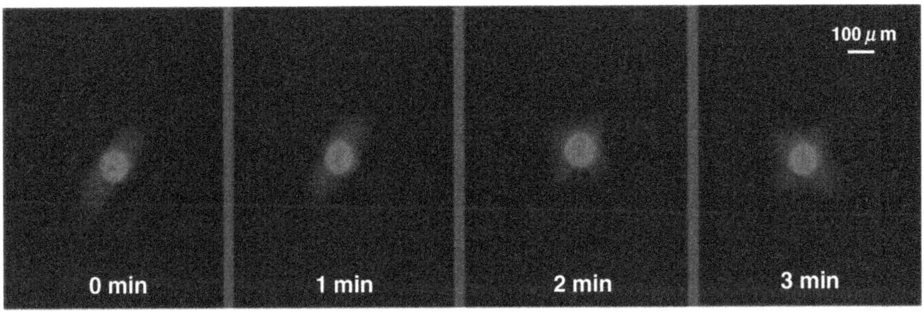

Fig. 9.21. Photographs of *blue-luminescence patterns* near the focus of a linear-polarized laser beam. The time shown in the figure is the duration after changing the polarization direction of the laser beam and irradiating the same location that was irradiated using the previous polarized laser beam

was focused inside the sample the spatial isotropy of the blue luminescence could be broken (Fig. 9.20b). The luminescence scattering increased along the direction of the pump polarization, while the circular shape of the pump beam remained unchanged. If they rotated the direction of the pump polarization by using a half-wave plate, the elongated pattern of the blue luminescence followed this rotation (Fig. 9.20c–e). The fact that the blue luminescence was elongated along the pump polarization indicated that some additional momentum was acquired by the photons along this direction. They believed that such transformation of the momentum could be caused by the photoelectrons moving along the direction of pump polarization. The photoelectrons with the anisotropic momentum distribution could be created via the multiphoton ionization of defects (two-photon ionization of Ge-ODC from the long-lived triplet state) by the linearly polarized light of high intensity.

Similarly but differently, Qiu et al. reported the observation of memorized polarization-dependent light scattering in a Eu^{2+}-doped fluoro-aluminate glass sample [2]. They observed anisotropic blue light scattering along the plane of the light polarization in the glass sample after the excitation of a focused 800 nm, 150 fs laser beam at a repetition rate of 200 kHz. The blue emission peaked at 460 nm was duo to the $5d \rightarrow 4f$ transition of Eu^{2+} ions. Calculations showed that the intensity of the blue luminescence at 460 nm was nearly proportional to the three units of power of the excitation power of the femtosecond laser. Therefore, the blue luminescence at 460 nm was due to three-photon absorption and subsequent relaxation from the $5d$ level to the ${}^8S_{7/2}$ ground state of the Eu^{2+} ions. When they changed the direction of the light polarization and irradiated the same location, as shown in Fig. 9.21, they observed an anisotropic blue-luminescence pattern identical to the original one at the beginning, but then observed a new pattern along the new direction of the light polarization while the original blue-luminescence pattern disappeared gradually with the passage of time.

In this case, the anisotropic light-scattering pattern was permanent and could be modified with another polarized beam. Therefore, the anisotropic light scattering in the Ge-doped silica glass cannot be directly used to explain the observed phenomenon in the Eu^{2+}-doped glass. They also measured electron-spin-resonance spectra in both types of glass under the same laser-irradiation conditions, and observed signals due to the permanently formed electron and hole trapping centers in them. Therefore, they suggested that the polarization-dependent permanent structure was induced during the intense laser irradiation. The focused femtosecond laser beam not only induced electron and hole trapping centers in the glass via multiphoton absorption and multiphoton ionization, but also acted as a driving force for inducing the distribution of induced defects, since a pair of electron and hole centers could be considered as a dipole. The driving force in the polarization direction was larger than that in the direction perpendicular to the light polarization; therefore, the permanent refractive-index fluctuations in the direction parallel to the light polarization were larger than those in the direction perpendicular to the light polarization. This structure resulted in stronger light scattering, e.g., Rayleigh scattering, in the direction of the light polarization because the scattering was proportional to the density or refractive-index fluctuations. The pump beam was scattered by the pump-beam-induced polarization-dependent permanent structure, and the scattered light excited the $^8S_{7/2}$ ground state to the $5d$ level of Eu^{2+} via three-photon absorption, finally resulting in the anisotropic blue luminescence. When the same location was irradiated in another direction of the light polarization, the previously induced structure was destroyed and a new polarization-dependent structure was created. Therefore, a blue-luminescence pattern could be observed along the new direction of the light polarization. Since this observed phenomenon was related to the permanent polarization-dependent structure.

Recently, Shimotsuma et al. observed polarization-dependent periodic nanostructures inside silica glass after irradiation by a focused beam of a femtosecond Ti:sapphire laser [40]. Backscattering electron images of the irradiated spot reveal a periodic structure (Fig. 9.22) of stripelike regions of 20 nm width with a low oxygen concentration (Fig. 9.23), which are aligned perpendicular to the laser polarization direction. These are the smallest embedded structures ever created by light. The period of self-organized grating structures can be controlled from 140 to 320 nm by the pulse energy and the number of irradiated pulses. The phenomenon is interpreted in terms of interference between the incident light field and the electric field of the bulk electron plasma wave, resulting in the periodic modulation of electron plasma concentration and the structural changes in glass. The polarization dependent nanograting is considered to be relevant to the polarization-dependent luminescence in glasses.

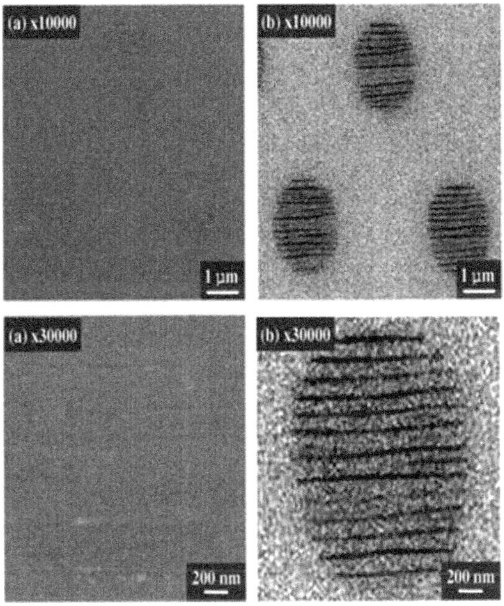

Fig. 9.22. (**a**) Secondary electron images of silica glass surface polished close to the depth of focal spot. (**b**) Light "fingerprints": Backscattering electron images of the same surface. The magnification of the upper and lower images is 10,000× and 30,000×, respectively

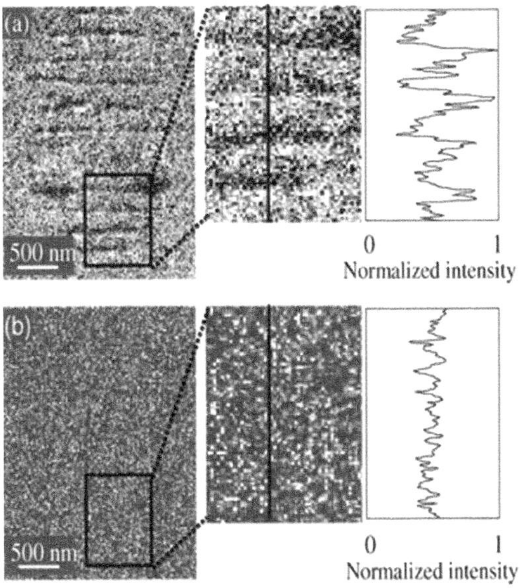

Fig. 9.23. Auger spectra maps and corresponding line scans of oxygen (**a**) and silicon (**b**) on same silica glass surface polished close to the depth of focal spot

9.6 Conclusion

We have introduced various luminescent phenomena in glasses, crystals and semiconductors induced by femtosecond laser, and discussed its mechanisms. As an excitation pump, femtosecond laser can induce efficient luminescence through multiphoton absorption processes. It can also induce the microstructural changes which results in different emission under the excitation of other pumping source. Femtosecond laser can be applied in biological imaging and optical storage with high density. The phosphorescence, particularly long-lasting phosphorescence has potential application in display and illumination, etc. The upconversion luminescence induced by infrared femtosecond laser will be useful in the fabrication of solid lasers working at wavelength of UV region and three-dimensional solid display. The polarization-dependent luminescence is a new phenomena, and more investigations are necessary to clarify this interesting phenomenon. We believe that femtosecond laser will open new possibilities in emission field.

Acknowledgements

This work was financially supported by National Natural Science Foundation of China (Grant No. 50672087), National Basic Research Program of China (2006CB806000b) and National High Technology Research and Development Program of China (2006AA03Z304). We also appreciate the support from the Program for Changjiang Scholars and Innovative Research Team in University.

References

1. G. Kazansky, H. Inouye, T. Mitsuyu, K. Miura, J. Qiu, K. Hirao, Phys. Rev. Lett. **82**, 2199 (1999)
2. J. Qiu, P.G. Kazansky, J. Si, K. Miura, T. Mitsuyu, K. Hirao, A.L. Gaeta, Appl. Phys. Lett. **77**, 1940 (2000)
3. Y. Dong, J. Xu, G. Zhao, C. Yan, G. Zhou, L. Su, L. Yang, J. Qiu, L. Lin, X. Liang, R. Li, Z. Xu, Q. Ren, Opt. Lett. **31**, 2175 (2006)
4. L. Yang, Y. Dong, D. Chen, C. Wang, N. Da, X. Jiang, C. Zhu, J. Qiu, Opt. Express **13**, 7893 (2005)
5. H. You, M. Nogami, Appl. Phys. Lett. **84**, 2076 (2004)
6. H. You, T. Hayakawa, M. Nogami, Appl. Phys. Lett. **85**, 3432 (2004)
7. X. Meng, K. Tanaka, S. Murai, K. Fujita, K. Miura, K. Hirao, Opt. Lett. **31**, 2867 (2006)
8. S. Zhang, B. Zhu, S. Zhou, S. Xu, J. Qiu, Opt. Express **15**, 6883 (2007)
9. B. Zhu, S. Zhang, S. Zhou, N. Jiang, J. Qiu, Opt. Lett. **32**, 653 (2007)
10. Y. Dong, J. Xu, G. Zhou, G. Zhao, M. Jie, L. Yang, L. Su, J. Qiu, W. Feng, L. Lin, Opt. Express **14**, 1899 (2006)
11. L. Yang, C. Wang, Y. Dong, N. Da, X. Hu, D. Chen, J. Qiu, Opt. Express **13**, 10157 (2005)

12. L. Yang, Y. Dong, D. Chen, C. Wang, X. Hu, N. Da, G. Zhao, J. Xu, X. Jiang, C. Zhu, J. Qiu, Opt. Express **14**, 243 (2006)
13. X. Wang, J. Song, H. Sun, Z. Xu, J. Qiu, Opt. Express **15**, 1384 (2007)
14. N.M. Dushkina, B. Ullrich, Opt. Eng. **41**, 2365 (2002)
15. A.L. Pan, R.B. Liu, B.S. Zou, Appl. Phys. Lett. **88**, 173102 (2006)
16. W.Z. Wu, Z.R. Zheng, W.L. Liu, J.P. Zhang, Y.X. Yan, Q.H. Jin, Y.Q. Yang, W.H. Su, Opt. Lett. **32**, 1174 (2007)
17. D.C. Dai, S.J. Xu, S.L. Shi, M.H. Xie, C.M. Che, Opt. Lett. **30**, 3377 (2005)
18. R. Prasanth, L.K. van Vugt, D.A.M. Vanmaekelbergh, H.C. Gerritsen, Appl. Phys. Lett. **88**, 181501 (2006)
19. Q. Li, S.J. Xu, G.Q. Li, D.C. Dai, C.M. Che, Appl. Phys. Lett. **89**, 011104 (2006)
20. H. Yang, S.J. Xu, Q. Li, J. Zhang, Appl. Phys. Lett. **88**, 161113 (2006)
21. C. Wu, C.H. Crouch, L. Zhao, E. Mazur, Appl. Phys. Lett. **81**, 1999 (2002)
22. J. Qiu, K. Miura, T. Suzuki, T. Mitsuyu, K. Hirao, Appl. Phys. Lett. **74**, 10 (1999)
23. J. Qiu, K. Kojima, K. Miura, T. Mitsuyu, K. Hirao, Opt. Lett. **24**, 786 (1999)
24. Q. Zhao, J. Qiu, X. Jiang, C. Zhao, C. Zhu, Opt. Express **12**, 4035 (2004)
25. J. Qiu, C. Zhu, T. Nakaya, J. Si, K. Hirao, Appl. Phys. Lett. **79**, 3567 (2001)
26. K. Miura, J. Qiu, S. Fujiwara, S. Sakaguchi, K. Hirao, Appl. Phys. Lett. **80**, 2263 (2002)
27. J. Qiu, Chem. Records **4**, 50 (2004)
28. Y. Watanabe, G. Namikawa, T. Onuki, K. Nishio, T. Tsuchiya, Appl. Phys. Lett. **78**, 2125 (2001)
29. A.V. Podlipensky, V. Grebenev, G. Seifert, H. Graener, J. Luminesc. **109**, 135 (2004)
30. T. Gleitsmann, T.M. Bernhardt, L. Wöste, Appl. Phys. A **82**, 125 (2006)
31. T. Gleitsmann, B. Stegemann, T.M. Bernhardt, Appl. Phys. Lett. **84**, 4050 (2004)
32. A. Takita, H. Yamamoto, Y. Hayasaki, N. Nishida, H. Misawa, Proc. SPIE, Bellingham, WA, **5662**, 722 (2004)
33. J. Qiu, K. Miura, H. Inouye, T. Mitsuyu, K. Hirao, J. Non-Cryst. Solids **244**, 185 (1999)
34. J. Qiu, A.L. Gaeta, K. Hirao, Chem. Phys. Lett. **333**, 236 (2001)
35. J. Qiu, K. Miura, H. Inouye, Y. Kondo, T. Mitsuyu, Appl. Phys. Lett. **73**, 1763 (1998)
36. J. Qiu, N. Kodama, M. Yamaga, K. Miura, T. Mitsuyu, K. Hirao, Appl. Opt. **38**, 7202 (1999)
37. X. Jiang, J. Qiu, Y. Fan, H. Hu, C. Zhu, J. Mater. Res. **18**, 616 (2003)
38. X. Jiang, J. Qiu, H. Zeng, C. Zhu, Chin. Phys. **12**, 1386 (2003)
39. J. Qiu, X. Jiang, C. Zhu, J. Si, C. Li, Q. Su, K. Hirao, Chem. Lett. **32**, 750 (2003)
40. Y. Shimotsuma, P.G. Kazansky, J. Qiu, K. Hirao, Phys. Rev. Lett. **91**, 247405 (2003)

10

Prospect of Laser-Driven X-Ray Lasers for Extension to Shorter Wavelengths

Yoshiaki Kato and Tetsuya Kawachi

Abstract. Several X-ray laser schemes which have possibilities for extending the present X-ray lasing wavelengths to shorter regions are reviewed. Some of these schemes may be worth reinvestigation using the ultrashort pulse, high intensity lasers that are available at present or in near future.

10.1 Introduction

Coherent X-ray science is rapidly evolving due to the progress in laser-driven plasma X-ray lasers [1], high-order harmonics generation [2] and extensive programs to develop free electron lasers in the X-ray region (X-FEL) [3–5]. The plasma X-ray lasers and high-order harmonics have various unique properties as coherent sources: they are compact, completely coherent in space and time, very short pulse duration of femto-sec to atto-sec regions, and their peak brightness is comparable to that of X-FEL. Their major limitation at present is that their wavelengths are limited to the soft X-ray region. Therefore one of the important issues in these X-ray sources is the extension to the shorter wavelengths, such as the water window region (2.33–4.36 nm) and into the sub-nm wavelengths. In this article, we focus on the laser driven plasma X-ray lasers and review the prospect for extension of the lasing to shorter wavelengths.

Since the successful demonstration of laser-driven plasma X-ray lasers in 1985 [6, 7], remarkable advances have been made in generation and application of the X-ray lasers [1]. With the collisional excitation X-ray lasers (CE-XRL), the lasing wavelength has been extended to 4.32 nm [8], the X-ray lasers have been downsized from the building size to the table-top size due to many inventions, such as multiple-pulse pumping [9, 10], transient collisional excitation [11], grazing incidence pumping [12], optical field ionization [13] and fast capillary discharge [14]. Also fully spatially coherent X-ray laser has been demonstrated by the oscillator-amplifier configuration with transverse pumping [15] and by injection seeding of high-order harmonics in the optical-field ionized low density plasma amplifier [16, 17].

The peak brightness of the plasma X-ray laser has reached 10^{26} photons/s/mm^2/mrad^2/0.1%BW [18] which is comparable to the expected brightness of the X-FELs under development. Therefore a part of the applications that are envisioned to be realized with X-FELs [19] will be studied with the present X-ray lasers. The X-ray laser applications that have been demonstrated include characterization of high density plasmas [20]; observation of the transient formation of microscopic domains of ferroelectric crystals [21], study of the electronic structure of metals with photoelectron spectroscopy [22], high resolution table-top microscopy [23] and explosion of rare gas clusters under intense X-ray irradiation [24].

This progress has been made almost exclusively with the CE-XRL, with which intense lasing is obtained very reliably under various pumping conditions. With reference to the energy level diagram of the nickel like ions shown in Fig. 10.1, the $4d$ state is pumped from the ground state of the $3d^{10}$ by monopole electron collisional excitation and population inversion is automatically created between the $4d$ and $4p$ levels since the $4p$ state rapidly decays to the ground state by radiative transition. In spite of the excellent property of the CE-XRL to automatically generate the population inversion, the possibility to extend the CE-XRL to shorter wavelengths is limited, because a large amount of energy is required to strip high-Z elements to the nickel like stage.

Various alternative X-ray laser schemes have been investigated over many years in order to achieve X-ray lasing at shorter wavelengths, with preliminary experimental investigations in some of these schemes. Owing to recent development of laser technology to generate ultrashort, very high intensity laser pulses, it is now possible to re-investigate some of these schemes under more ideal conditions than previous experiments. For example, the Ti:sapphire laser at the Advanced Photon Research Center of JAEA delivers laser pulses with

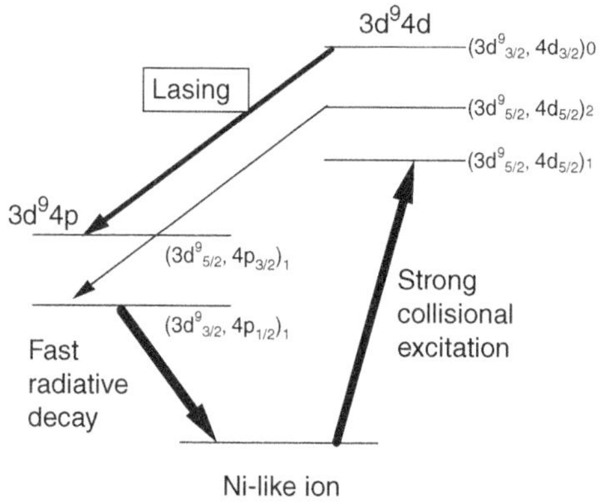

Fig. 10.1. Schematic energy level diagram of Ni-like ions. Typical *lasing lines* are obtained in the $4d \rightarrow 4p$ transitions of $J = 0 \rightarrow 1$ and $J = 2 \rightarrow 1$

~20 fs pulse duration, 100 TW-1 PW peak power [25] and a very high contrast of over 10^9 [26]. Also various schemes have been developed recently to generate high intensity X-ray and electron pulses which may be applicable to X-ray laser pumping. In view of these recent technical advances, intension of this article is to revisit some of the X-ray laser schemes that have possibility to achieve X-ray lasing at shorter wavelengths.

10.2 Photo Pumping by Use of Resonance Lines

In the photo-pumping scheme, spectral line emission from particular ions is absorbed by different element ions to create the population inversion in the latter. The success of this scheme as an X-ray laser depends upon exact spectral coincidence between the emission line and the absorption line. The widths of the spectral lines of ions in plasma, that are due mainly to the Doppler broadening and the Stark broadening, are typically $\sim\Delta\tilde{\lambda}\lambda < 0.01\%$. Therefore high resolution spectroscopic studies are required to find the good pair of the "emitter" and "absorber" ions.

In the early 1990s, the atomic physics group at the Lawrence Livermore National Laboratory (LLNL) has determined the accurate wavelengths of many spectral lines of highly charged ions by use of the electron beam ion trap (EBIT) [27–29]. The data obtained by the EBIT group, in collaboration with the X-ray laser specialists at LLNL, was used to find the candidate pairs of ions for photo-pumping X-ray lasers. Figure 10.2 shows an example where the hydrogen-like magnesium is an emitter and the neon-like germanium is

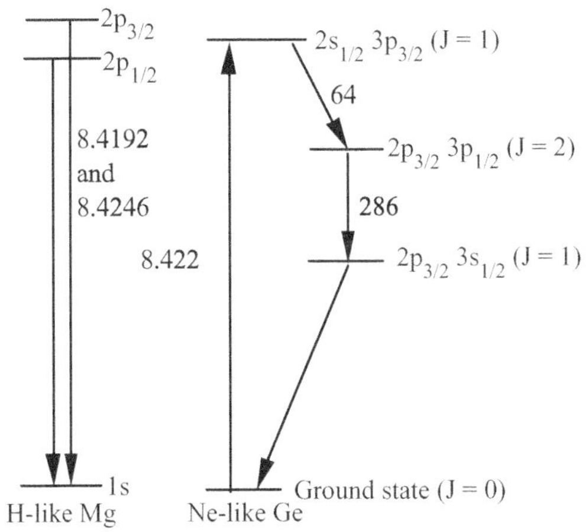

Fig. 10.2. The energy level diagram of the H-like Mg and the Ne-like Ge ions, where the wavelengths of the radiative transitions are shown in unit of Ångstrom. The Ly-α line of H-like Mg is resonantly absorbed by the $2s-3p$ transition of Ne-like Ge, and the lasing at 6.4 nm is expected (from [28])

an absorber [28]. In the following, we use the notation to describe the highly charged ions as H-like Mg, Ne-like Ge, etc. In Fig. 10.2, the Ly-α, $2p \rightarrow 1s$, emission line of the H-like Mg is used to pump the Ne-like Ge ions from the ground state to the excited level of $\overline{2s_{1/2}}3p_{1/2}$ $(J = 1)$ where the upper bar denotes the vacancy state. The population inversion is generated between $\overline{2s_{1/2}}3p_{1/2}$ $(J = 1)$ and $\overline{2p_{3/2}}3p_{1/2}$ $(J = 2)$. The expected lasing wavelength is 6.4 nm, which is three times shorter than the typical lasing wavelength of the Ne-like Ge collisional-excitation X-ray laser (19.6 nm).

Owing to the extensive investigation at LLNL, many pairs of candidate ions for the photo-pumping X-ray laser have been found. For example, the $3d_{3/2}$–$4f_{5/2}$ transition of the Ni-like ions with the atomic number of $(2Z + 4)$ can be used to pump the $2p_{1/2}$–$4d_{3/2}$ transition of the Ne-like ions with the atomic number of Z. In the pair of the Ni-like Pt (emitter) and the Ne-like Rb (absorber), the difference of the energy is only 0.4 eV for the transition energy of 2,512 eV. In other emitter-absorber combinations, the Ly-α line of the H-like Na and the $2s_{1/2}$–$4p_{3/2}$ line of the Ne-like Co differ by 0.1 eV at 1,237 eV, and the $1s$–$2p$ line of He-like Ar and the $3p_{1/2}$–$5d_{3/2}$ line of Ne-like Y differ by 0.4 eV at 3,140 eV [30].

In the photo-pumping scheme, the emitter and the absorber ions should be located as close as possible so that the pumping emission reaches to the absorber ions efficiently. At the same time, the electron temperature should be high for the emitter to increase the emissivity of the ions, whereas the lower temperature is preferred for the absorber ions to avoid the "thermal" population in the lower lasing level, which reduces the amplification gain. In the early 1990s, this scheme has been studied extensively using the laser drivers available at that time, but the appropriate plasma condition for the optimum abundances of the emitter and the absorber ions could not be realized. However, by using the ultrashort pulse, high power lasers, it is now possible to better control the ion abundance in the plasma with the appropriate electron temperature by using the optical field ionization (OFI) process. With OFI, it is possible to realize the condition where a high temperature plasma and a low temperature plasmas are located at a short distance each other by independently controlling the polarizations of the pumping lasers for the emitter and the absorber ions.

10.3 Population Inversion Involving Atomic Inner-Shell Vacancy States

Consider an atom with an electronic vacancy in an inner shell, which is called as an inner-shell vacancy atom. Schematic figure of this atom is shown in Fig. 10.3. This atom is unstable and decays very rapidly through spontaneous transition processes such as the radiative decay and the autoionization process. In the case of the radiative decay, the vacancy is filled with by an electron in the outer-shell, and the vacancy moves from the inner-shell to the outer-shell.

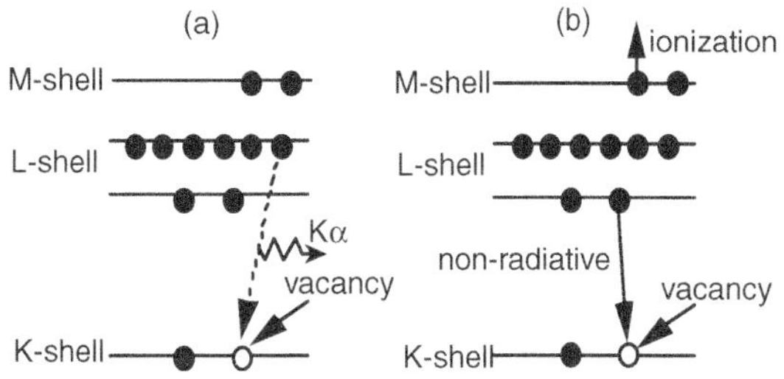

Fig. 10.3. A schematic diagram of an ion with a K-shell vacancy, where solid circle is the electron and open circle is the vacancy. The vacancy is filled by an L-shell electron; (**a**) with K-α line emission, and (**b**) with ionization of an outer-shell electron (autoionization)

Here we denote the vacancy in the K-shell, L-shell and M-shell as K_l, L_l, and M_l, respectively, where the index l corresponds to the angular momentum quantum number of the vacancy. For example, the atom with the $2p$ vacancy in the L-shell is denoted as L_2. If we can realize the population inversion between the inner-shell vacancies levels such as $n(L_2)/g(L_2) > n(M_1)/g(M_1)$, the spectral line corresponding to the $L_2{-}M_1$ transition will be amplified, where $n(p)$ and $g(p)$ represents the population density and the statistical weight of the level p, respectively.

The X-ray laser involving the inner-shell vacancy has two advantages compared with the CE-XRL in which the excited states of the valence electrons are used for population inversion. The first advantage is that, with the inner-shell ionization scheme, the photon energy of the lasing transition with respect to the pumping energy is much larger than that of the CE laser. This is because only the ionization energy of an inner-shell electron of a neutral atom is necessary to create the population inversion in the case of the inner-shell ionization scheme, whereas with the CE laser, a sum of the ionization potential energies of all the ions up the neon-like or nickel-like stage is required. For example, the 4.3 nm nickel-like tungsten laser requires more than 20 keV energy to ionize the tungsten atom up to the nickel-like ion, whereas the 4.1 nm X-ray laser using a K_1 vacancy of a neutral carbon atom requires only ~0.32 keV to ionize the carbon.

The second advantage of this scheme is that it will generate ultra-short duration X-ray laser pulses. Since the lifetime of the inner-shell vacancy is determined mainly by the probability of the autoionization process, which is typically of the order of 10^{13}–$10^{14}\,\mathrm{s}^{-1}$, duration of the X-ray lasers involving the inner-shell vacancy is expected to be several 10 fs. This value is much shorter than the typical duration of the TCE X-ray lasers of 1–10 ps. It also means that in order to generate substantial population inversion in the inner-shell vacancy, the pumping duration should be less than ~10 fs. For the fast pumping, uses of the energetic electron pulse and the intense X-ray pulse have been proposed. In the following these methods are reviewed and examine the possibility of realization of the inner-shell ionization X-ray laser.

10.3.1 Energetic Electron Pulse Pumping

Let us assume that an atom is collisionally ionized by a very short-duration pulse of mono-energetic electrons. The relative populations of the produced states are determined by the electron impact collisional ionization cross sections. Population inversion is not produced by the electron impact ionization, since the collisional ionization cross section for the outer-shell electrons is larger than that of the inner-shell electrons. However, after this initial distribution, the populations of the outer-shell and inner-shell vacancy states change rapidly through the radiative and nonradiative transitions. As the nonradiative decay processes, we consider the Auger decay and the Coster–Krönig decay processes. The Auger decay process is the n-changing transition, where the inner shell vacancy is filled with an outer shell electron of a different principal quantum number through the nonradiative transition accompanied with ionization of an outer shell electron. Whereas the Coster–Krönig decay process is the l-changing transition, where the vacancy is filled with an electron of the same principal quantum number, accompanied with ionization of an outer shell electron.

Typically the rate of the Coster–Krönig process is much faster than that of the Auger process. Furthermore since the rate of the Coster–Krönig process of a certain shell vacancy is significantly larger than that of next deeper shell vacancy, the population of the outer-shell vacancy decreases rapidly by changing its ionic stage compared with that in the inner-shell vacancy. This implies that it may be possible to generate the population inversion between the levels with the inner-shell and the outer-shell vacancies. This possibility has been proposed by Kim et al. [31], where they have surveyed the radiative and nonradiative (Auger and Coster–Krönig) rates of K, L, M and N-shells for virtually all the elements up to $Z = 90$ and calculated the expected population inversion density for several elements. Table 10.1 shows the population

Table 10.1. Characteristics of the inner-shell transitions in tested atoms. The decay rate is for the upper state

Transition	Tested atoms Z range	(Z)	λ (nm)	Decay rate (fs^{-1})	D ratio	P ratio	N_{inv} ($10^{19}\,\text{cm}^{-3}$)
$L_1 \rightarrow M_3$	52–75	Cs(55)	2.63	6.68	1.2	46	$<10^{-10}$
$L_2 \rightarrow M_1$	20–90	Ti(22)	3.09	0.37	13.5	11.8	1.1
$L_3 \rightarrow M_1$	20–90	Ti(22)	3.15	0.37	13.5	11.8	0.5
$L_3 \rightarrow N_5$	65–85	Er(68)	0.15	7.13	1.3	288	$<10^{-10}$
$M_4 \rightarrow N_2$	44–90	Sn(50)	3.10	0.79	28.5	26	0.012

The D and P ratios are the ratios of the decay and the collisional ionization rates, respectively, of the lower state to that of the upper state. The inversion densities were calculated for the electron pulse with duration of 10 fs FWHM and the maximum density of $10^{21}\,\text{cm}^{-3}$ (from [31])

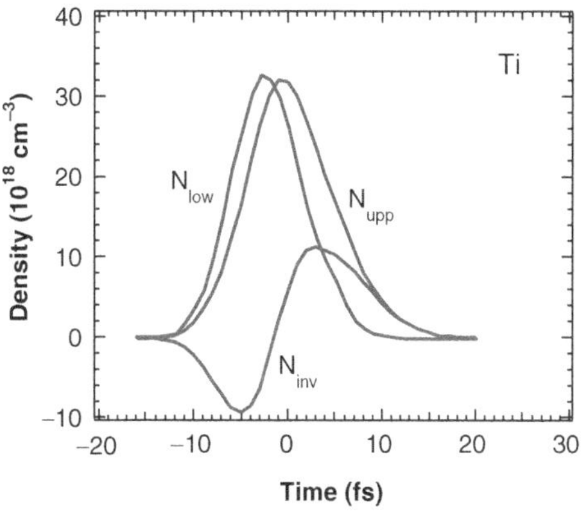

Fig. 10.4. Temporal profiles of the upper and lower lasing levels together with the population inversion density. Here the Ti is pumped by an electron pulse with the duration of 10 fs (FWHM) and the density of 10^{21} cm^{-3} (from [31])

inversion densities (N_{inv}) and wavelengths of the X-ray lasers due to some of candidate transitions.

Figure 10.4 shows the calculated temporal behavior of the populations of the upper and the lower lasing levels together with the population inversion density, when titanium ($Z = 23$) is pumped by an electron pulse with 10 fs duration (full width at half maximum), 1 keV energy and the density of 10^{21} cm^{-3} [31]. It should be noted that the pumping rate (or the collisional-ionization rate) should be less than the destruction rate of the vacancy of the lower lasing level in this scheme, since the population inversion can not be generated initially.

This idea has been expanded by employing more sophisticated cross sections of the inner-shell ionization processes [32], and the amplification gain was investigated for the neutral Cu target [33]. The calculated result indicates that the substantial gain ($g > 10\,\text{cm}^{-1}$) can be obtained in the transition of $M_1 \rightarrow L_3$ at a wavelength of 1.53 nm with a long lifetime (more than 70 fs) if the electron temperature and the flux are high enough (~100 keV and 10^{19} cm^{-3} s^{-1}).

Generation of relativistic electron pulses has been studied intensively in the laser-plasma interaction studies [34–36]. Recently generation of an intense directional electron beam has been demonstrated, where a gold cone with a copper wire was irradiated by a sub-ps CPA Nd:glass laser pulse with the intensity of more than 10^{19} W cm^{-2} [37]. An electron current with the current density of around TA cm^{-2} and the average energy of 3.3 MeV has propagated through the copper wire. The current density obtained in this experiment may be sufficient for the lasing in the $M_2 \rightarrow L_3$ transition of the inner-shell ionized copper ions.

10.3.2 X-Ray Pulse Pumping

Pumping by Continuum X-Ray

The XUV laser using the inner-shell vacancy was first proposed in 1967 [38], where the sodium atom is irradiated by the photon beam at a wavelength of around 20 nm. Figures 10.5 and 6 show, respectively, the schematic energy level diagram of the Na^+ ion and the photo-ionization cross sections of the $2p$ and $3s$ electrons of the Na atom. In this scheme, substantial $2p$ inner-shell vacancy ($2p^5 3s$ in Fig. 10.5) could be generated, since the photo-ionization

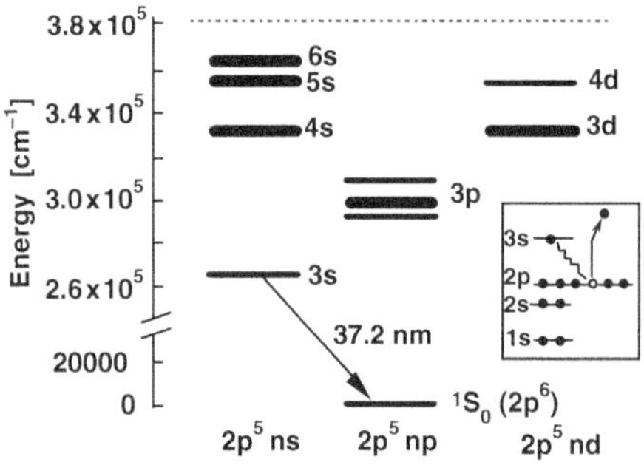

Fig. 10.5. The energy level diagram of a Na^+ ion which has the ground state of $2p^6$. The $2p^5 3s$ state is generated by photo-ionization of a $2p$ electron from the Na atom ($2p^6 3s$) (from [41])

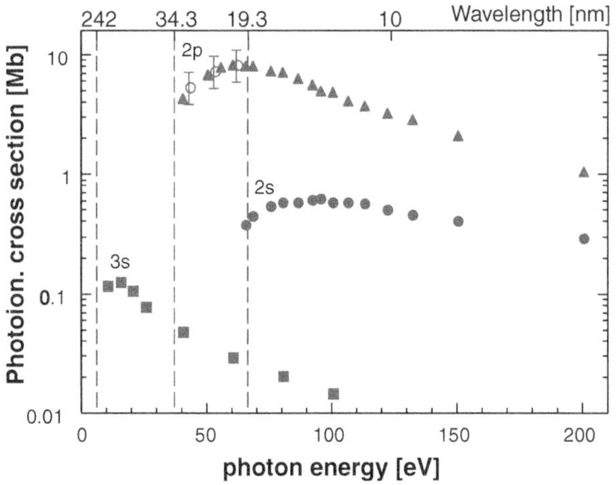

Fig. 10.6. The photo-ionization cross sections of the $3s, 2p$ and $2s$ states of Na atom. The photo-ionization cross section of the $2p$ electron is larger by two orders of magnitude than that of the $3s$ electron for the pumping photons of >35 eV (from [41])

cross section of the $2p$ electron is larger than that of the $3s$ electron for the photon energies higher than 35 eV as shown in Fig. 10.6. As a result, the population inversion will be generated for the $2p^5 3s$–$2p^6$ transition of the Na^+ ions at the wavelength of 37.2 nm. Required pumping intensity to obtain substantial amplification gain was estimated to be $\sim 10^{11}$ W cm^{-2}. This scheme has been evaluated in detail for the case of ultra-fast continuum X-ray pumping [39].

Several years ago, this Na scheme was experimentally tested by using the collisional-excitation X-ray lasers as a pump source at the wavelengths of 19.6 nm (Ne-like Ge laser) [40] and 21.2 nm (Ne-like Zn laser) [41]. However, the amplification has not been verified, possibly due to insufficient intensity of these X-ray laser beams, typically order of 10^{10} W cm^{-2} in the Na vapor.

The inner-shell vacancy X-ray lasers with X-ray pumping operating at shorter wavelength regions have been theoretically studied. The first one is the same scheme as with the electron pulse pumping [31], except the electron pulse pumping is replaced by the continuum or monochromatic X-ray pulse pumping [42]. Figure 10.7 shows a schematic energy level diagram of the Ca atom and the Ca ions. Consider that the Ca atom is irradiated by the continuum X-ray pulse. If proper spectral filters are used to select the spectral regions for ionization mainly of the L-shell electrons, the population inversion in the $L_2 \rightarrow M_1$ and $L_3 \rightarrow M_1$ transitions at a wavelength of $\sim$4.1 nm can be generated. This population inversion will increase by the fast de-population of the M-shell vacancy (the lower lasing level) into the Ca^{2+} ion via the Coster–Krönig process. The duration of the population inversion is determined by generation of the M_1 state due to collisional ionization of the valence electrons of the Ca atom by the photo-ionized electrons, which is about 10 fs.

The second example is the pumping of a carbon foil by an ultra-short duration continuum X-ray pulse [43]. The energy level diagram for this

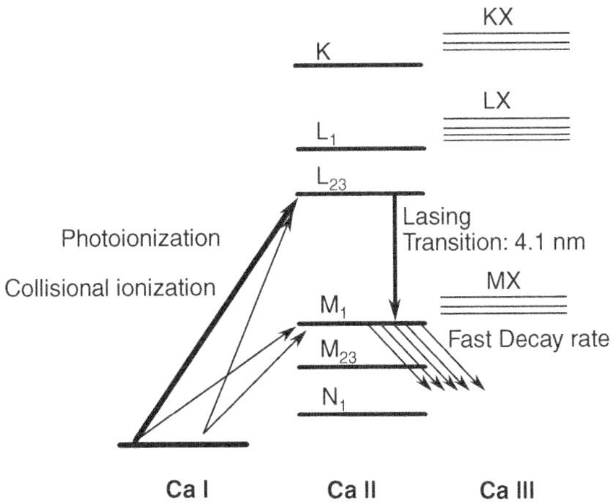

Fig. 10.7. A simplified energy level diagram of the inner-shell vacancy states of Ca (from [42])

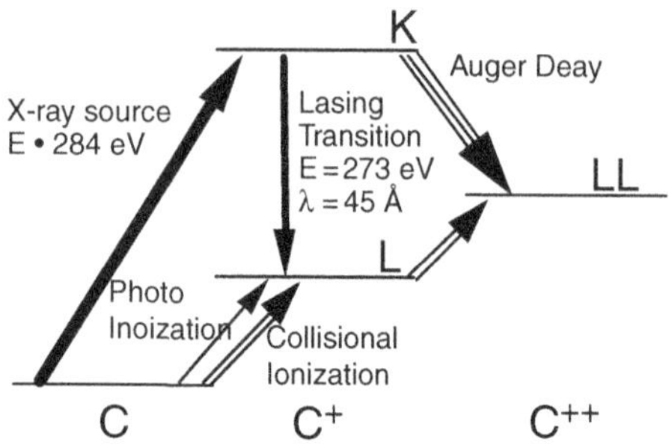

Fig. 10.8. A simplified energy level diagram of carbon atom and ions. A high-energy X-ray pulse can preferentially photo-ionize an inner-shell electron creating a K-shell vacancy and lasing due to the $K-L$ transition is expected (from [43])

scheme is shown in Fig. 10.8, where the $1s$ electron is selectively ionized by an X-ray pulse with the energy higher than the K-shell absorption edge (e.g., $h\nu > 284\,\text{eV}$). Since the photo-ionization cross section for the production of the K-shell vacancy is larger than that for the valence electron ($2p$ electron), population inversion can be generated transiently between the K-shell vacancy and the ground state of the singly ionized carbon. The upper-level has the fast Auger decay rate of the order of $10^{13}\,\text{s}^{-1}$, whereas the lower level population increases due to collisional ionization of the neutral carbon by the photo-ionized electrons. Since the pumping speed should be faster than these destruction processes, the pumping rate required for this scheme is shorter than 20 fs.

Figure 10.9 shows an example of the target geometry for this scheme. The target consists of three-layers; a thin high-Z layer, a metallic high-pass filter and a low density carbon. The high Z substrate, which works as a continuum X-ray emitter, is irradiated by a high intensity laser pulse. It is known that intense continuum X-ray is generated from laser-driven high-Z elements [44, 45] with the demonstrated conversion efficiency (the ratio of the output X-ray energy to the incident laser energy) of 5%. After cutting the long wavelength part of the continuum radiation with the metallic filter, the photons with $h\nu > 280\,\text{eV}$ irradiate the carbon target. Since the distance between the emitter and the carbon target is only around $0.5\,\mu\text{m}$, good coupling is achieved for pumping the carbon. It has been shown by calculation that the amplification gain of $\sim 15\,\text{cm}^{-1}$ is expected over 60 fs duration, when a gold foil is irradiated with a laser pulse of 1 J energy and 40 fs duration focused to a line of 10 mm length and $10\,\mu\text{m}$ width. It should be noted that in this scheme, the X-ray photon flux required for pumping is 10^{29} photons $\text{s}^{-1}\,\text{cm}^{-2}$ to obtain the gain coefficient of $10\,\text{cm}^{-1}$.

The success of the inner-shell photo-ionization scheme depends upon the generation of the ultra-short intense X-ray pulse. The required characteristics for the X-ray pulse are (1) quasi-monochromatic and (2) fast rise time of less than $\sim$20 fs. The first condition is important to increase the gain duration since the high energy component, which is difficult to cut by a filter, is not

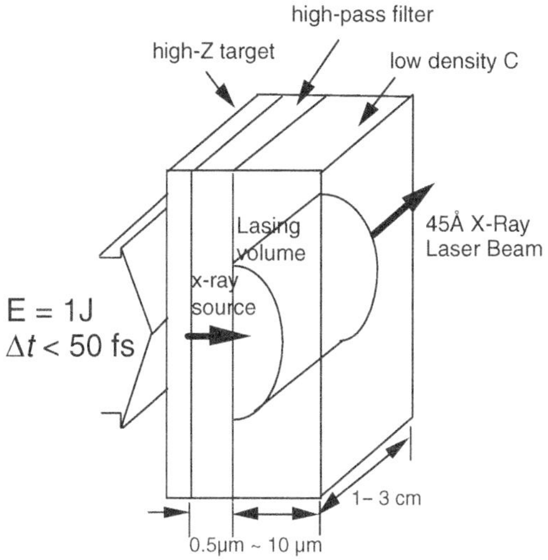

Fig. 10.9. A schematic layout of the experimental configuration. A high-Z target (emitter) is irradiated by an ultra-short laser pulse. The continuum X-ray spectrum appropriate for the pumping is selected by the high-pass filter and absorbed by a low-density carbon creating the K-shell vacancy. The wavelength of the K–L transition is 4.5 nm (from [43])

only ineffective to create the inner-shell vacancy in the upper lasing level, but also it generates the photo-ionized electrons which produce larger population of the lower-lasing level, resulting in suppression of the population inversion density. The second condition is required to generate the population inversion by competing with the rapid decay processes such as the Auger decay of the upper-lasing level and the increase in the population of the lower-lasing level by collisional ionization by photo-ionized electrons.

The X-ray sources which may fulfill these two features have been studied experimentally and theoretically. Larmor radiation or nonlinear Thomson scattering is one of the candidates. This radiation originates from the relativistic nonlinear motion of the electrons quivered by high intensity laser field to produce high-order harmonic radiations [46–48]. A collimated continuum soft X-ray radiation with the peak photon energy of 0.15 keV has been generated by using a 30 fs/1.5 J Ti:sapphire pulse [49]. Another candidate of the light source is the directional (cone angle of 0.5 mrad) continuum X-ray beam with keV photon energies, which is generated when a linearly polarized 30 fs/1 J laser pulse is injected into a He gas jet [50] where ion channeling with the length of over 1 mm is formed due to self-focusing of the incident laser pulse. This X-ray emission is due to the following processes: an electron bunch, which is accelerated by the wake field of the laser pulse, oscillates radially by the strong radial electrostatic field formed by the electron density depression right behind the laser pulse, resulting in generation of directional synchrotron radiation. Since the research on these X-ray sources has just started, further experimental and theoretical investigations are necessary to use them for the X-ray laser pumping.

Since the duration of the inner-shell scheme is very short ($\sim$ a few 10 fs), traveling wave pumping which matches very accurately to the propagation of the X-ray pulse is necessary. For example, traveling wave pumping with the speed higher than $0.9997c$ is necessary, where c is the speed of light in vacuum, for the gain duration of 10 fs and the gain length of 10 mm, when we assume the speed of the X-ray laser pulse in the gain medium is the same with c. This precise traveling wave pump may be realized by combination of a holographically grooved grating with a pair of cylindrical mirrors [51]. Since this optical system is rather complex, development of more simplified system for precise traveling wave pumping is one of the important issues of the inner-shell X-ray lasers.

Pumping By Bound-Bound Transition

Inner-shell pumping with the X-rays of discrete spectra has been also investigated. In this case, the pumping sources are the strong X-ray radiations such as the resonance lines as He-α, Ly-α, and K-α. Intense K-α line radiation has been observed from wide variety of Z-elements [52–55]. Predicted conversion efficiency of the K-α line (ratio of output energy of Kα to the input laser energy) reaches 10^{-4}/str [55], and close to the theoretical conversion efficiency has been obtained in several experiments [53,56]. Since the wavelength of the K-α line of an atom with nuclear charge Z is slightly shorter than the absorption K-edge of other atoms with smaller nuclear charges, a wide variety of pump and absorber combinations becomes possible. For example, the Cu Kα line can be used for the inner-shell photo-pumping of Co. Therefore, this scheme enables wavelength scaling of the X-ray laser [57].

10.4 Alternative Pumping Schemes

Several years ago, strong X-ray emission with the wavelength shorter than 0.3 nm due to $2p$–$3d$ transitions originating from L-shell hollow-atoms of $Xe^{27+}\sim Xe^{37+}$ ions (Fig. 10.10) has been reported [58]. In this experiment, a laser beam of 248 nm wavelength and 500 fs duration from a hybrid Ti:sapphire/KrF laser was focused to Xe clusters with the intensity of 10^{18}–10^{19}W cm^{-2}. The observed emissions were ascribed to the generation of $2p$ vacancies, where the $2p$ electrons were ionized due to electron impact by the electrons which were photo-ionized from an outer orbit and accelerated by the laser electric field to a sufficient energy to eject the $2p$ electrons. Fully relativistic classical model for the trajectory of the photo-ionized electrons showed that the photo-ionized electrons from the $4p$ bound electrons could return to their origin with the energy of 10–20 keV under the intensity of 10^{18}–10^{19} W cm^{-2} of 248 nm laser, which was sufficient to ionize the $2p$ bound electrons.

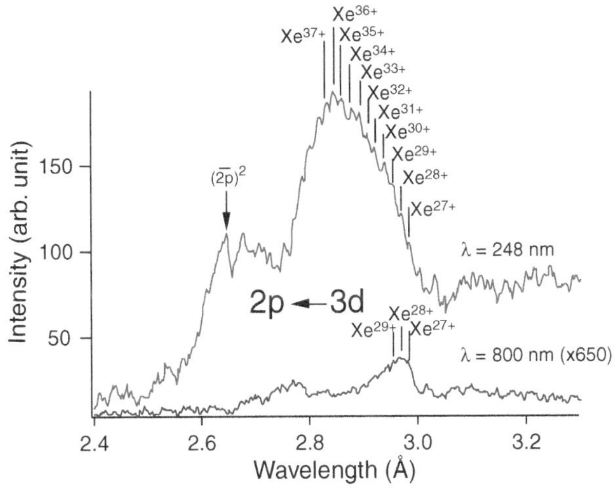

Fig. 10.10. The Xe L-shell ($3d \rightarrow 2p$) X-ray spectra emitted by 5–20 atom Xe clusters pumped by ultraviolet (248 nm) and infrared (800 nm) sub-picosecond laser pulses. The charge state designations are the results of analysis based on the Cowan code (from [58])

They have further shown in the succeeding theoretical paper [59] that the photo-ionized $4p$ electrons kept the wave function characteristics of the $4p$ bound electrons within a half cycle of the 248 nm laser (=0.8 fs), e.g., the spatial size, the shapes of the radial and angular components and the phase information. The $4p$ electrons have selectively ionized the $2p$ electrons since the $4p$ wave functions had larger overlap with the $2p$ wave functions than that of the $3p$ electrons.

A possibility of population inversion in the $3d - 2p$ transitions in highly charged Xe ions with double $2p$ electron holes has been also indicated from the experimental spectra [59, 60]. Further measurements of the properties of these X-ray emissions such as beam divergence, exponential intensity growth with the gain length and coherence are required to confirm the amplification.

Another interesting possibility is frequency up-conversion from a visible laser to the X-ray region called as laser X-ray satellite, which was first proposed in 1970s [61]. With reference to Fig. 10.11, we consider He-like ions where B is the ground state $1s^2$, B* is $1s2s\ ^1S_0$, and K is $1snp\ ^1P$ (n = 2, 3, 4, ...). Since the $B^* \rightarrow B$ transition is optically forbidden, the B* level becomes metastable. When a high intensity laser pulse interacts with this ion, the B* state is coupled with the laser photons and forms a virtual state. Since the wave-function of the virtual state is described as a sum over the contribution from the $1snp\ ^1P_1$ levels (n = 2, 3, 4, ...), it has certain spontaneous transition probability to the ground state. As the result, the satellite lines with the angular frequencies of $\omega' = \omega_{BB^*} + \omega_{las}$ and $\omega'' = \omega_{BB^*} - \omega_{las}$ will be emitted. From the probability of this process, which is calculated from the

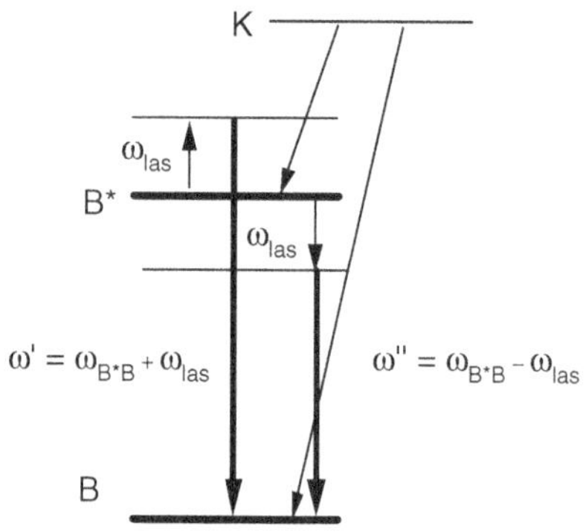

Fig. 10.11. A schematic energy level diagram of He-like ion. B is the ground state and B* is the 2^1S_0 metastable state. $\omega\omega_{las}$ is the angular frequency of the incident laser pulse, and ω_{BB^*} is that of the transition of B$^* \rightarrow$ B

Raman scattering cross sections, the required laser intensity is $\sim 10^{16}$ W cm^{-2}. Several years ago, the laser X-ray satellite has been observed near the resonance line of the He-like Mg $1s2p\,{}^1P_1 \rightarrow 1s^2\,{}^1S_0$ in laser-produced magnesium plasma [62]. In this experiment, the plasma was generated by a 400 fs duration, 1 J energy glass laser, and the laser X-ray satellite emission was obtained only when the laser pulse existed.

Here we consider the recombining plasma. If we choose the appropriate plasma condition, we may realize the population inversion between the metastable state and the ground state although the lifetime may not be very long due to collisional de-excitation. Under such a condition, by injecting an ultra-short laser pulse with the intensity of 10^{16} W cm^{-2} to the plasma, we may obtain the signal of the stimulated Raman amplification at the angular frequency of $\omega' = \omega_{BB^*} + \omega_{las}$ and $\omega'' = \omega_{BB^*} - \omega_{las}$. This scheme has advantages, compared with ordinary recombining plasma laser involving the ground state [63, 64], that the lifetime of the population inversion is longer and the gain duration can be controlled by changing the intensity of the laser.

10.5 Summary

We have reviewed various X-ray laser schemes which have possibilities for extending the present X-ray lasing wavelength to shorter regions. These schemes require pumping with ultrashort duration X-ray or particle beam pulses since the gain durations for the allowed transitions in the X-ray region are very short. Considering that the high-intensity, ultrashort pulse lasers are now available worldwide, it is now possible to re-investigate the various schemes which have been only partially investigated with the lasers of longer pulse durations. These studies will lead to developing laboratory size coherent

X-ray sources, which will open innovative applications in science, industry and medicine.

References

1. P. Jeaglé, Coherent sources of XUV radiation: soft X-ray lasers and high-order harmonics generation, *Springer Series in Optical Sciences* (Springer, New York, 2005)
2. S. Watanabe, K. Midorikawa (ed.) Ultrafast Optics V, Springer Series in Optical Sciences, Springer, New York (2007)
3. Linac Coherent Light Source, Stanford Linear Accelerator Center, http://www-ssrl.slac.stanford.edu/lcls/
4. The European X-Ray Laser Project XFEL, http://xfel.desy.de/
5. RIKEN JASRI SPring-8 Joint Project for XFEL, http://www.riken.jp/XFEL/eng/index.html
6. D.L. Matthews, P. Hagelstein et al., Demonstration of a soft x-ray amplifier. Phys. Rev. Lett. **54**, 110–113 (1985)
7. S. Suckewer, C.H. Skinner et al., Amplification of stimulated soft x-ray emission in a confined plasma column. Phys. Rev. Lett. **55**, 1753–1756 (1985)
8. B.J. MacGowan, S. Maxon S et al., Demonstration of x-ray amplifiers near the carbon K edge. Phys. Rev. Lett. **65**, 420–423 (1990)
9. H. Daido, S. Ninomiya et al., Efficient soft x-ray lasing at 6 to 8 nm with nickel-like lanthanide ions. Phys. Rev. Lett. **75**, 1074–1077 (1995)
10. J. Zhang, A.G. MacPhee et al., Demonstration of saturation in a Ni-like Ag x-ray laser at 14 nm. Phys. Rev. Lett. **78**, 3856–3859 (1997)
11. P.V. Nickles, V.N. Shlyaptsev et al., Short pulse x-ray laser at 32.6 nm based on transient gain in Ne-like titanium. Phys. Rev. Lett. **78**, 2748–2751 (1997)
12. R. Keenan, J. Dunn et al., High-repetition-rate grazing-incidence pumped x-ray laser operation at 18.9 nm. Phys. Rev. Lett. **94**, 103901 (2005)
13. B.E. Lemoff, G.Y. Yin et al., Demonstration of a 10-Hz femtosecond-pulse-driven xuv laser at 41.8 nm in Xe IX. Phys. Rev. Lett. **74**, 1574–1577 (1995)
14. B.R. Benware, Macchietto et al., Demonstration of a high average power tabletop soft x-ray laser. Phys. Rev. Lett. **81**, 5804–5807 (1998)
15. M. Nishikino, M. Tanaka et al., Demonstration of a soft-x-ray laser at 13.9 nm with full spatial coherence. Phys. Rev. A **68**, 061802(R) (2003)
16. Ph Zeitoun, G. Falvre et al., A high-intensity highly coherent soft x-ray femtosecond laser seeded by a high harmonics beam. Nature **431**, 426–429 (2004)
17. N. Hasegawa, T. Kawachi et al., Direct measurement of the temporal profile of the amplification gain of the transient collisional excitation neonlike manganese x-ray laser medium. Phys. Rev. A **76**, 043805 (2007)
18. B. Rus, T. Mocek et al., Multimillijoule, highly coherent x-ray laser at 21 nm operating in deep saturation through double-pass amplification. Phys. Rev. A **66**, 063806 (2002)
19. C. Pellegrini, J. Stöhr, X-ray free-electron lasers principles, properties and applications. Nucl. Instr. Meth. Phys. Res. Sect. A **500**, 33–40 (2003)
20. J. Filevich, J.J. Rocca et al., Observation of a multiply ionized plasma with index of refraction greater the one. Phys. Rev. Lett. **94**, 035005 (2005)

21. R.Z. Tai, K. Namikawa et al., Picosecond snapshot of the speckles from ferroelectric $BaTiO_3$ by means of x-ray lasers. Phys. Rev. Lett. **89**, 257602 (2002)
22. A.J. Nelson, J. Dunn et al., X-ray laser induced photoelectron spectroscopy for single-state measurements. Appl. Phys. Lett. **85**, 6290–6292 (2004)
23. G. Vaschenko, C. Brewer et al., Sub-38 nm resolution tabletop microscopy with 13 nm wavelength laser light. Opt. Lett. **31**, 1214–1216 (2006)
24. S. Namba, N. Hasegawa et al., Enhancement of double Auger decay probability in xenon clusters irradiated with a soft x-ray laser pulse. Phys. Rev. Lett. **99**, 043004 (2007)
25. M. Aoyama, K. Yamakawa et al., 0.85PW, 33-fs Ti:sapphire laser. Opt. Lett. **28**, 1594–1596 (2003)
26. H. Kiriyama, M. Mori et al., High-energy, high-contrast, multiterawatt laser pulses by optical parametric chirped-pulse amplification. Opt. Lett. **32**, 2315–2317 (2007)
27. S.R. Elliott, P. Biersdorfer, J. Nilsen, Measurement of line overlap for resonant photopumping of transitions in neonlike ions by nickel-like ions. Phys. Rev. A **47**, 1403–1406 (1993)
28. J. Nilsen, P. Biersdorfer et al., Measurement of the Ly-α Mg resonance with the $2s \rightarrow 3p$ Ne-like Ge line. Phys. Rev. A **50**, 2143–2149 (1994)
29. S.R. Elliott, P. Biersdorfer, B.J. MacGowan, J. Nilsen, Measurements of line overlap for resonant spoiling of x-ray lasing transitions in nickel-like tungsten. Phys. Rev. A **52**, 2689–2692 (1995)
30. S.R. Elliott, P. Biersdorfer, J. Nilsen, EBIT X-ray spectroscopy studies for applications to photo-pumped X-ray lasers. Proc. X-ray Lasers, eds by D.C. Eder, D.L. Matthews, AIP-CP **332**, 307–311 (1994)
31. D. Kim, C. Toth, C.P.J. Barty, Population inversion between atomic inner-shell vacancy states created by electron-impact ionization and Coster–Kronig decay. Phys. Rev. A **59**, R4129–R4132 (1999)
32. H. Deutsch, D. Margreiter, T. Mark, A semiclassical approach to the calculation of absolute innershell electron impact ionization cross sections. Z. Phys. D **29**, 31–37 (1994)
33. L.M. Upcraft, Gain calculations for inner-shell lasing by electron collisional ionization. Proceedings Eighth International Conference on X-ray Lasers, ed by J.J. Rocca, J. Dunn, S. Suckewer, AIP CP-**641**, 349–353 (2002)
34. U. Teubner, I. Uschmann et al., Absorption and hot electron production by high intensity femtosecond uv-laser pulses in solid targets. Phys. Rev. E **54**, 4167–4177 (1996)
35. F.N. Beg, A.R. Bell et al., A study of picosecond laser solid interactions up to 10^{19} Wcm^2. Phys. Plasmas **4**, 447–452 (1997)
36. B. Luther-Davies, A. Perry, K.A. Nugent, $K\alpha$ emission measurements and superthermal electron transport in layered laser-irradiated disk targets. Phys. Rev. A **35**, 4306–4313 (1987)
37. R. Kodama, Y. Sentoku et al., Plasma devices to guide and collimate a high density of MeV electons. Nature 432, **7020**, 1005–1008 (2004)
38. M.A. Duguay, P.M. Rentzipis, Some approaches to vacuum uv and x-ray lasers. Appl. Phys. Lett. **10**, 350–352 (1967)
39. K. Moribayashi, A. Sasaki, T. Tajima, Ultrafast x-ray processes with hollow atoms. Phys. Rev. **A58**, 2007–2015 (1998)

40. S. Meyer, T. Menzel et al., Investigations towards the realization of an inner-shell x-ray laser in Na. Proceedings X-ray lasers 1996, ed by S. Svanberg, C.G. Wahlstrom, IOP **151**, 173–175 (1996)
41. S. Meyer, C. Reinhardt et al., Investigations towards the realization of an inner-shell x-ray laser in Na, Proceedings of X-ray lasers 1998, ed by Y. Kato, H. Takuma, H. Daido, IOP **159**, 313–316 (1998)
42. D. Kim, S.H. Son et al., Gain characteristics of inner-shell photoionization-pumped $L_{23}M_1$ transition in Ca. Phys. Rev. A **63**, 023806 (2001)
43. S.J. Moon, D.C. Eder, Theoretical investigation of an ultrashort-pulse coherent x-ray source at 45 Å. Phys. Rev. A **57**, 1391–1394 (1998)
44. H.C. Kapteyn, Photoionization-pumped x-ray lasers using ultrashort-pulse excitation. Appl. Opt. **31**, 4931–4935 (1992)
45. G.L. Stobel, D.C. Eder et al., Innershell photoionized x-ray laser schemes. Proceedings SPIE, ed by H.A. Baldis **1860**, 157–166 (1993)
46. Y. Ueshima, A. Sasaki, T. Tajima T, Laser larmor x-ray radiation from low-Z matter, laser part. Beams **17**, 45–58 (1999)
47. B. Sheng et al., High order harmonic generation due to nonlinear Thomson scattering. Opt. Commun. **136**, 239–242 (1997)
48. S.K. Ride, E. Esarey, M. Baine, Thomson scattering of intense lasers from electron beams at arbitrary interaction angles. Phys. Rev. E **52**, 5425–5442 (1995)
49. K. Ta Phuoc, A. Rousse et al., X-ray radiation from nonlinear Thomson scattering of an intense femtosecond laser on relativistic electrons in a helium plasma. Phys. Rev. Lett. **91**, 195001 (2003)
50. A. Rousse, K. Ta Phuoc, Production of a keV x-ray beam from synchrotron radiation in relativistic laser-plasma interaction. Phys. Rev. Lett. **93**, 135005 (2004)
51. S.J. Moon, F.A. Weber et al., Advances toward inner-shell photo-ionization x-ray lasing at 45 Å. Proc. X-ray Lasers 2002, ed. by J.J. Rocca, J. Dunn, S. Suckewer, AIP-CP641: 342–348 (2002)
52. A. Rousse et al., Efficient $K\alpha$ x-ray source from femtosecond laser-produced plasmas. Phys. Rev. E **50**, 2200–2207 (1994)
53. D.C. Eder, G. Pretzler, E. Fill et al., Spatial characteristics of $K\alpha$ radiation from weakly relativistic laser plasmas. Appl. Phys. B: Laser Opt. **70**, 211–217 (2000)
54. S. Bastiani et al., Experimental study of the interaction of subpicosecond laser pulses with solid targets of varying initial scale lengths. Phys. Rev. E **56**, 7179–7185 (1997)
55. D. Salzmann, Ch Reich, I. UschmannI, E. Forster E, Theory of $K\alpha$ generation by femtosecond laser-produced hot electrons in thin foils. Phys. Rev. E **65**, 036402 (2002)
56. Ch Reich, I. Uschmann et al., Spatial characteristics of $K\alpha$ x-ray emission from relativistic femtosecond laser plasmas. Phys. Rev. E **68**, 056408 (2003)
57. G. Pretzler, Th Schlegel, E. Fill, D.C. Eder, Hot-electron generation in copper and photopumping of cobalt. Phys. Rev. E **62**, 5618–5623 (2000)
58. A. McPherson, B.D. Thompson, A.B. Borisov, Multiphoton-induced x-ray emission at 4–5 keV from Xe atoms with multiple core vacancies. Nature **370**, 631–634 (1994)
59. W.A. Schroeder, T.R. Nelson et al., An efficient, selective collisional ejection mechanism for inner-shell population inversion in laser-driven plasmas. J. Phys. B: At. Mol. Opt. Phys. **34**, 297–319 (2001)
60. A.B. Borisov, X. Song et al., Ultrabright multikilovolt coherent tunable x-ray source at $\lambda \sim 2.71$–2.93 Å. J. Phys. B: At. Mol. Opt. Phys. **36**, 3433–3456 (2003)

61. A. Vinogradov, E.A. Yukov, Influence of two-photon processes on the x-ray emission spectrum of a laser plasma Sov. J. Quant. Electron **3**, 163–164 (1973)
62. S.A. Pikuz, A. Maksimchuk, D. Umstadtler et al., Observation of laser satellites in a plasma produced by a femtosecond laser pulse. JETP Lett. **66**, 480–486 (1997)
63. Y.K. Nagata, K. Midorikawa et al., Soft-x-ray amplification of the Lyman-α transition by optical-field-induced ionization. Phys. Rev. Lett. **71**, 3774–3777 (1993)
64. D.V. Korobkin, C.H. Nam et al., Demonstration of soft x-ray lasing to ground state in Li III. Phys. Rev. Lett. **77**, 5206–5209 (1996)

11

Femtosecond Laser Applications in Micro/Nano Science and Technology: Nonlinear Effects in Photonic Crystal Fibers, Femtosecond Laser-Induced Forward Transfer, and Femtosecond Laser Manipulation System for Biology

Ching-Yue Wang, Qi-Rong Xing, Yan-Feng Li, Ming-Lie Hu, Li Yang, Ji-Xian Gong, Wei Jia, and Lu Chai

Abstract. A review of our recent work on applications of femtosecond laser in micro/nano science and technology is presented. In particular, we first discuss enhanced nonlinear effects generated by unamplified femtosecond pulse propagation in photonic crystal fibers, including supercontinuum generation, frequency conversion and soliton self-frequency shift. Following this, we show microdroplet deposition of copper thin film by laser-induced forward transfer using amplified femtosecond pulses. Finally, a femtosecond laser manipulation system for biological research is demonstrated, on which we have achieved femtosecond laser optical trapping, two-photon excitation, and femtosecond laser-induced cell fusion.

11.1 Introduction

The advances of femtosecond (fs, 10^{-15}s) laser technology [1–3] have been so rapid that today sub-30 fs pulses can be routinely generated from femtosecond oscillators and sub-50 fs pulses can be routinely obtained from femtosecond amplifiers. These femtosecond light sources are now commercially available.

Femtosecond light pulses are characterized by their short duration in the time domain, broad width in the spectral domain, and high peak power in intensity; these features make femtosecond light sources a useful tool in time, spectral, and spatial resolved applications in a wide range of fields and subjects [4–7]. The feature of extremely high peak power with low average power makes femtosecond laser pulses very attractive in applications in light-matter interaction [8–10].

Although the applications of femtosecond lasers are diverse and numerous, in this chapter we will just review our latest results on the applications of femtosecond laser technology in micro/nano science and technology. By micro/nano science and technology, we mean the study of materials having structures on the micro-/nanometer scale and biological cells of micrometer dimensions. The structure of this chapter is organized as follows: Enhanced nonlinear effects of femtosecond laser pulses in photonic crystal fibers are discussed in Sect. 11.2, including supercontinuum generation, frequency conversion, and soliton self-frequency shift. Then, microdroplet deposition of copper thin film by femtosecond laser-induced forward transfer is presented in Sect. 11.3. Section 11.4 is devoted to a demonstration of a femtosecond laser manipulation system for biological research, on which femtosecond laser optical trapping, two-photon excitation, and femtosecond laser-induced cell fusion are performed. Finally, a conclusion is drawn in Sect. 11.5.

11.2 Enhanced Nonlinear Effects of Femtosecond Laser Pulses in Photonic Crystal Fibers

11.2.1 Introduction

Photonic crystal fibers (PCFs) are a new class of single-material optical fibers which usually have wavelength-scale air holes running down the entire fiber length [11–13]. A typical scanning electron microscopy (SEM) image of the PCF is shown in Fig. 11.1. The idea behind PCFs dates back to the concept of photonic crystals, periodic dielectric media proposed in as early as 1987 by Yablonovitch [14] and John [15]. It was proposed that in analogy with electronic bandgaps, photonic crystals can possess photonic bandgaps for photons by light scattering from the periodic dielectric structures. Photons with frequency falling within the bandgap are forbidden to propagate in the periodic

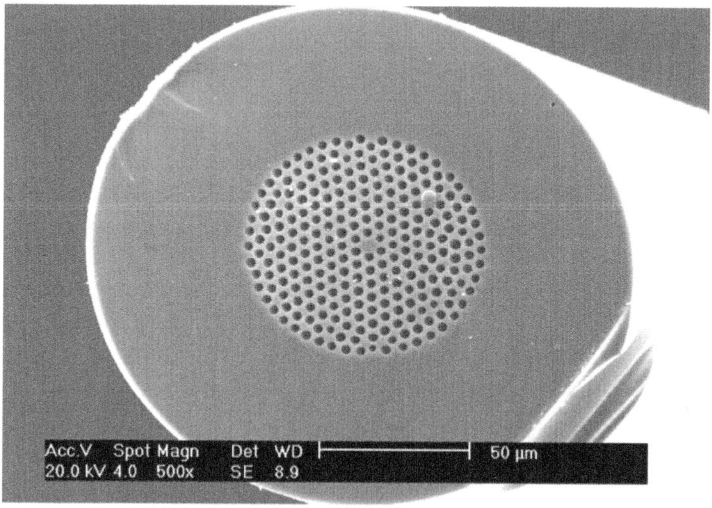

Fig. 11.1. SEM image of an index-guiding PCF

medium. Light can be localized at a defect or propagate down a defect in an otherwise periodic structure, and thus the behavior of light can be controlled in photonic crystals as electrons in semiconductors [16].

The first working PCF as a two-dimensional waveguide was reported in 1996 [17]. This fiber, however, was not based on the bandgap effect (the first PCF based on the bandgap effect was not demonstrated until 1998 [18]), but on modified total internal reflection because the fiber has a core with a higher refractive index than that of the air-hole interspersed cladding [19]. Fibers of this type are also known as holey fibers or microstructure fibers, from which some intriguing properties different from and even superior to those of conventional optical fibers (COFs) have been found and used to advantage [11–13]. PCFs are currently a subject of increasing scientific and technological interest in the field of fiber optics. In this section, we will only concentrate on the index-guiding type.

It has been shown by intensive research activities that many novel properties unimaginable with COFs can result from the holey structure in PCFs, which include, among other things, endlessly single mode operation [19], anomalous dispersion at shorter wavelengths than in COFs [20], small or large mode areas [21,22], and high birefringence [23].

In a COF the normalized frequency V must be below 2.405 in order for the fiber to be single mode. A similar value V_{eff} can be defined for a PCF with a strongly wavelength-dependent cladding index replacing the nearly wavelength-independent cladding index in V. V_{eff} tends to approach a constant value when the wavelength λ is in the high frequency limit ($\lambda \rightarrow 0$), and thus below a certain ratio of the air hole diameter D to the air hole pitch Λ (about $D/\Lambda \leq 0.4$) [13] the fiber remains single mode. This is because more light is confined in silica regions when $\lambda \rightarrow 0$ and the effective cladding index can counteract the decrease of λ.

The total dispersion in a silica-based step-index fiber is normal below 1.3 μm, whereas the strong waveguide dispersion resulting from tight light confinement and the large index difference between air and the fiber material in a PCF can cancel the large negative dispersion of silica glass at visible wavelengths [20]. Therefore, anomalous dispersion is possible below 1.3 μm in PCFs. Furthermore, the dispersion of PCFs can be designed by varying the air hole spacing and air filling fraction in the cladding [24–26] to such an unprecedented extent that the propagation of ultrashort light pulses in these fibers can be more easily controlled [27,28].

The commonly used stack-and-draw fabrication process [17] allows PCFs with tailorable mode areas to be readily fabricated. For instance, PCFs with an extremely small core can be fabricated with enhanced nonlinearity [21,28]. PCFs with enhanced nonlinearity and tailorable dispersion are ideally suited as nonlinear media [29–31], as will be exemplified by our experimental results in the following on three nonlinear effects in PCFs conducted with femtosecond light pulses: supercontinuum generation (SG), frequency conversion, and soliton self-frequency shift (SSFS) effects.

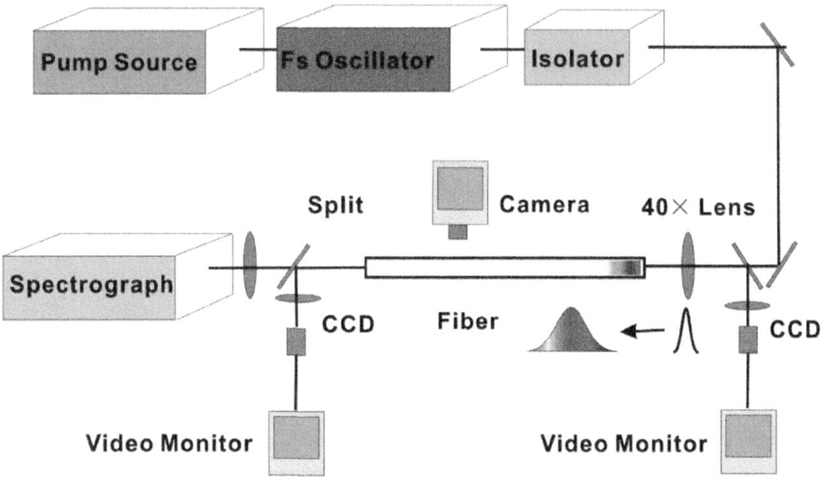

Fig. 11.2. Schematic of the experimental setup for nonlinear effects in PCFs

11.2.2 Nonlinear Effects in Photonic Crystal Fibers

Experimental Setup

The typical setup used in our experiments is depicted in Fig. 11.2, where femtosecond light pulses of about 30 fs with an average power of 600–800 mW are generated from a home-made self-mode locking Ti:sapphire oscillator operating around 800 nm. Depending on the particular system used, the pulse repetition rate is either 80 or 100 MHz. A Faraday isolator is used to prevent retroreflected radiation from disturbing the operation of the laser oscillator. The femtosecond pulses are coupled into the PCF with an objective lens (typically ×40), and the coupling and the output are monitored by CCD cameras. The spectral properties of the output pulses are measured by an optical spectrum analyzer.

Supercontinuum Generation

The first observation of supercontinuum generation, a phenomenon of dramatic spectral broadening of optical pulses and thereby potentially octave-spanning output, in PCFs was reported by Ranka et al. in 2000 [32] and SG have subsequently been studied intensively [33, 34]. Effects like group-velocity dispersion, third and higher-order dispersion, self-phase modulation, cross-phase modulation, four-wave-mixing (FWM), stimulated Raman scattering, higher-order soliton fission, and self-steeping all can come into play, see [34] for a review SG processes in PCFs. Here, we just show some typical results on SG we got by propagating femtosecond light pulses in PCFs.

Figure 11.3 shows the observed spectral broadening [35] at different pump powers by coupling pulses of 30 fs at 800 nm into a PCF provided by University

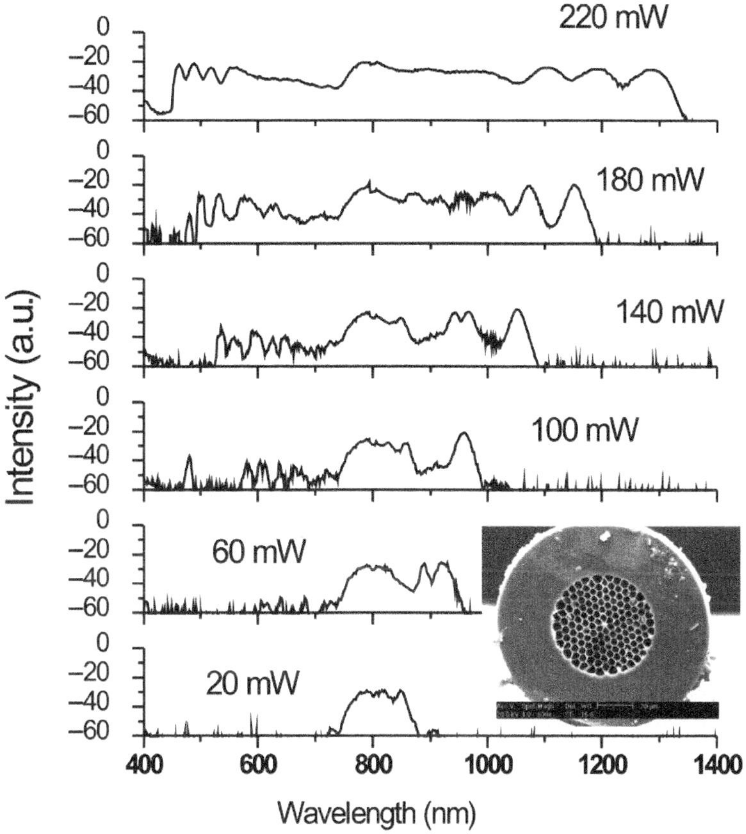

Fig. 11.3. Output spectra in a PCF as shown in the inset at different pump powers of 30 fs input pulses at 800 nm

of Bath, an SEM image of which is shown in the inset. This fiber has a zero-dispersion wavelength at 730 nm, and therefore the pulses fall within the anomalous dispersion regime and the role of anomalous dispersion and FWM is evident. With the increase of the pump power, red-shifted solitonic features are clearly observable and phase-matched nonsolitonic components extend the output spectra towards the blue side. Meanwhile, FWM generated spectral components are able to smooth the spectrum [36, 37].

The different phase-matching processes depend crucially on the fiber dispersion and hence the wavelength of the input pulses. By tuning the wavelength of the pump pulses, we can control those processes and further shape the output spectrum. We find that the further the operating wavelength is tuned from the zero-dispersion of the fiber, the broader the output spectrum. This is due to the fact that when the wavelength of the input pulses are tuned further from the zero-dispersion of the PCF, the nonsolitonic components generated in the soliton-fission processes are more blue-shifted.

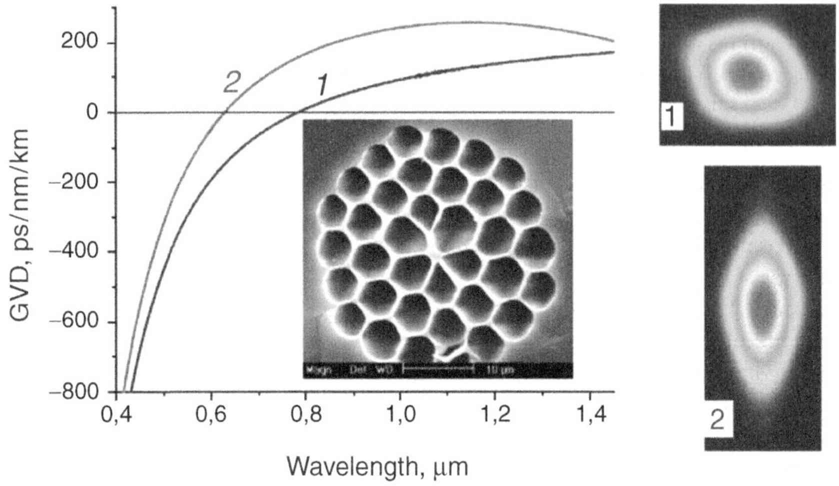

Fig. 11.4. GVD curves and field intensity profiles of the two modes (labeled as 1 and 2, respectively) supported by the PCF with a comma-shaped core (shown in the inset)

PCFs with only twofold rotational symmetry are highly birefringent [23, 38]. The fundamental modes along the two polarization axes are no longer degenerate and they have different dispersion properties, and thus SG processes can be further enriched by the combined effects of birefringence and nonlinearity. Here, we show how SG is tuned in a high-index-step PCF with a comma-shaped core [39]; the fiber is shown in Fig. 11.4 along with the GVD curves and the field intensity profiles of the two modes supported.

The zero dispersion wavelengths are calculated to be 783 and 630 nm for modes of type 1 and type 2, respectively. As a result, pump pulses of 800 nm fall in the anomalous dispersion regime, no matter whether modes of the first type or the second type are excited. Two different physical mechanisms are identified as being responsible for the different features in the output supercontinuum spectra: modes of type 1 generate a smooth spectrum covering a wavelength range from 450 to 1,400 nm, whereas a supercontinuum with an enhanced short-wavelength wing, dominated by intense spectral lines centered at 400–450 nm is emitted in modes of the second type, as shown in Fig. 11.5. When modes of type 1 are excited, the depletion of the pump field is caused by FWM around the zero dispersion wavelength which plays an important role in the initial stages of SG. When modes of the second type are excited, phase matching processes involving the red-shifted soliton and dispersive wave lead to the pronounced short-wavelength wing. By displacing the input end of the fiber with respect to the laser beam in the transverse direction, we can further switch the two regimes of supercontinuum generation and the two types of output spectra.

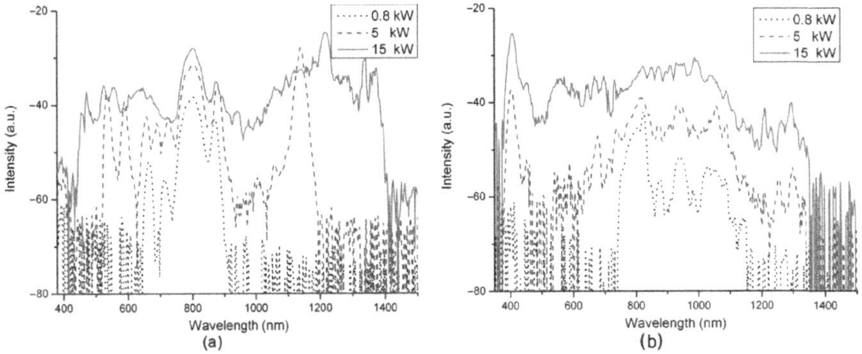

Fig. 11.5. Output spectra generated by 30 fs laser pulses coupled into the first (**a**) and second (**b**) type mode of the PCF as shown in Fig. 11.4. The initial peak power of laser pulses is 0.8, 5, and 15 kW, respectively

Frequency Conversion

In addition to FWM, another important nonlinear effect that plays an important role in transforming input femtosecond light pulses into anti-stokes frequency components in PCFs is the soliton fission process [36,37]. The phase-matching condition of these two processes is highly dispersion-sensitive and polarization-sensitive. What is more, the phase-matching condition can be mode-selective. In this subsection, we will show our work on the generation of frequency-tunable anti-stokes components by propagating fs pulses down highly nonlinear birefringent PCFs.

We can generate intense blue-shifted anti-stokes components centered at 490 and 510 nm, respectively, at the output of a 10-cm birefringent PCF by polarizing the pump laser pulses of 35 fs at 820 nm along the fast or slow axis of the fiber core, as shown in Fig. 11.6a, b [40, 41]. The fiber core (shown in the inset in Fig. 11.6a) has an elliptical shape ($2.0\,\mu m \times 1.6\,\mu m$), leading to dramatically different GVD properties of the two eigenmodes of the PCF. The frequencies of the anti-stokes lines can be further controlled by the pump power. The extra polarization-separating devices added after the fiber output in Fig. 11.2 allow us to selectively observe the change of the two-color frequency conversion by accurately polarizing the pump field with respect to one of the principal axes of the elliptically deformed fiber core [42]. Not only can we excite either of the two eigenmodes centered at 490 and 510 nm, but we are able to excite both modes of the fundamental doublet by polarizing the pump filed at 45° with respect to the fiber axis. These results are shown in Fig. 11.6c.

Highly nonlinear PCFs with large air holes can support higher-order modes which are often involved in phase-matched frequency conversion processes [43]. This mode-selective vectorial nonlinear frequency transformation process can

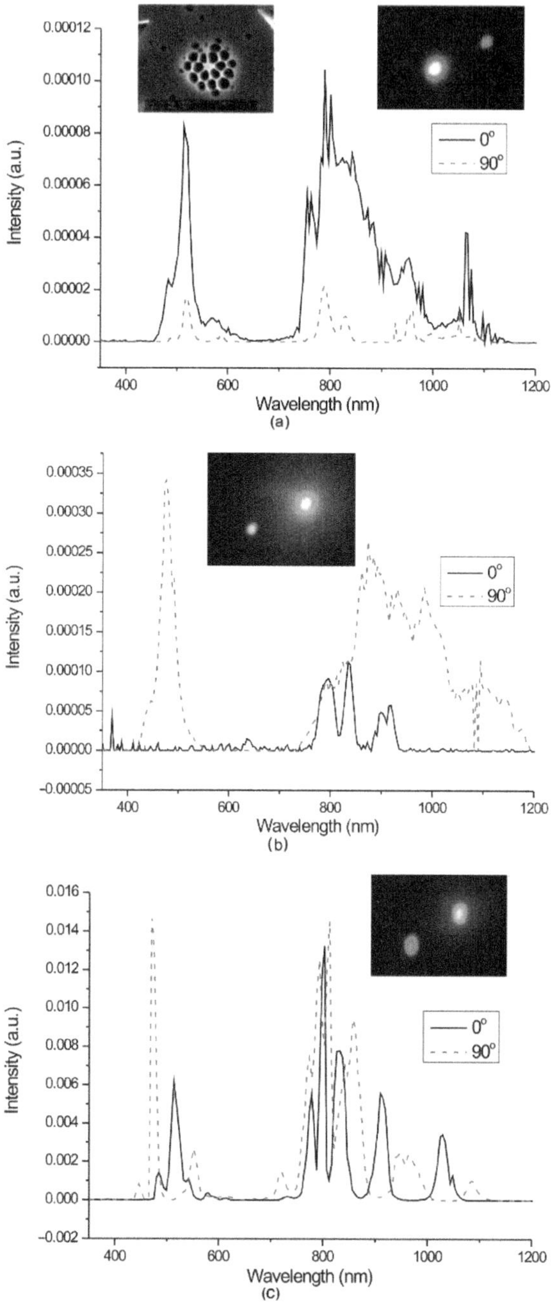

Fig. 11.6. Polarization-demultiplexed anti-stokes line generation in a birefringent PCF. The pump field is polarized at an angle of (**a**) 90°, (**b**) 0°, and (**c**) 45° with respect to the x-axis of the fiber. The output beam images are shown in the insets for the output radiation polarized along the y-axis (0°) and the x-axis (90°)

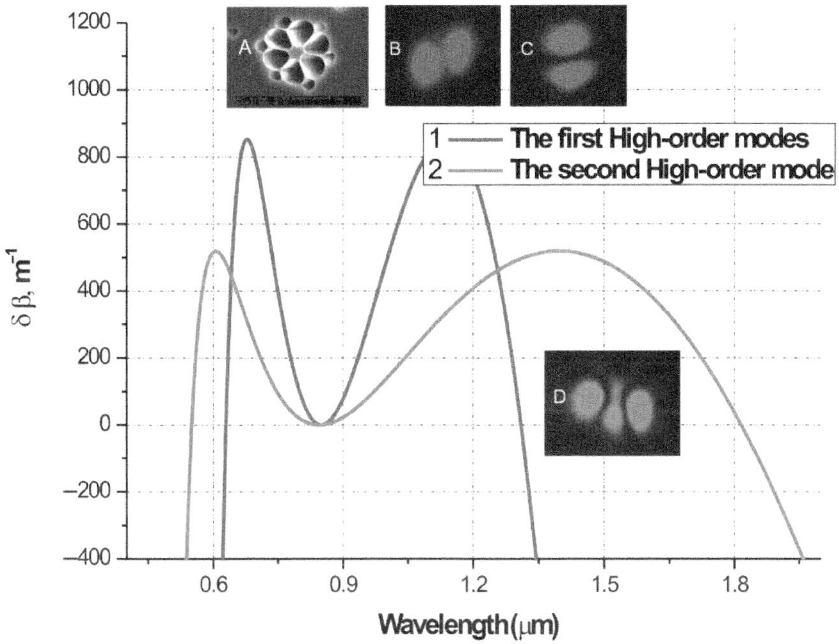

Fig. 11.7. Phase matching diagram for a PCF shown in inset A for (1) the first doublet of higher order modes (mode patterns B and C) and (2) the next higher-order mode (mode pattern D)

be controlled by varying the tilt angle of the input beam and rotating the polarization of the input field, as we will show below.

By coupling 30 fs Ti: sapphire laser pulses into a birefringent PCF (shown in inset A in Fig. 11.7), either the fundamental doublet of the fiber (mode patterns B and C in Fig. 11.7) or the second multiplet (mode pattern D in Fig. 11.7) can be generated by coupling the pump light into the PCF at an angle of 10° or 18° relative to the fiber axis, respectively [44, 45]. The output spectra of both the fundamental mode and the higher-order mode can be further tuned by the polarization of the pump field, see Fig. 11.8.

Finally, in addition to the generation of the usual fundamental fiber mode, similar nonlinear-optical transformation techniques can be used to produce a high-quality spectrally isolated hollow beam in a higher order mode of a PCF [46, 47]. The far-field image of the hollow-beam output features the perfect sixth-fold rotational symmetry of the PCF structure, as demonstrated in Fig. 11.9. The frequency of the hollow beam can be further varied by changing the input beam parameters, having potential applications for the guiding and trapping of atoms and creation of all-fiber optical tweezers, a subject of study in Sect. 11.4.

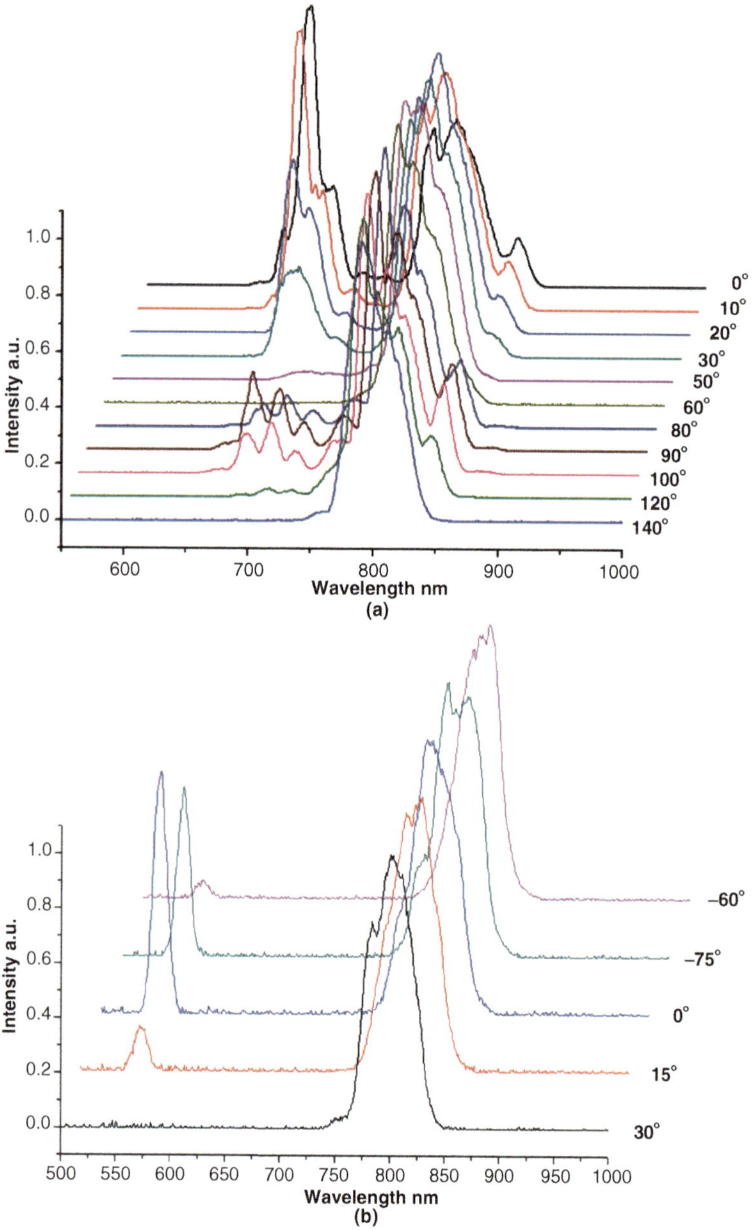

Fig. 11.8. Output spectra at the output of the PCF in Fig. 11.7 measured as a function of the angle of the polarization vector of the linearly polarized pump field relative to the fast axis of the fiber. The pump field is coupled into the fiber at an angle of (**a**) 10° and (**b**) 18° with respect to the fiber axis

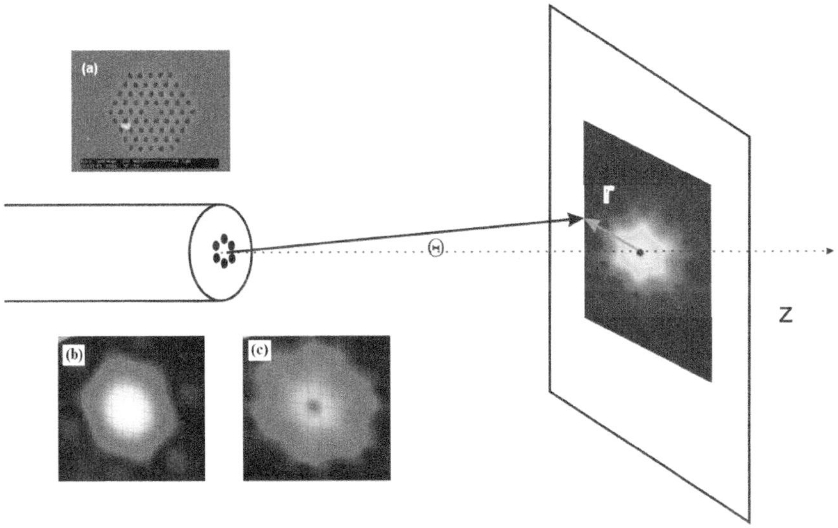

Fig. 11.9. A fundamental fiber mode (**b**) and a hollow beam (**c**) generated by nonlinear-optical transformation of Ti:sapphire laser pulses in a higher-order mode of a PCF (**a**)

Soliton Effects

Solitons, special kinds of wave packets that can propagate undistorted over long distances, are of both fundamental interest and practical application significance in nonlinear fiber optics and fiber optic communications [48]. Anomalous GVD can be realized in the visible and near-infrared wavelengths in PCFs, thus allowing solitons to occur at wavelengths previously inaccessible, covering the spectral range from 0.78 to 1.677 μm [49–53].

In spit of the fact that solitons have been shown to play a significant role in supercontinuum generation and frequency conversion processes in the previous two subsections, distinct solitonic features can be observed by carefully choosing the fiber and pulse parameters. The evolution of SSFS is shown in Fig. 11.10 when the relative polarization of the 30 fs pulses at 820 nm coupled into the same PCF as in Fig. 11.6a is changed [54]. When the input light is polarized along either of the two polarization axes, different features are clearly observable due to the birefringence of the fiber [55].

11.2.3 Application Aspects of Nonlinear Effects

The nonlinear effects in PCFs as demonstrated so far can be so efficient that they are finding an increasingly wider rang of applications in fiber optics and other fields. In particular, octave-spanning supercontinua generated from PCFs have contributed to breakthroughs in precision frequency spectroscopy [56, 57] and carrier-envelope phase control of femtosecond mode-locked lasers [58, 59]. They are also useful in optical coherence tomography

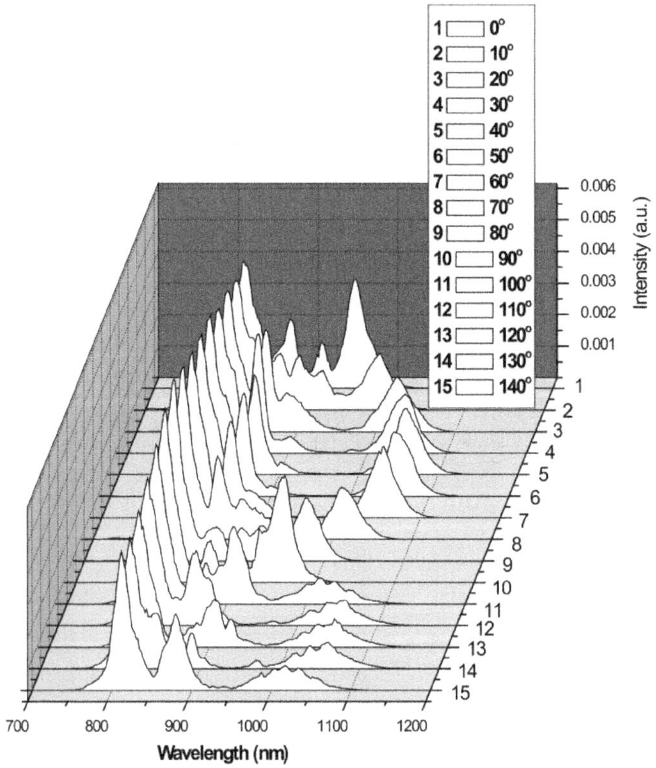

Fig. 11.10. Soliton features for different polarizations of the linearly polarized input field with respect to the x-axis of the fiber shown in the inset in Fig. 11.6a. The initial pulse duration is 30 fs and the initial pulse energy is 0.5 nJ

[60,61], and other spectroscopic and microscopic applications [62–64]. Besides, the supercontinua can be compressed to yield ultrashort light pulses [65,66] or act as WDM sources for communications [67].

Secondly, PCFs are highly efficient frequency converters, allowing the generation of frequency-tunable ultrashort pulses, which are important sources of tunable radiation for applications in spectroscopy, photochemistry, optical metrology, photobiology, biomedicine, quantum optics and so on [68–70]. Solitons generated from PCFs are not only potential sources for communications, but they also provide a good means for the generation of tunable light sources [50,71–73].

Finally, other nonlinear effects not covered here like self-phase modulation, cross-phase modulation, stimulated Raman scattering can all be utilized to make PCF-based components such as amplifiers, lasers, gratings, signal-processing units in fiber optics and fiber optic communications [74,75].

11.3 Microdroplet Deposition of Copper Film by Femtosecond Laser-Induced Forward Transfer

Many investigations on the interaction of ultrashort laser pulses with solid materials have shown that the heat affected zones are minimized during the ablation of materials because the high laser energy of the pulses are delivered in such a short period of time that the material is heated to very high temperatures, leading to direct evaporation. Micromachining with ultrashort laser pulses can offer precise laser-induced breakdown threshold with reduced laser fluence and good repeatability; the resulting structures are characterized by being free of burrs and having minimal damage [10, 76]. These advantages make femtosecond laser pulses very desirable in a wide range of micromachining situations where micrometer and sub-micrometer feature sizes are required [77]. In this section, we will show the advantages of femtosecond pulses in deposition of copper film by laser-induced forward transfer.

In the laser-induced forward transfer (LIFT) process, thin film material from a donor substrate is selectively removed by laser pulses and transferred in the form of micron-sized dots onto an acceptor substrate placed nearby. Since the first performance by Bohandy et al. [78, 79] to achieve direct writing of 50 μm-wide Cu lines by using single 15 ns excimer laser pulses of 193 nm radiation, the LIFT has been studied extensively as a method of additive micropattening in recent years [80, 81]. The ability to deposit submicron-scale features using picosecond and femtosecond laser pulses has been demonstrated [82–84].

The experimental setup used for the femtosecond LIFT process [85] is schematically shown in Fig. 11.11, where the light source is a commercial femtosecond laser amplifier-based workstation (UMW-2110i, Clark-MXR Inc.)

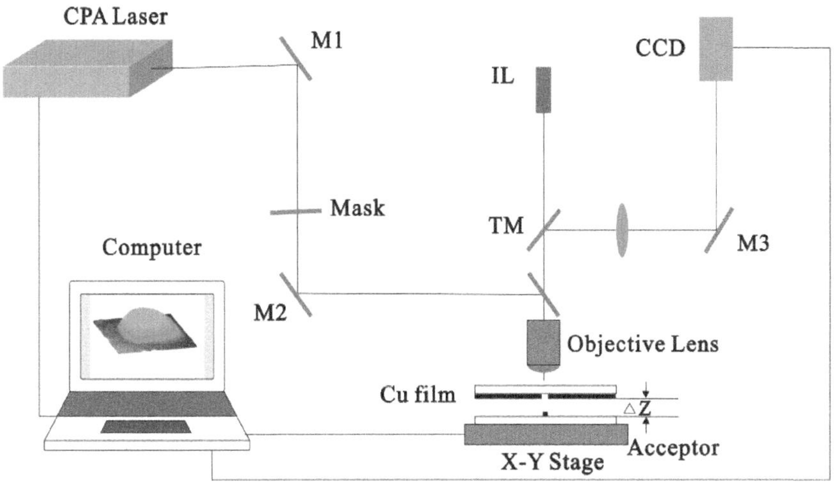

Fig. 11.11. Experimental setup for femtosecond laser-induce forward transfer

delivering pulses of 148 fs at 775 nm. A copper thin film of ~80 nm deposited on a quartz substrate 3 mm in thickness serves as the donor. Another quartz substrate of the same specifications and thickness is used as the acceptor. The samples are mounted on a motorized $x - y$ precision stage with a resolution of 10 nm. The laser beam is focused on the Cu thin film (donor) by a 5× objective, leading to a focus around 8 μm. All the experiments are performed in contact mode ($\Delta z = 0$) with single fs laser pulses in atmosphere at room temperature.

When the laser pulse energy is below ~6 μJ, no transfer of dots is observed. Using laser energies from 6.5 to 20 μJ, copper dots with diameters ranging from 3 to 15 μm are transferred. Three typical morphologies corresponding to three different pulse energies have been identified by the SEM and atomic force microscopy (AFM) images in Figs. 11.12 and 13, respectively. The formation of these morphologies is discussed in the following.

Based on the experimental observations, the physical mechanism responsible for the droplet formation, we believe, can be described as follows. In femtosecond LIFT processes the thin film is heated only at a constrained interface (between the donor film and the substrate) rather than the free surface, see Fig. 11.14 for illustration.

Femtosecond laser pulses inject energy so quickly into the thin film that the temperature of the constrained interface is increased to the melting point within several picoseconds [86], so short a period of time that normal boiling can not happen in the melted thin film because there is no enough time to generate heterogeneity bubbles [87]. Instead, phase explosion happens. Plasma with high temperature and high pressure is formed at the constrained interface with the free surface of the film melted in the meantime, and the melted film is pushed out of the surface by the pressure of the plasma generated at the interface. Metal in liquid phase has very high surface tension and the pushing force of the plasma is moderate when the laser pulse energy is low, so the melted material is transferred and deposited onto the acceptor substrate in the form of liquid droplets. This conclusion is supported by the fact that the diameter of the transferred copper droplets is around 2–3 μm and the biggest height is about 1 μm by AFM detection, while a circular film region of ~8 μm in diameter has been removed completely and a hole-like crater in the donor substrate is formed. Thus for laser energies slightly above the transfer threshold, the transferred dots display good symmetry and a smooth surface, as shown for a pulse energy of 6.5 μJ in Figs. 11.12a and 13a, where droplets of 2–3 μm in diameter can be deposited in a repeatable way. We see that island-like structures with a thinned down central part are formed as a result of the rising pulse energy (Figs. 11.12b and 13b). In this situation the thickness and pressure of the plasma in the inside surface of the film are both increased to cause the material to be transferred in a mixture of liquid and gas phases. Large kinetic energy from the very high phase explosive strike makes the surface of the acceptor super-hot to reduce the viscosity coefficient in the process of deposition. The inhomogeneous intensity distribution of the

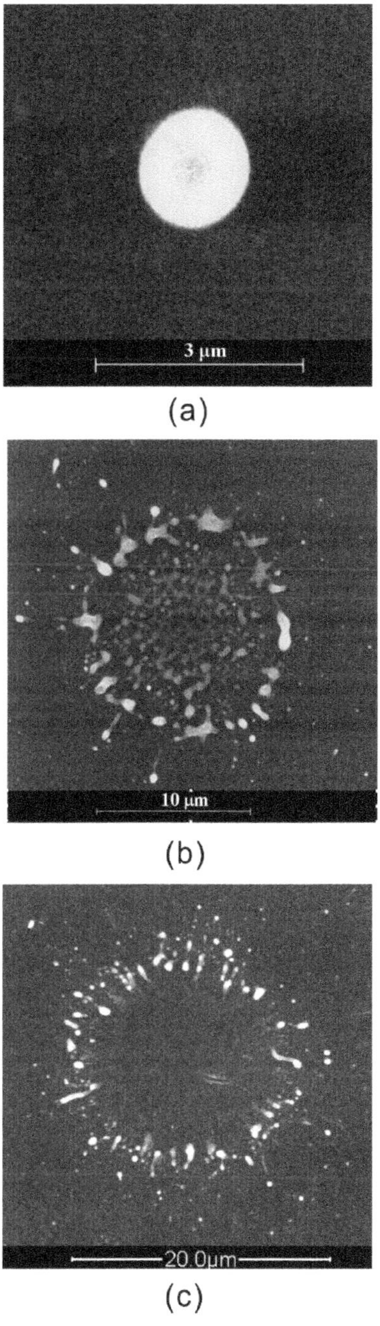

Fig. 11.12. SEM images of deposited Cu dots by femtosecond LIFT for laser pulse energy (**a**) 6.5 μJ, (**b**) 11 μJ, and (**c**) 16 μJ, respectively. (Reprinted with permission from [85]. Copyright 2006, American Institute of Physics.)

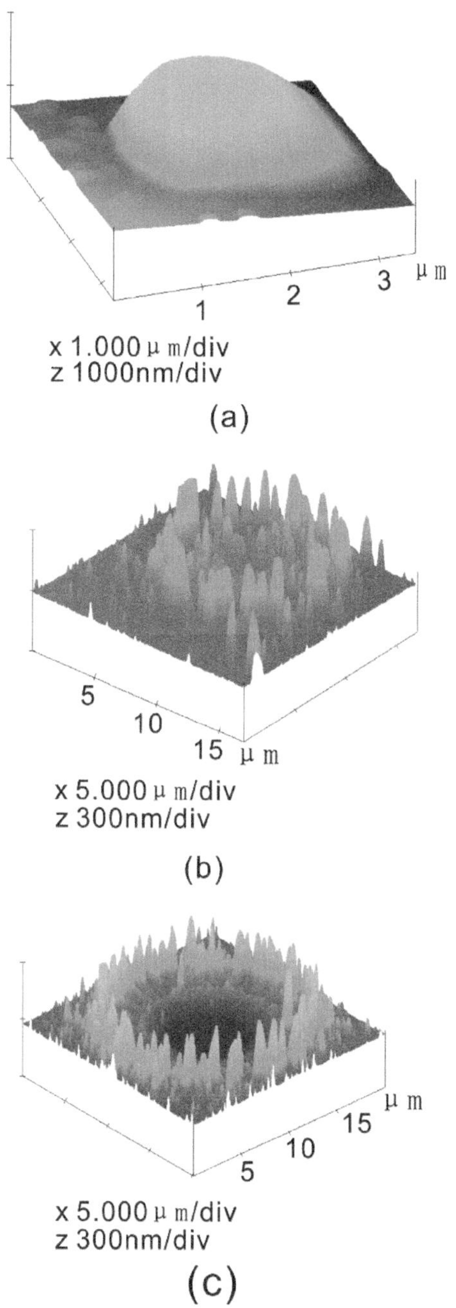

Fig. 11.13. AFM images of deposited Cu dots by femtosecond LIFT for laser pulse energy (**a**) 6.5 μJ, (**b**) 11 μJ, and (**c**) 16 μJ, respectively. (Reprinted with permission from [85]. Copyright 2006, American Institute of Physics.)

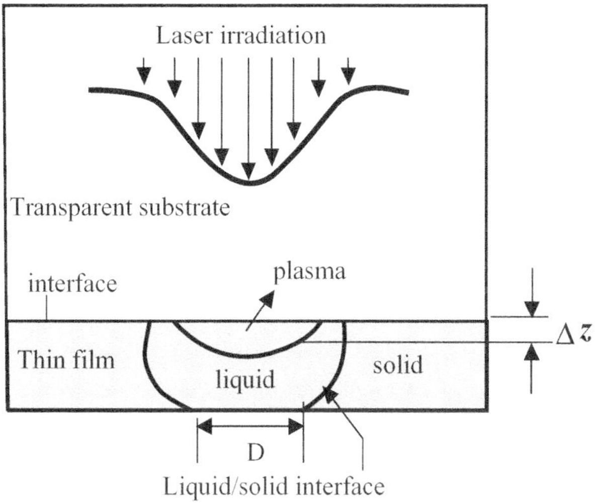

Fig. 11.14. Schematic diagram of heating of the donor film in femotosecond LIFT. (Reprinted with permission from [85]. Copyright 2006, American Institute of Physics.)

Gaussian beam leads to the sticking coefficient being minimal at the center of the irradiation. Therefore, a crater-like morphology is formed at the center of the deposited dots [88], and this effect becomes more prominent with a further increase of the pulse energy, see, for instance, the case of pulse energy of 16 μJ in Figs. 11.12c and 13c. For this pulse energy, the very big shock causes a wide spread of the deposited material during the collision between the transferred particles and the acceptor substrate, and a ring-like morphology around a central crater is clearly visible.

Thus, we have shown that copper microdroplets can be deposited onto a quartz substrate by LIFT using femtosecond laser pulses, where a different mechanism from microdroplet transfer by nanosecond LIFT is described. Droplet formation is a result of the blow-off of the molten film from the quartz substrate by a compressive stress of plasma when the free surface is melted. This femtosecond laser-induced forward transfer process can be very useful in various applications such as rapid prototyping of microelectronic circuits for research applications and additive micropattening.

11.4 Femtosecond Laser Manipulation System for Biological Research

11.4.1 Introduction

Optical tweezers, based on the optical gradient forces as first demonstrated by Ashkin et al. [89], provide a non-invasive method for manipulating biological cells and even individual molecules. Optical tweezers have been widely used in basic research in life sciences, such as single cell molecular biology [90] and the mechanics of single DNA molecules [91].

So far, to realize trapping in three dimensions virtually all optical tweezers used in biology are based on continuous-wave (CW) laser sources with powers of only a few mW. Nevertheless, ultra-high temporal [92] and spatial [93] resolutions are really desirable in biological studies, and these resolutions can be acquired by using femtosecond laser pulses. In most cases, femtosecond laser pulses serve as a scalpel or an excitation source only [94–97] because they can supply impressively high peak powers albeit with a tiny single pulse energy. In the optical trapping context, second harmonic generation from optically trapped nonlinear particles were demonstrated [98] using a femtosecond laser, and in situ control of two-photon fluorescence from dye-doped polymer microspheres were performed on femtosecond optical tweezers [99]. We are led by the promise shown by femtosecond lasers applied in the biological field to propose femtosecond laser tweezers for biological research [100]. Subsequently, our numerical modeling and theoretical analysis predict that femtosecond optical tweezers are entirely feasible [101]. Finally, the feasibility of femtosecond optical tweezers has been demonstrated [102].

In this section, we shall present our optical manipulation system based on a femtosecond laser that functions both as optical tweezers and as a scalpel. Stable optical trapping of human red blood cells (RBCs) is demonstrated, and the trapping capability of femtosecond laser tweezers is compared with CW laser tweezers [102]. On our system, we have also performed experiments for photodynamic therapy based on two-photon excitation using liver cancer cells as the specimen [102], and for cell fusion induced by femtosecond laser [103].

11.4.2 Experimental Setup

Our optical manipulation system (see Fig. 11.15) comprises a commercial inverted research microscope (IX71, Olympus) and a homemade Ti:sapphire laser which delivers 40 fs pulses at 810 nm with a 100 MHz repetition rate. The femtosecond laser can be conveniently switched between the CW and the mode-locking (ML) modes. Passing through a number of lenses and mirrors, the femtosecond laser beam is focused on the sample plane by the microscope objective. The average power of the femtosecond laser beam is tuned by the rotatable polarizer P (half plate denoted in Fig. 11.15). After the polarizer, we set a mechanical shutter to control the time slot of the femtosecond laser beam which is sent into the microscope at the left side port of the microscope. The distance between lenses L_1 and L_2 is finely adjusted by mounting lens L_1 on a z-direction (the optical axial direction) translation stage. The focal lengths of lenses L_1 and L_2 are 150 mm and 180 mm, respectively. Using another two flat mirrors, M_3 and M_4, the laser beam is introduced into the microscope through the left side port originally provided by the manufacturer for imaging purposes. After passing a transmission cube C, the laser beam is fed into a

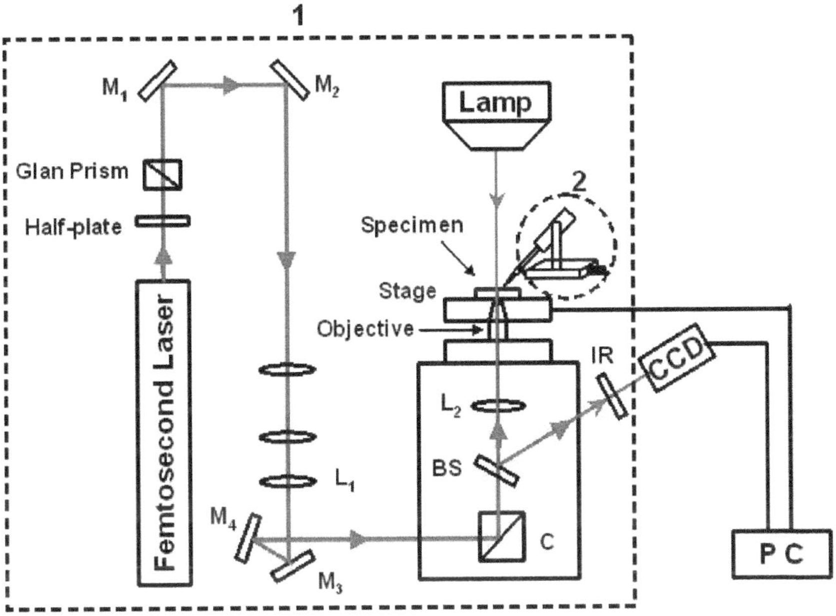

Fig. 11.15. Schematic of the optical manipulation system: L_1, L_2, lenses; M_1, M_2, M_3, M_4, mirrors; C, transmission cube; BS, beam splitter; IR, IR blocking filter

100× objective without immersion fluid (NEA100, NA = 1.25 Olympus). The specimen stage is a two-dimensional ($x-y$) translation stage (M423, Newport, CA) controlled by two linear actuators (T28, Zaber, Canada) with a minimum step of 0.2 μm. A CCD camera on the trinocular tube and the halogen lamp of the microscope make real-time observation of the trapped targets possible. The actuators and the CCD camera are connected to a computer, which with a specially designed program controls the movement of the specimen stage and monitors the trapping process. An IR blocking filter is placed into the optical path of the monitor for eye and camera protection.

11.4.3 The Optical Trap

The specimen used in our experiments is human RBCs. RBCs have a flat, biconcave disc-like shape with a diameter of about 5 μm. Untrapped RBCs suspended in the sodium chloride solution are observed to be freely squirming, as shown in Fig. 11.16a. Then, the laser is set to run in ML mode to produce an average power of 1.6 mW at the ×100 lens focus. We can scan a target RBC into the laser spot by carefully moving the specimen stage. As soon as the target RBC is close enough to the trap, it is abruptly dragged to the center of the trap, and thus trapped. Set upright by the optical force, the trapped cell takes a dumbbell-like shape as shown in Fig. 11.16b. Once trapped, even if the chamber is moved at a suitable speed, the cell will be held steady, as in

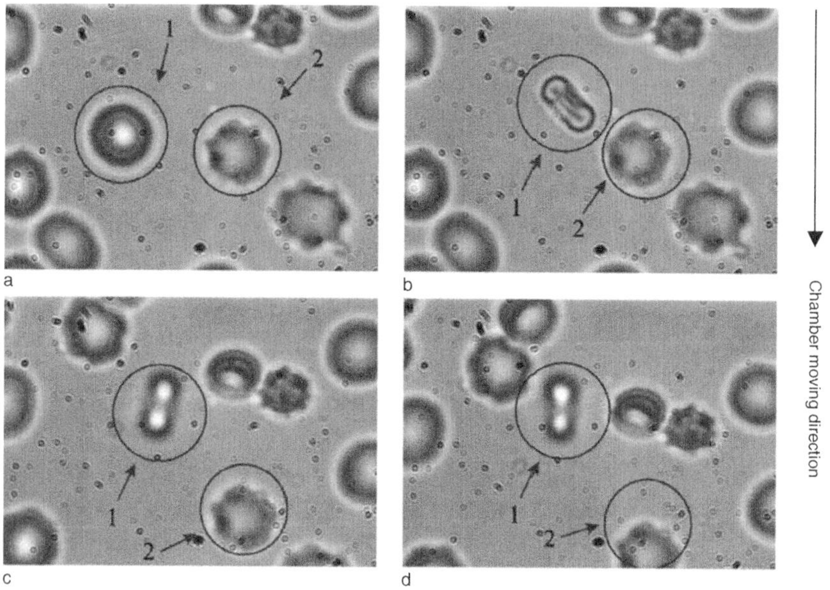

Fig. 11.16. Sequential video frames showing a human RBC tranpped with femtosecond laser pulses with an average power of 5 mW. The *arrows* point to (1) the target cell and (2) the reference cell. (**a**) untrapped target cell; (**b**)–(**d**) trapped cell. The fact that the target cell is stably trapped can be observed from the change of the relative position between the trapped cell and the reference cell, as shown in (**b**)–(**d**). (Reprinted with permission from [102], pp. 358–363. Copyright 2005, Elsevier.)

Fig. 11.16c, d. When the laser beam is blocked, the trapped cell can return to its original state. No evident damage to the target cell is found after it has been trapped for tens of minutes.

A similar experimental method to that by Agate et al. [99] is performed to compare the optical trapping capability between CW and fs lasers using RBCs again as the specimen. First we define the escape speed of the trapped targets as the critical velocity at which the laser tweezers fail to trap. Then, we determine the trapping capability of the system by measuring the escape speed vs. the average trapping power at the $\times 100$ lens focus The higher the escape speed, the stronger the trapping capability of the optical tweezers. Speed control is realized by moving the specimen stage with respect to the trap via the computer-controlled $x - y$ translation stage.

The escape speed of the trapped cell vs. the average laser power at the $\times 100$ lens focus in ML mode is first recorded, followed by the escape speed of the CW laser tweezers obtained by switching the laser to the CW mode. From the experimental results shown in Fig. 11.17, we can see the following three points: (1) almost the same minimum trapping power (corresponding to the near zero escape speed) is found for both the CW and the femtosecond

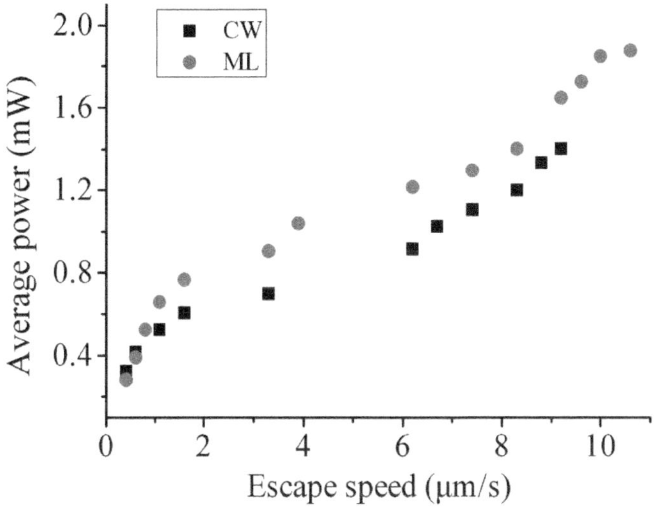

Fig. 11.17. The dependence of the escape speeds of the trapped cell on the average powers incident on the target RBC. *Filled squares* and *circles* are for CW and femtosecond laser tweezers, respectively. (Reprinted with permission from [102], pp. 358–363. Copyright 2005, Elsevier.)

modes; (2) when the escape speed is increased, the trapping power of the femtosecond laser tweezers is rising a little faster than that of the CW laser tweezers; and (3) the average trapping power of the femtosecond laser tweezers is always slightly above that of the CW laser tweezers. These obvious features of the experimental results can be understood by noting that CW light exerts a continuous force on the cell, whereas the force by the femtosecond pulses is only interim. During the intervals where there is no force resulting from femtosecond light, the trapped cell will tend to escape from the trap, requiring more power from the femtosecond laser to pull it back.

11.4.4 Two-Photon Excitation

As an example of using fs laser light as a scalpel to kill target cancer cells similar to Liu et al. [104], two-photon photodynamic therapy has been performed on our system [102] to take advantage of the impressively high peak powers offered by femtosecond laser pulses [94–97]. A newly synthesized chlorine derivative (CDP) [105] is used as the photosensitizer in the experiments, and the specimen is Bel-7402 liver cancer cells. An NEA-10× Olympus lens is used instead of an NEA 100× to obtain a large working distance. We examine the cell-damaging effects of the two-photon induced photodynamic therapy under three conditions: (1) Cells without CDP are illuminated by femtosecond light for 1 min; (2) Cells with CDP are illuminated by femtosecond light for 1 min; and (3) Cells with CDP are illuminated

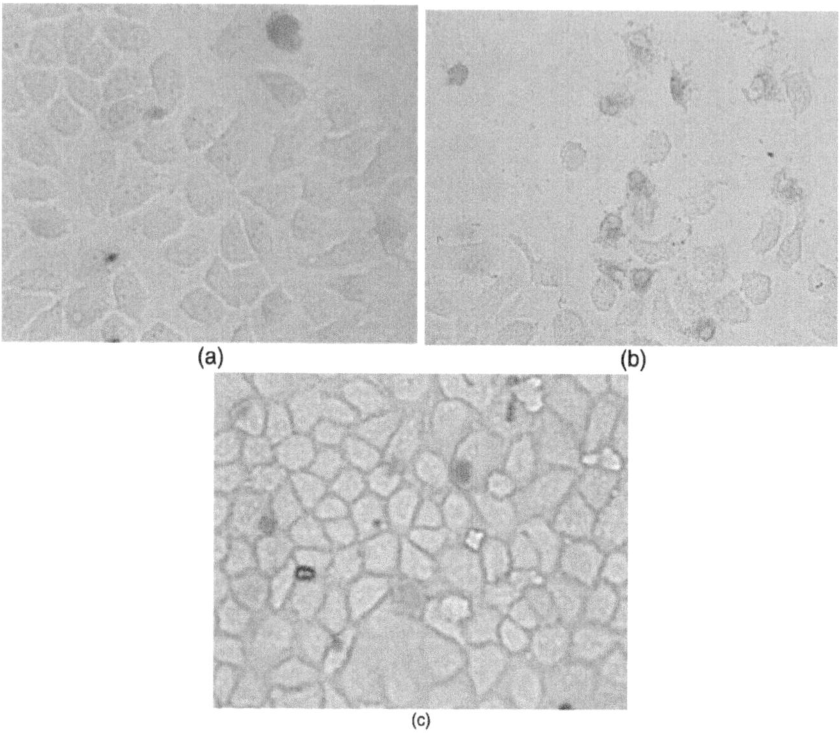

Fig. 11.18. Images of Eosin stained liver cancer cells. (a) Cells without CDP illuminated by femtosecond pulses; (b) Cells with CDP illuminated by femtosecond pulses; (c) Cells with CDP illuminated by CW light. (Reprinted with permission from [102], pp. 358–363. Copyright 2005, Elsevier.)

by CW light for 1 min. The same average laser power of 5 mW for CW or ML modes is used under the above three conditions. The viable and nonviable cells after illumination are identified by Eosin Y with a concentration of 0.2%: Dead cells can be dyed by Eosin Y, while viable cells can not. The results in Fig. 11.18a demonstrate that the femtosecond laser does not kill the cells without CDP. As shown in Fig. 11.18b, there are dyed cells and undyed cells. The dyed cells are those with CDP illuminated by femtosecond light, and the undyed cells are those not illuminated. It is shown in Fig. 11.18c that the cells with CDP illuminated by CW laser are not killed, proof that CW laser is not able to excite the CDP effect to kill the target cells. The dyed cells shown in Fig. 11.18a, c may be naturally dead. These experimental results demonstrate that biological research at single cell level based on nonlinear processes can be realized on our system.

11.4.5 Femtosecond Laser-Induced Cell Fusion

In the experiments on femtosecond laser-induced cell fusion [103], the biological samples are *Phaffia rhodozyma*, a sort of red yeast, which are used as one of the main microorganisms in natural astaxanthin production and have attracted attention throughout the world. The *Phaffia rhodozyma*, provided by Institute of Hydrobiology of Chinese Academy of Science, is maintained at 22 °C in a yeast malt medium. For a detailed description of the culture methods see, for example [106].

The cells in logarithmic stage are harvested by centrifugation at 6,000 rpm for 10 min and then resolved in the enzyme (Glucanex® 200G, Novozyme) solution to give about 1×10^7 cells mL^{-1}. The reaction mixture is then incubated at 22 °C with reciprocal shaking at 90 rpm for 16 h. After incubation, the suspension is washed with osmotic solution to remove the enzyme.

A Petri dish with a hole at the bottom and covered with a thin glass is prepared to form the manipulation chamber. A suspension of *Phaffia rhodozyma*

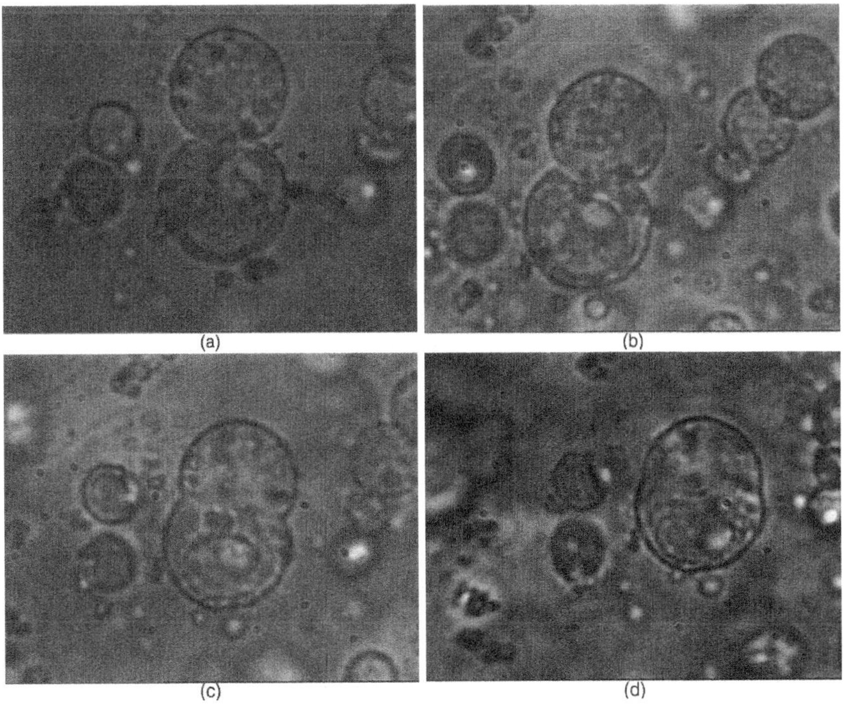

Fig. 11.19. Laser-induced fusion of *Phaffia rhodozyma* protoplasts. Photographs of video frames of the cell fusion are shown. (**a**) A pair of close contacted protoplasts prepared for fusion; (**b**) the cells start to be fused 23 min later after being hit by femtosecond laser pulses; (**c**) about 90 min post-fusion; (**d**) the two cells are almost fused in 160 min

protoplasts containing 10% of polyethylene glycol (PEG) and 0.02 mol L^{-1} of $CaCl_2$ is prepared and placed into the manipulation chamber. The chamber is then put under the microscope and the room temperature is kept at approximately 23 °C using air-conditioning. The area-contact of the target cells is obtained by the effect of PEG, with the procedure monitored with the CCD camera.

Photographs selected from video frames of the cell fusion are shown in Fig. 11.19. A pair of cells in close contact are selected, and then the laser beam is targeted to hit the contiguous area (contact surface) of the two cells. The contact surface is irradiated by femtosecond laser beam with 55 mW for 0.25 s. The effect of the cell fusion is visible about 20 min later. The fusion process is relatively fast at the beginning and then slows down when the time is close to when the hybrid cell has been round again. The course of cells fusion lasts for 160 min and finally the two cells merges into one large cell.

11.4.6 Summary

We have developed an optical manipulation platform, which has performed a number of functions like optical trapping, two-photon excitation-based photodynamic study and cell fusion using femtosecond laser pulses only. Such an integrated system will be of extreme value for biological research.

11.5 Conclusion

In this chapter, we have reviewed the work done by our group on applying femtosecond laser technology to materials and biological cells on the micro-/nanometer scale. The interaction between femtosecond laser light with photonic crystal fibers is first discussed, and we show that photonic crystal fibers are very efficient nonlinear media for unamplified femtosecond laser pulses, using supercontinuum generation, frequency conversion, and soliton self-frequency shift as examples. Next, it is shown that copper microdroplets of 2–3 μm can be removed from a donor film with a thickness of ~80 nm and deposited onto a quartz substrate by the laser-induced forward transfer using amplified single femtosecond laser pulses. Finally, a femtosecond laser-based optical manipulation system is developed, which allows a number of functions to be simultaneously realized for biological research such as optical trapping, two-photon excitation-based photodynamic study and cell fusion by femtosecond laser sources only. We believe that these initial investigations can shed important light on the interaction between femtosecond laser with materials and biological cells, and pave the way for further applications.

Acknowledgements

The authors are indebted to Professor Aleksei M. Zheltikov and his group at M.V. Lomonosov Moscow State University, and Professor Lantian Hou and his group at Yanshan University for providing the fibers and fruitful collaboration on the photonic crystal fiber project. We would also like to thank Professor Jonathan Knight of University of Bath and Fiberhome Telecommunication Technology Co. Ltd for providing fiber samples. The work described in this chapter was supported by the National Key Basic Research Special Foundation (Grant Nos. 2007CB310408 and 2006CB806002), the National High Technology Development Program of China (Grant No. 2007AA03Z447), the National Natural Science Foundation of China (Grant Nos. 60838004 and 60678012), the Key Project of the Ministry of Education of China (Grant No. 108032), the Natural Science Foundation of Tianjin (Grant Nos. 043103911 and 013800311), and Project of Invited Scholar of Key Laboratory of Optoelectronics Information and Technical Science, Ministry of Education (Grant No. B06006), Tianjin University.

References

1. J. Diels, W. Rudolph, *Ultrashort Laser Pulse Phenomena: Fundamentals, Techniques, and Applications on a Femtosecond Time Scale* (Academic Press, London, 1996)
2. M.E. Fermann, A. Galvanauskas, G. Sucha (eds), *Ultrafast Lasers: Technology and Applications* (Marcel Dekker, New York, 2003)
3. U. Keller, Nature **424**, 831 (2003)
4. A.H. Zewail, J. Phys. Chem. **100**, 12701 (1996)
5. S. Baker, J.S. Robinson, C.A. Haworth, H. Teng, R.A. Smith, C.C. Chirilă, M. Lein, J.W.G. Tisch, J.P. Marangos, Science **312**, 424 (2006)
6. V.M. Axt, T. Kuhn, Rep. Prog. Phys. **67**, 433 (2004)
7. B. Lavorel, H. Tran, E. Hertz, O. Faucher, P. Joubert, M. Motzkus, T. Buckup, T. Lang, H. Skenderovi, G. Knopp, P. Beaud, H.M. Frey, C. R. Physique **5**, 215 (2004)
8. G.A. Mourou, C. R. Acad. Sci. Paris, t. 2, Série IV **2**, 1407 (2001)
9. D. Umstadter, J. Phys. D **36**, R151 (2003)
10. X. Liu, D. Du, G. Mourou, IEEE J. Quant. Electron. **33**, 1706 (1997)
11. P. Russell, Science **299**, 358 (2003)
12. J.C. Knight, Nature **424**, 847 (2003)
13. A. Bjarklev, J. Broeng, A.S. Bjarklev, *Photonic Crystal Fibres* (Kluwer, Dordecht, 2003)
14. E. Yablonovitch, Phys. Rev. Lett. **58**, 2059 (1987)
15. S. John, Phys. Rev. Lett. **58**, 2486 (1987)
16. J.D. Joannopoulos, R.D. Meade, J.N. Winn, *Photonic Crystals: Molding the Flow of Light* (Princeton University Press, Princeton, NJ, 1995)
17. J.C. Knight, T.A. Birks, P.St.J. Russell, D.M. Atkin, Opt. Lett. **21**, 1547 (1996)
18. J.C. Knight, J. Broeng, T.A. Birks, P.St.J. Russell, Science **282**, 1476 (1998)
19. T.A. Birks, J.C. Knight, P.St.J. Russell, Opt. Lett. **22**, 961 (1997)
20. J.C. Knight, J. Arriaga, T.A. Birks, A. Ortigosa-Blanch, W.J. Wadsworth, P.St.J. Russell, IEEE Photon. Technol. Lett. **12**, 807 (2000)
21. N.G.R. Broderick, T.M. Monro, P.J. Bennett, D.J. Richardson, Opt. Lett. **24**, 1395 (1999)

22. J.C. Knight, T.A. Birks, R.F. Cregan, P.St.J. Russell, J.-P. de Sandro, Electron. Lett. **34**, 1347 (1998)
23. A. Ortigosa-Blanch, J.C. Knight, W.J. Wadsworth, J. Arriaga, B.J. Mangan, T.A. Birks, P.St.J. Russell, Opt. Lett. **25**, 1325 (2000)
24. A. Ferrando, E. Silvestre, P. Andrés, J.J. Miret, M.V. Andrés, Opt. Express **9**, 687 (2001)
25. W.H. Reeves, J.C. Knight, P.St.J. Russell, P.J. Roberts, Opt. Express **10**, 609 (2002)
26. P.J. Roberts, B.J. Mangan, H. Sabert, F. Couny, T.A. Birks, J.C. Knight, P.St.J. Russell, J. Opt. Fiber Commun. Rep. **2**, 435 (2005)
27. W.H. Reeves, D.V. Skryabin, F. Biancalana, J.C. Knight, P.St.J. Russell, F.G. Omenetto, A. Efimov, A.J. Taylor, Nature **424**, 511 (2003)
28. D.V. Skryabin, F. Luan, J.C. Knight, P.St.J. Russell, Science **301**, 1705 (2003)
29. K.P. Hansen, J. Opt. Fiber Commun. Rep. **2**, 226 (2005)
30. LI Yan-feng, HU Ming-lie, CHAI Lu, WANG Ching-yue, Front. Phys. China **1**, 160 (2006)
31. A. Zheltikov, J. Opt. A **8**, S47 (2006)
32. J.K. Ranka, R.S. Windeler, A.J. Stentz, Opt. Lett. **25**, 25 (2000)
33. A. Zheltikov (ed.), Appl. Phys. B **77**, 143 (2003)
34. J.M. Dudley, G. Genty, S. Coen, Rev. Mod. Phys. **78**, 1135 (2006)
35. HU Ming-Lie, WANG Qing-Yue, LI Yan-Feng, WANG Zhuan, ZHANG Zhi-Gang, CHAI Lu, ZHANG Ruo-Bing, Acta Phys. Sin. **53**, 4243 (2004) (in Chinese)
36. A.V. Husakou, J. Herrmann, Phys. Rev. Lett. **87**, 203901 (2001)
37. J. Herrmann, U. Griebner, N. Zhavoronkov, A. Husakou, D. Nickel, J.C. Knight, W.J. Wadsworth, P.St.J. Russell, G. Korn, Phys. Rev. Lett. **88**, 173901 (2002)
38. M.J. Steel, T.P. White, C. Martin de Sterke, R.C. McPhedran, L.C. Botten, Opt. Lett. **26**, 488 (2001)
39. M. Hu, C. Wang, Y. Li, L. Chai, A. Zheltikov, Opt. Express **14**, 1942 (2006)
40. M. Hu, C. Wang, L. Chai, A. Zheltikov, Opt. Express **12**, 1932 (2004)
41. M. Hu, C. Wang, L. Chai, Y. Li, K.V. Dukel'skii, A.V. Khokhlov, V.S. Shevandin, Yu.N. Kondrat'ev, A.M. Zheltikov, Laser Phys. Lett. **1**, 299 (2004)
42. M. Hu, C. Wang, Y. Li, L. Chai, A. Zheltikov, Opt. Express **13**, 5947 (2005)
43. A. Efimov, A.J. Taylor, F.G. Omenetto, J.C. Knight, W.J. Wadsworth, P.St.J. Russell, Opt. Express **11**, 910 (2003)
44. M. Hu, C.-Y. Wang, Y. Li, Z.Wang, L. Chai, Y.N. Kondrat'ev, C. Sibilia, A.M. Zheltikov, Appl. Phys. B **79**, 805 (2004)
45. M. Hu, C. Wang, Y. Li, Z. Wang, L. Chai, A.M. Zheltikov, IEEE Photon. Technol. Lett. **17**, 630 (2005)
46. M. Hu, C. Wang, Y. Song, Y. Li, L. Chai, E.E. Serebryannikov, A.M. Zheltikov, Opt. Express 14, 4128 (2006)
47. M. Hu, C. Wang, E.E. Serebryannikov, Y. Song, Y. Li, L. Chai, K.V. Dukel'skii, A.V. Khokhlov, V.S. Shevandin, Yu.N. Kondrat'ev, A.M. Zheltikov, Laser Phys. Lett. **3**, 306 (2006)
48. G.P. Agrawal, *Nonlinear Fiber Optics*, 3rd edn. (Academic Press, London, 2001)
49. X. Liu, C. Xu, W.H. Knox, J.K. Chandalia, B.J. Eggleton, S.G. Kosinski, R.S. Windeler, Opt. Lett. **26**, 358 (2001)

50. N. Nishizawa, Y. Ito, T. Goto, Jpn J. Appl. Phys. **42**, 449 (2003)
51. F. Druon, N. Sanner, G. Lucas-Leclin, P. Georges, K.P. Hansen, A. Petersson, Appl. Opt. **42**, 6768 (2003)
52. K.S. Abedin, F. Kubota, Opt. Lett. **28**, 1760 (2003)
53. D.T. Reid, I.G. Cormack, W.J. Wadsworth, J.C. Knight, P.St.J. Russell, J. Mod. Opt. **49**, 757 (2002)
54. E.E. Serebryannikov, M. Hu, Y. Li, C. Wang, Z. Wang, L. Chai, A.M. Zheltikov, JETP Lett. **81**, 487 (2005)
55. E.E. Serebryannikov, M. Hu, C. Wang, Y. Li, Z. Wang, L. Chai, X. Ren, K.V. Dukel'skii, A.V. Khokhlov, V.S. Shevandin, Yu. N. Kondrat'ev, A.M. Zheltikov, J. Raman Spectrosc. **37**, 416 (2006)
56. S.A. Diddams, D.J. Jones, J. Ye, S.T. Cundiff, J.L. Hall, J.K. Ranka, R.S. Windeler, R. Holzwarth, Th. Udem, T.W. Hänsch, Phys. Rev. Lett. **84**, 5102 (2000)
57. R. Holzwarth, Th. Udem, T.W. Hänsch, J.C. Knight, W.J. Wadsworth, P.St.J. Russell, Phys. Rev. Lett. **85**, 2264 (2000)
58. D.J. Jones, S.A. Diddams, J.K. Ranka, A. Stentz, R.S. Windeler, J.L. Hall, S.T. Cundiff, Science **288**, 635 (2000)
59. F.W. Helbing, G. Steinmeyer, U. Keller, IEEE J. Sel. Top. Quant. Electron. **9**, 1030 (2003)
60. I. Hartl, X.D. Li, C. Chudoba, R.K. Ghanta, T.H. Ko, J.G. Fujimoto, J.K. Ranka, R.S. Windeler, Opt. Lett. **26**, 608 (2001)
61. B. Povazay, K. Bizheva, A. Unterhuber, B. Hermann, H. Sattmann, A.F. Fercher, W. Drexler, A. Apolonski, W.J. Wadsworth, J.C. Knight, P.St.J. Russell, M. Vetterlein, E. Scherzer, Opt. Lett. **27**, 1800 (2002)
62. A.B. Fedotov, P. Zhou, A.P. Tarasevitch, K.V. Dukel'skii, Yu.N. Kondrat'ev, V.S. Shevandin, V.B. Smirnov, D. von der Linde, A.M. Zheltikov, J. Raman Spectrosc. **33**, 888 (2002)
63. H. Kano, H. Hamaguchi, Appl. Phys. Lett. **85**, 4298 (2004)
64. T. Nagahara, K. Imura, H. Okamoto, Rev. Sci. Instrum. **75**, 4528 (2004)
65. G. Chang, T.B. Norris, H.G. Winful, Opt. Lett. **28**, 546 (2003)
66. B. Schenkel, R. Paschotta, U. Keller, J. Opt. Soc. Am. B **22**, 687 (2005)
67. Z. Yusoff, P. Petropoulos, K. Furusawa, T.M. Monro, D.J. Richardson, IEEE Photon. Technol. Lett. **15**, 1689 (2003)
68. A.M. Zheltikov, Laser Phys. Lett. 2004, **1**, 220 (2004)
69. S.O. Konorov, A.M. Zheltikov, Opt. Express **11**, 2440 (2003)
70. S. Lorenz, Ch. Silberhorn, N. Korolkova, R.S. Windeler, G. Leuchs, Appl. Phys. B **73**, 855 (2001)
71. K.S. Abedin, T. Miyazaki, F. Kubota, IEEE Photon. Technol. Lett. **16**, 1119 (2004)
72. H. Lim, J. Buckley, A. Chong, F.W. Wise, Electron. Lett. **40**, 1523 (2004)
73. G. Genty, M. Lehtonen, H. Ludvigsen, Opt. Lett. **30**, 756 (2005)
74. K. Hougaard, F.D. Nielsen, J. Opt. Fiber Commun. Rep. **1**, 63 (2004)
75. J.H. Lee, IEICE Trans. Electron. **E88-C**, 327 (2005)
76. S. Nolte, G. Kamlage, F. Korte, T. Bauer, T. Wagner, A. Ostendorf, C. Fallnich, H. Welling, Adv. Engin. Mat. **2**, 23 (2000)
77. S. Kawata, H. Sun, T. Tanaka, K. Takada, Nature **412**, 697 (2001)
78. J. Bohandy, B.F. Kim, F.J. Adrian, J. Appl. Phys. **60**, 1538 (1986)
79. J. Bohandy, B.F. Kim, F.J. Adrian, A.N. Jette, J. Appl. Phys. **63**, 1158 (1988)

80. A. Piqué, D.B. Chrisey, R.C.Y. Auyeung, J. Fitz-Gerald, H.D. Wu, R.A. McGill, S. Lakeou, P.K. Wu, V. Nguyen, M. Duignan, Appl. Phys. A **69**, S279 (1999)
81. I. Zergioti, D.G. Papazoglou, A. Karaiskou, N.A. Vainos, C. Fotakis, Appl. Surf. Sci. **197–198**, 868 (2002)
82. I. Zergioti, S. Mailis, N.A. Vainos, C. Fotakis, S. Chen, C.P. Grigoropoulos, Appl. Surf. Sci. **127–129**, 601 (1998)
83. I. Zergioti, S. Mailis, N.A. Vainos, P. Papakonstantinou, C. Kalpouzos, C.P. Grigoropoulos, C. Fotakis, Appl. Phys. A **66**, 579 (1998)
84. P. Papakonstantinou, N.A. Vainos, C. Fotakis, Appl. Surf. Sci. **151**, 159 (1999)
85. L. Yang, C. Wang, X. Ni, Z. Wang, W. Jia, L. Chai, Appl. Phys. Lett. **89**, 161110 (2006)
86. R.W. Schoenlein, W.Z. Lin, J.G. Fujimoto, G.L. Eesley, Phys. Rev. Lett. **58**, 1680 (1987)
87. R. Kelly, A. Miotello, Appl. Surf. Sci. **96–98**, 205 (1996)
88. B. Tan, K. Venkatakrishnan, K.G. Tok, Appl. Surf. Sci. **207**, 365 (2003)
89. A. Ashkin, J.M. Dziedzic, J.E. Bjorkholm, S. Chu, Opt. Lett. **11**, 288 (1986)
90. K. Schnütze, G. Pösl, G. Lahr, Cell Mol. Biol. **44**, 735 (1998)
91. C. Bustamante, Z. Bryant, S.B. Smith, Nature **421**, 423 (2003)
92. R. Nakamura, Y. Kanematsu, M. Kumauchi, N. Hamada, F. Tokunaga, J. Lumin. **102–103**, 21 (2003)
93. K. König, I. Riemann, W. Fritzsche, Opt. Lett. **26**, 819 (2001)
94. K. König, Hisochem. Cell Biol. **114**, 79 (2000)
95. D. Morrish, X. Gan, M. Gu, Opt. Express **12**, 4198 (2004)
96. W. Watanabe, N. Arakawa, S. Matsunaga, T. Higashi, K. Fukui, K. Isobe, K. Itoh, Opt. Express **12**, 4203 (2004)
97. M. Goksör, J. Enger, D. Hanstorp, Appl. Opt. **43**, 4831 (2004)
98. L. Malmqvist, H.M. Hertz, Appl. Opt. **34**, 3392 (1995)
99. B. Agate, C. Brown, W. Sibbett, K. Dholakia, Opt. Express **12**, 3011 (2004)
100. XING Qi-rong, MAO Fang-lin, LI Yan-feng, Wei He-ying, CHAI Lu, WANG Qing-yue, J. Optoelectron. Laser **13**, 102 (2002) (in Chinese)
101. Q. Xing, F. Mao, L. Chai, Q. Wang, Opt. Laser Technol. **36**, 635 (2004)
102. F. Mao, Q. Xing, K. Wang, L. Lang, Z. Wang, L. Chai, Q. Wang, Opt. Commun. **256**, 358 (2005)
103. J. Gong, X. Zhao, Q. Xing, F. Li, H. Li, Y. Li, L. Chai, Q. Wang, A. Zheltikov, Appl. Phys. Lett, **92**, 093901 (2008)
104. J. Liu, Y.W. Zhao, J.Q. Zhao, A.D. Xia, L.J. Jiang, S. Wu, L. Ma, Y.Q. Dong, Y.H. Gu, J. Photochem. Photobiol. B **68**, 156 (2002)
105. P.D. Zhao, P. Chen, G.D. Tang, G.L. Zhang, W.J. Chen, Chem. Phys. Lett. **390**, 41 (2004)
106. Q. Dong, X. Zhao, Catal. Today **98**, 537 (2004)

Index

Above threshold dissociation (ATD), 67
Above-threshold ionization (ATI)
absolute phase
evidence, 21
krypton atom, 20, 21
mirror and inversion symmetry, 19
asymmetric ionization
energy-integrated photoelectron yield, 21, 24
pulse duration, 22, 23
time-dependent Schrodinger equation (TDSE), 23
energy-resolved spectra
phase change, 25
typical structure, 25
xenon photoelectron spectrum, 26
high-energy (plateau) electrons
absolute phases, 33, 34
cutoff energy, 32
phasemeter, 32
pulse duration, 31
time-dependent Schrodinger equation (TDSE), 33
low-energy (direct) electrons
asymmetry, 30
cutoff energy, 28
stereo-TOF spectrometer, 20
Adiabaticity, 138
Attosecond waveform
pulse duration extraction, 150
spectral and temporal response, 150
TDSE, 150
temporal light confinement, 150
temporal pulse sequence, 152
Auger decay process, 220

Beta-barium borate (BBO), 161
Bond hardening (BH)
See, Vibrational trapping (VT), 67
Bond-softening (BS) mechanism, 67

Cell-damaging effects, 253
Charge resonance enhanced ionization (CREI), 69
Classical dynamics
axicon (radially polarized) laser fields, 128, 129
center-of-mass motion, 127
electron acceleration
energy gain, 130
scaled field amplitude, 129
Collisional excitation X-ray lasers (CE-XRL), 215
Colored conical emissions
fs pulses, 174
FWHM, 175
pump intensity, 175
seeded amplification
beam divergence, 176
carrier envelope phase stabilization, 179
fundamental-wave (FW) pulse, 176
FWHM, 176
photon-momentum conservation, 176
spatiotemporal modulational instability, 176

space and time degrees of freedom, 175
spatiotemporal modulational instability, 173, 174
three-photon processes, 173
Conventional optical fibers (COFs), 235
Coster–Krönig decay process, 220
Coulomb explosions
fs pump-probe setup, 59
KER, 58
momentum distributions, 58

Dirac equation
adaptive-grid, 117
algorithms, 115
FFT, 115
Hamiltonian, 116
matrix formalism, 116
Dynamical reaction theory
Arnold Web
characteristics, 97
IVR, 96
nonlinear resonances, 96
stable and unstable manifolds, 97
NHIM
definition, 95
essence, 95
stable and unstable manifolds, 95
TS concept, 96

Fast Fourier Transforms (FFT), 115
Femtosecond laser
active ions valence state change, 206
glass sample composition, 199
photoluminescence spectraum, 200
photoreduction, 200, 201
signal-to-noise ratio, 202
Sm^{3+} ions doped alumina borate glass, 199
long-lasting phosphorescence
calcium aluminosilicate glasses, 207
mechanism, 206
oxygen-deficient Ge-doped silica glass, 207
Tb^{3+} ions doped ZBLAN fluoride glass, 206
multiphoton excited upconversion luminescence
advantages, 187
annealing, 198
crystals, 193, 195, 196
inorganic glasses, 187, 190
mechanisms, 186
micro-crystallites and rare-earth ions, 192
semiconductors, 196–199
polarization-dependent luminescence
electron-spin-resonance spectrum, 210
Eu^{2+}-doped fluoro-aluminate glass, 209
Ge-ODC, 208
refractive-index fluctuations, 210
Femtosecond laser technology
copper film microdroplet deposition
LIFT process, 245, 246
morphology, 246
quartz substrate, 249
optical manipulation system
cell fusion, 255, 256
experimental setup, 250
optical trap, 251, 252
optical tweezer, 250
two-photon excitation, 253, 254
PCF
cladding index, 235
COF, 235
light behavior, 234
nonlinear effects, 236, 238, 239, 243
scanning electron microscopy, 234
silica glass, 235
stack-and-draw fabrication process, 235
Fractional behavior
Arnold web, 104
HCN isomerization, 105
Maxwell's demon, 106
NHIM, 104
origin, 103
reaction process, 106
survival probability, 104
typical model Hamiltonian, 103
Full width at half maximum (FWHM), 175, 176, 180

Germanium oxygen deficient centers (Ge-ODC), 208
Gouy effect, 20

H_2^+, 45, 67
Hohg Xuv source
higher order harmonic generation, 135
intense isolated attosecond bursts emission, 137
parameters, 138

Intensity-difference spectra (IDS), 72
Intramolecular vibrational-energy redistribution (IVR)
acetylene
analysis, 98
symmetry, 98
Arnold Web, 98, 99
Dunham expansion, 98
Kellman's constant, 101
nondiagonal couplings, 99
Ion dissociation
angular distribution, 75
KER distribution, 75
Ionization
Coulomb explosion channel, 79
internuclear distance, 78
KER distribution, 82
two-dimensional velocity distribution, 80
Ionization vs. dissociation
branching ratio, 85
IDS, 84

Keldysh parameter, 42, 138
Kinematically complete, 71
Kinetic energy release (KER), 58, 71

Laser-driven X-ray lasers
alternative pumping schemes
frequency up-conversion, 227
recombining plasma, 228
photo-pumping scheme, 217, 218
population inversion
energetic electron pulse pumping, 220, 221
ionization energy, 219
pumping duration, 219
shell vacancy, 218, 219
X-ray pulse pumping, 222–226
Laser-induced forward transfer (LIFT) process, 245

Molecular hydrogen ion (H_2^+)
charge resonance-enhanced ionization, 55
Coulomb explosions, 58, 59
electronic and nuclear dynamics, 50, 51
electrostatic potential, 46
ionization rate, 54, 56
nonadiabatic multielectron dynamics, 57
nonadiabatic transitions and ionization probability, 53
nuclear dynamics and ionization dynamics, 48
time-dependent adiabatic states, 45
time-dependent eigenfunctions, 46
time-dependent Schrödinger equation, 44
Multi-photon process, 67
Multiphoton excited upconversion luminescence
crystals
emission spectrum, 193
infrared to ultraviolet and visible, 195
log–log relationship, 196
three photon induced fluorescence, 193
inorganic glasses
absorption spectrum, 188
Gaussian mode, 188
Ge-doped silica glass, 188
near-infrared to visible upconversion fluorescence, 188, 191
photoluminescence spectrum, 188
preparation, 189
three-photon absorption cross section, 191
transition-metal ions potential, 191
mechanisms, 186
micro-crystallites and rare-earth ions
Eu^{3+}-doped glass–ceramics, 192
half-width at middle maximum, 192
second-harmonic generation, 192
rare-earth doped material (Eu^{2+} CaF_2), 186
semiconductors
broad visible (VIS) band, 197
power index, 198

two-photon absorption, 198
two-photon photoluminescence (TPL), 198

Non-linear split-mirror volume autocorrelator
helium gas jet harmonic radiation, 146
ionization signal, 147
Non-linear XUV processes
adiabaticity, 138
direct double ionization, 142
multiphoton ionization rate, 139
two-XUV-photon ionization, 139, 140
X-ray spectral regime, 145, 146
Normally Hyperbolic Invariant Manifolds (NHIMs), 94

One electron molecular ion, (H_2^+)
CREI, 69
intense laser fields
dissociation, 73–75, 77
energy spectrum, 71
ionization, 77, 78, 80
ionization vs. dissociation, 81, 83–85
KER, 71
parameters, 69
potential energy curve (PEC), 68, 69
One electron molecular ion, (H_2^+)
intense laser fields
Intensity-difference spectra (IDS), 72
Optical field ionization (OFI) process, 218
Optical manipulation system
cell fusion, 255
experimental setup, 250
optical trap
escape speed *vs.* average laser power, 252
human red blood cells (RBCs), 251
sodium chloride, 251
trapping capability and power, 252
two-photon excitation, 253

Phase-dependent interferences
argon photoelectron spectra, 35, 36
Photo-pumping scheme
emitter and absorber ion pairs, 217, 218
OFI process, 218
population inversion, 218
spectral lines, 217
Photonic crystal fibers (PCFs)
cladding index, 235
COF, 235
light behavior, 234
nonlinear effects
application aspects, 243, 244
frequency conversion, 239
soliton effects, 243
supercontinuum generation, 236, 238
scanning electron microscopy, 234
silica glass, 235
stack-and-draw fabrication process, 235
Polyatomic molecule (C_{60})
classical MD method, 61
lifetime, 61
normal mode analysis, 61
structural stability, 61
time dependent adiabatic states, 59
vibrational dynamics, 62
Population inversion
energetic electron pulse pumping
current density, 221
nonradiative decay processes, 220
temporal behavior, 221
pumping duration, 219
shell vacancy, 218, 219
X-ray pulse pumping
bound-bound transition, 226
carbon foil, 224
characteristics, 224
energy level diagram, 223
inner-shell vacancy, 222
photo-ionization cross section, 223
target geometry, 224
Potential energy curves (PEC), 67

Quadratic nonlinear media
colored conical emissions
fs pulses, 174
FWHM, 175
pump intensity, 175
seeded amplification, 176, 177

space and time degrees of freedom, 175
spatiotemporal modulational instability, 173, 174
three-photon processes, 173
second harmonic (SH) control beam
optical quantum coherent control, 173
tunable relative phase shift, 172
two-dimensional multicolored up-converted parametric amplification, 173
two-dimensional pattern suppression, 172
two-dimensional multicolored transverse arrays
BBO crystal, 161
beam divergence, 163
degrees of freedom, 164
spatial beam breakup, 164
Ti-Sapphire system, 161
wave-packet breakup, 162
two-dimensional transient gratings
beating signal, 167
diffraction and amplification, 169
high-peak-power optical pulse, 171
phase coherence, 170
phase difference, 167
spectral interference, 168
super-continuum pulse, 164, 167
synchronous coherent beam, 164
two-dimensional multicolored up-converted parametric amplification, 164
Quantum dynamics
bound electrons
electronic ground-state wave function, 122
multi-photon regime, 121
over-the-barrier regime, 121
ponderomotive potential, 121
probability density, 126
tunneling-regime, 121
Dirac equation, 115–117
free electron, 119
free electronic wave packet, 119, 120
Gaussian distribution, 120
Volkov states, 118
Lorentz-invariance, 115
Second harmonic (SH) control beam
optical quantum coherent control, 173
tunable relative phase shift, 172
two-dimensional multicolored up-converted parametric amplification, 173
two-dimensional pattern suppression, 172
Statistical reaction theory
assumptions, 93
limitations, 94
rate constant, 94
transition states (TS), 94
Strong-field photoionization, *see* Above-threshold ionization (ATI)
Symmetric multiprocessing environments (SMP), 116

Time dependent Schrödinger equation (TDSE), 142, 150
Time of flight ion mass spectra, 140, 141
Time-of flight (TOF) technique, 71
Time-of-flight spectra, 86
Transition state concept
definition, 95
limitations, 107–109
statistical reaction theory, 93, 94
Transmission grating interferometer, 148
Two-dimensional multicolored transverse arrays
BBO crystal, 161
beam divergence, 163
degrees of freedom, 164
spatial beam breakup, 164
Ti-Sapphire system, 161
wave-packet breakup, 162
Two-dimensional transient gratings
beating signal, 167
high-peak-power optical pulse, 171
phase coherence, 170
phase difference, 167
spectral interference, 168
super-continuum pulse, 164, 167
synchronous coherent beam, 164

two-dimensional multicolored up-converted parametric amplification, 164
Two-XUV-photon ionization (TXUVPI)
attosecond pulse train, 139
photoelectron energy, 141

Vibrational trapping (VT), 67

Wavepacket dynamics
Molecular hydrogen ion (H_2^+)
adiabatic states and nonadiabatic coupling, 46, 48, 50
adiabatic tunnel ionization and nonadiabatic ionization, 54, 55, 57
Coulomb explosions, 58, 59
electronic and nuclear dynamics, 50
time-dependent Schrödinger equation, 44, 45
Polyatomic molecule (C_{60}), 59, 61, 63

X-ray pulse pumping
bound-bound transition, 226
carbon foil, 223
characteristics, 224
energy level diagram, 223, 224
inner-shell vacancy, 222
photo-ionization cross section, 222
sodium atom, 222
target geometry, 224
XUV auto- and cross-correlators
non-linear split-mirror volume autocorrelator, 146, 147
transmission grating interferometer, 148

GPSR Compliance

The European Union's (EU) General Product Safety Regulation (GPSR) is a set of rules that requires consumer products to be safe and our obligations to ensure this.

If you have any concerns about our products, you can contact us on ProductSafety@springernature.com

In case Publisher is established outside the EU, the EU authorized representative is:

Springer Nature Customer Service Center GmbH
Europaplatz 3
69115 Heidelberg, Germany

Batch number: 10838015

Printed by Printforce, the Netherlands